Cattle in the Backlands

[ROBERT W. WILCOX]

Cattle in the Backlands

MATO GROSSO AND THE EVOLUTION OF
RANCHING IN THE BRAZILIAN TROPICS

University of Texas Press AUSTIN

Requests for permission to reproduce material from this work should be
sent to:
Permissions
University of Texas Press
P.O. Box 7819
Austin, TX 78713-7819
http://utpress.utexas.edu/index.php/rp-form

⊗ The paper used in this book meets the minimum requirements of
ANSI/NISO Z39.48-1992 (R1997) (Permanence of Paper).

LIBRARY OF CONGRESS CATALOGING-IN-PUBLICATION DATA

Names: Wilcox, Robert W., author.
Title: Cattle in the Backlands : Mato Grosso and the evolution of ranching in
    the Brazilian tropics / Robert W. Wilcox.
Description: First edition. | Austin : University of Texas Press, 2017. |
    Includes bibliographical references and index.
Identifiers: LCCN 2016020979 | ISBN 978-1-4773-1114-1 (cloth : alk. paper) |
    ISBN 978-1-4773-1174-5 (pbk. : alk. paper) | ISBN 978-1-4773-1115-8
    (library e-book) | ISBN 978-1-4773-1116-5 (non-library e-book)
Subjects: LCSH: Ranching—Brazil—Mato Grosso (State)—History—
    19th century.| Ranching—Brazil—Mato Grosso (State)—History—
    20th century. | Cattle—Technological innovations—Brazil—Mato Grosso
    (State)—History—19th century. | Cattle—Technological innovations—
    Brazil—Mato Grosso (State)—History—20th century. | Beef cattle—
    Environmental aspects—Brazil—Mato Grosso (State)—History. | Mato
    Grosso (Brazil : State)—Economic conditions—19th century. | Mato
    Grosso (Brazil : State)—Economic conditions—20th century.
Classification: LCC SF196.B6 W55 2017 | DDC 636.2/01098172—dc23
LC record available at https://lccn.loc.gov/2016020979

doi:10.7560/311141

In memory of my parents, Roy and Lois, curious supporters of my sometimes perplexing journeys into and across the American tropics.

Contents

Selected Timeline for Cattle Ranching in Mato Grosso, 1580s–1980

1580s   Establishment of Spanish Jesuit missions in territory that would become Mato Grosso; cattle reportedly part of the missions

1718   Discovery of gold and founding of Cuiabá

1729   First record of nonmission cattle in Mato Grosso

1750   Treaty of Madrid

1767   Expulsion of Jesuits from Spanish territories

1770s–1780s   First *fazendas reais* (royal ranches) established; with independence, became *fazendas nacionaes*

1778   Founding of Albuquerque (Corumbá) and a fort at Miranda

1822   Brazilian independence

1820s   First reports of the existence of Jacobina ranch in the northern Pantanal

1830s–1840s   First ranching settlers in southern Mato Grosso, including the Pantanal

1838   Founding of Paranaíba

1850s–1860s   Military agricultural colonies established in various locations

1850   National land law; revised in 1854; abolition of African slave trade

1850s   First reports of mal das cadeiras (surra) in Pantanal of Mato Grosso; first reported in Brazil in 1820s on Marajó Island

1857   Shipping by way of the Paraguay River opened to Brazilian vessels

1858   First mention of export of Mato Grosso cattle to other states (Rio de Janeiro)

1864–1870   Paraguayan War; extensive depopulation and dispersal of people and cattle

1870s   Beginnings of resettlement of the region by former settlers, ex-soldiers, and others; direct fluvial shipping reestablished on Paraguay River

1870s–1880s   Reestablishment of national military agricultural colonies

1872   Founding of Campo Grande

1873–1874   Establishment of Descalvados jerky factory near Jacobina

1881   Purchase of Descalvados for the production of beef bouillon by Jaime Cibils y Buxareo

1882   Beginnings of the erva mate (tea) industry with concession granted to Thomaz Larangeira

1888   Abolition of slavery

1889   Overthrow of the emperor and declaration of Brazil as a republic; provinces become states

1890   Lloyd Brasileiro shipping company formed by the Brazilian government

1891   Banco Rio e Mato Grosso established

1892   First Mato Grosso state land law; founding of Aquidauana

1893   Beginning of the organized importation of zebu cattle from India to Minas Gerais

1894   State rental of more than five million hectares in the south to erva company Mate Larangeira

1895   Purchase of Descalvados by a Belgian syndicate; first appearance of foot-and-mouth disease in Brazil, in Minas Gerais

1895–1910   Gaúcho immigration into Mato Grosso

1899–1906   Civil war in Mato Grosso

c. 1900   First introduction of barbed wire into Mato Grosso; first reports of formal horse breeding in Mato Grosso highlands

1903   Formal reserve of 370,000 hectares granted to Kadiwéu peoples by the Mato Grosso state government

1904–1905   Lands encompassing 5,400 hectares in two villages granted to Terena peoples by the state government

1904–1907   Establishment of formal cattle trails to São Paulo by Major Cecílio and Francisco Tibiriçá

1907   Mato Grosso state homestead law; first charqueadas (jerky factories) established in southern Mato Grosso

1908   Construction begun in Mato Grosso on the Estrada de Ferro Noroeste do Brasil (EFNOB)

1910   Creation of the Serviço de Proteção aos Indios (SPI) by the federal government

1911   Purchase of Descalvados and other ranching properties by Percival Farquhar syndicate, Brazil Land, Cattle and Packing Company, more than 1.8 million hectares total

1912–1913   State agricultural school and *campo de demonstração* (experimental ranch) established near Cuiabá

1914   Completion of the EFNOB from the Paraná River to near the Paraguay River; a link to Porto Esperança on the Paraguay River near Corumbá completed in 1917; first frigoríficos (packing plants) established in Brazil (São Paulo); first exports of frozen and chilled beef

1915   Official founding of Três Lagoas

1916   Branch of the Bank of Brazil in Corumbá established; other branches followed in Campo Grande and Três Lagoas

1918   First organized experiments with planted pasture in Mato Grosso

1919   Creation of a cattle market and fair (*feira de gado*) in Três Lagoas

1921   Rinderpest outbreak in São Paulo; contained to São Paulo stockyards

1924   Model ranch (*fazenda modelo*) established at Campo Grande with federal and state cooperation; did not actually function until 1936

1928   Bridge completed over Paraná River at Três Lagoas

1930   Military coup and revolution led by Getúlio Vargas

1936   Indubrasil officially recognized as a distinct breed

1937–1945   Estado Novo dictatorship

1938   National zebu genealogical registry authorized

1947   Bridge completed over Paraguay River at Porto Esperança

1955   Illegal import of 100 zebu cattle (Gir) through Corumbá

1960   First permanent frigorífico in Mato Grosso built at Campo Grande; decade of the decline of Mato Grosso charqueadas

1962   Last legal importation of zebu cattle (Nelore) into Brazil

1964–1985   Military dictatorship

1970s–1980s   Increased investment in the Center West cattle sector, including introduction of new grasses such as brachiaria; development of soybean agriculture in the Center West, including the Mato Grossos

1973   Creation of the government agricultural research corporation, Empresa Brasileira de Pesquisa Agropecuária (EMBRAPA)

1977   Inauguration of EMBRAPA Gado de Corte outside Campo Grande

1979   Division of Mato Grosso into two states, Mato Grosso and Mato Grosso do Sul

MAP 1. *Geography and Environments of Mato Grosso do Sul. Map created by Jones Maps & Diagrams, Ltd., Delta, British Columbia.*

MAP 2. *Regions of the Pantanal. Map created by Jones Maps & Diagrams, Ltd., Delta, British Columbia.*

MAP 3. *Cattle in Historical Southern Mato Grosso. Map created by Jones Maps &*
Diagrams, Ltd., Delta, British Columbia.

Preface and Acknowledgments

When I first embarked on my tentative journey into cattle country, I imagined myself engaging in Frontier History. Indeed, the introduction to my dissertation, the seed from which this study germinated, concentrated extensively on the frontier thesis of Frederick Jackson Turner, his critics, and more recent studies of the development of the frontier and pioneer fronts in Brazil. My impression at the time was that ranching was a characteristic of the frontier and as such represented the very essence of human occupation of so-called empty spaces. Naturally, this also included discussion of the environment and the impact of ranching on the land, but I was not yet convinced of the central place of the environment in the story. I have since come to realize how much the story of cattle on the land is one that includes not only economic development, migration, land struggles, technologies, and uneven labor relations but also the role of humans in their ecosystems, most particularly their impact on those ecologies along with how much those environments determined settlement and human action.

This perspective was influenced by noting how much of what I was reading in the archives of the Arquivo Público do Estado de Mato Grosso (APMT) and elsewhere—government reports, scientific studies of cattle raising, observations of actors on the ground—was similar to what I saw when I stepped out of the air conditioned rooms and into the humid air of Cuiabá. The region was undergoing rapid transformation that was reported daily in the news media, inside and outside Brazil. Chico Mendes had been murdered not long before I began my research, and his story and that of Brazil's Amazon of the 1980s and 1990s seemed to reflect what I was finding in the archives. It gradually dawned on me that I was engaging in a form of living history, and the environment was a major player. Nonetheless, for several years I let my work languish as I pursued other priorities, in a way risking the opportunity to tell the story while it still had a strong relationship to the headlines. Sadly, that connection is still deeply relevant, as cattle ranching continues to be a major player in the occupation and ecological transformation of the Brazilian interior. It seems the need to tell the story of ranching's early days is even more urgent today than it was years ago, and it is in this spirit that I write this book, in humble recognition of those who have lived and continue to live the experience today. To them goes all the credit for any recognition my book may receive, to me the responsibility for its shortcomings and errors of interpretation.

There are far too many people to thank for all the help I have received over the many years of this project; for those not mentioned here, I hope it is understood that this book could not have been shaped without you. I must, however, specifically thank some groups and individuals for their unflagging interest in my research and for helping me access important sources and understand better the world of tropical ranching. First of all, I am deeply indebted to the past and present archivists and other employees in the APMT and the Núcleo de Documentação e Informação Histórica Regional (NDIHR), both located in Cuiabá. The NDIHR is part of the Universidade Federal de Mato Grosso (UFMT), where a number of professors and students expressed deep curiosity in my work. I am particularly grateful for the support and friendship of Fernando Tadeu de Miranda Borges, who befriended me when we were both lowly graduate researchers; he is now a prestigious professor at UFMT and was recently inducted into the Academia Mato-Grossense de Letras, a truly outstanding honor. The late Dr. Lenine Póvoas also was unreservedly welcoming, especially inviting my wife and me into his home during our first foray into the region. Others were infinitely helpful in the Arquivo Nacional, the Biblioteca Nacional, and the Instituto Histórico e Geográfico Brasileiro, all in Rio de Janeiro, and in various other archives and libraries in Rio and São Paulo and in Asunción, Paraguay. In addition, the staffs at the Baker Library at the Harvard Business School, at the Manuscripts and Archives section of Yale University Library, and at the Southwest Collection/Special Collections Library of Texas Tech University bent over backward to help me find details of United States investment in Mato Grosso cattle ranching.

In Mato Grosso do Sul, the center of most of the region's cattle sector during the period under study, I thank the staff at the Arquivo Público do Estado de Mato Grosso do Sul and the Biblioteca Pública Isais Paim in Campo Grande. Most crucial were the support and friendship of Valmir Batista Corrêa and Lúcia Salsa Corrêa, who gave me open access to their extensive personal library when I was in Corumbá (later they moved the library intact to Campo Grande). Over many years they accumulated countless books, manuscripts, and pamphlets and applied what they learned to their excellent research; their library remains a treasure trove for present and future researchers. Others in Mato Grosso do Sul who have offered their tireless and good-humored help and with whom I have forged lasting friendships include Gilmar Arruda, Paulo Cimó Queiroz, Eudes Fernando Leite, Dr. Alain Moreau, and José Luiz Amador (Zé Fisgo).

With no personal experience in ranching, I owe a tremendous debt to those who raise cattle in the region for their interest, support, and at times patient tolerance of my many, no doubt obvious questions about the business and its historical practice. They include Renato Alves Ribeiro and his family, the late Dr. Paulo Coelho Machado, Abílio Leite de Barros, the late Clóvis de Barros, Dr. Cassio Leite de Barros, and Sra. Elza Dória Passos. At the Empresa Brasileira de Pesquisa Agropecuária (EMBRAPA), in Campo Grande and Corumbá, I am indebted to Dr. Afonso Simões Corrêa and Dr. Arnildo Pott, who clarified my concerns and doubts about the science of ranching in tropical regions. I also would like to thank head librarian Elane de Souza Salles at EMBRAPA Gado de Corte in Campo Grande, who greatly facilitated my access to the many books and journals in the center's library.

Elsewhere, I am most grateful for the continued interest and encouragement of my many colleagues at Northern Kentucky University and in the fields of Brazilian and environmental history. Again, there are too many to list here, but I would like to express my deep appreciation for the continual support over many years of John Soluri, Stephen Bell, and Sterling Evans in the United States; Shawn Van Ausdal, Claudia Leal, and Stefania Gallini in Colombia; Reinaldo Funes Monzote in Cuba; and Lise Sedrez and José Augusto Pádua in Rio de Janeiro. They are all truly exceptional colleagues.

I would be remiss if I didn't thank the Social Sciences and Humanities Research Council of Canada for its generosity in awarding me two doctoral fellowships, New York University for the Dean's Dissertation Fellowship and the Leo B. Gershoy Development Fellowship, and

Northern Kenucky University for various grants, fellowships, and sabbaticals. Without such financial support this study could not have come to fruition.

I have been blessed with very patient and understanding editors at the University of Texas Press. Thanks especially to Casey Kittrell, Lynne Chapman, and Lorraine Atherton, who guided me through the publishing process with good-natured forbearance. And my gratitude to Lester Jones and Lisa Rivero for their excellent work on the maps and the index. You are all the best!

On a personal level, words are not sufficient to express my gratitude for the guidance and support of the late, great Warren Dean, who was my dissertation advisor and friend during my graduate school experience. Plenty has been said about Warren's contributions to the field of Latin American environmental history and his empathy for and support of young, naive grad students, which needs no repeating here. Without question, I could not have had a better advisor; Warren was a highly intelligent and dedicated professor, with a delicately subtle sense of humor, who made my graduate experience much more than tolerable, indeed enjoyable. He remains close to my and my wife's hearts.

Finally, my family deserves my sincerest thanks and love for their patient support over far too many years. Katia and Dan, you are constantly the key to my success on so many levels. Mil gracias.

Cattle in the Backlands

The Paradox of Tropical Ranching

Cattle are raised [according] to the law of nature, released into the fields [campos], subject to bad weather, entrusted to public honor; there is no stabling; forage is acquired in the wild [campos], no matter the season; [only] one or two ranchers have fencing, causing frequent rustling among neighbors; existing breeds, perhaps the same as those of colonial times, have not been improved through selective breeding.

MATO GROSSO VICE PRESIDENT JOSÉ JOAQUIM RAMOS
FERREIRA, 1887

Mato Grosso Vice President José Joaquim Ramos Ferreira's 1887 description of the weaknesses of ranching in his remote province illustrates a theme repeated by numerous legislators in Brazil and much of Latin America over the late nineteenth century and well into the twentieth. Rudimentary ranching was perceived to be a drag on regional and national economies, and frequently, local ranchers were considered obstacles to the introduction of so-called rational development, which promised to transform backward provinces and triumphantly lead the nation out of its centuries-old lethargy and into the modern era. This premise has dominated the global development discourse since the nineteenth century, and in Latin America ranching has played a key role in the multiple debates that ensued.[1]

Responses to such critiques came in fits and starts throughout the continent, yet efforts to address apparent limitations eventually led to economic development that assured ranching's indisputable place as a key driver of agricultural expansion. That was especially the case in

Brazil, today a major provider of beef to areas of the world as diverse as Russia, the Middle East, Hong Kong, and Venezuela. As suggested by Ramos Ferreira's evaluation, the road to such prosperity was slow and uneven, while the kind of development proposed for remote regions like Mato Grosso often was inappropriate to local economic, social, and environmental conditions. These apparent contradictions in the conversion of ranching from "backward" to "modern" are the focus of this study.

WHY RANCHING?

The spectacular expansion of ranching into tropical forest and savanna in the Americas over the past few decades, with widespread social and environmental consequences, has attracted considerable attention and concern around the globe. Indeed, the casual visitor to Brazil cannot help but notice the huge numbers of cattle grazing the nation's fields. According to the Instituto Brasileiro de Geografia e Estatística (IBGE), in 2011 the nation boasted the second largest bovine population in the world, after India, at almost 213 million head. In the same year, the combined cattle populations of two states, Mato Grosso and Mato Grosso do Sul, stood at more than 50 million, and another 43.2 million head were pastured in the Norte, or Amazon Basin.[2] These figures are reflected in national exports of beef and beef products, which topped 1.9 million tons in 2014, second only to India.[3] At the same time, intense national and international criticism has been leveled at ranching for its prominent role in social and environmental disruption, particularly in the Amazon. While this expansion has generated heated debate over the role of ranching in development today, what has largely been missing is a closer examination of the historical trajectory of the cattle business from an allegedly neglected sector in the nineteenth century into the vibrant, if controversial, economic pole of development in the twenty-first. Crucial to this transformation are the multiple forces involved in the expansion of cattle ranching into tropical zones, especially those influencing perceptions and priorities.[4]

Many readers will know that the dominant literature on ranching in Latin America has tended to focus on frontiers. Although I recognize this as a productive approach to understanding social and political relations, I believe that in several ways such attention has distracted from actual day-to-day activities in so-called frontier zones.[5] The emphasis on state formation and violence in remote regions has tended to ignore the subtleties of relations not only between human beings but also between humans, animals, and the land. The ambiguities of establishing and running a ranch in an area with what appear to be favorable natural conditions but with uncertain future opportunities meant that economic results were irregular, sometimes spectacularly successful, often a difficult struggle. At the same time, ranching development did not occur in historical isolation, nor was it linear. To address these issues more closely, I draw primarily on the literatures of geography and the agricultural, social, and environmental history of Latin America and Brazil, and argue that the expansion of cattle was more disorganized than generally assumed, largely because of a multitude of contradictory factors inherent in ranching. Using the far western state of Mato Grosso as the model, often referred to in the literature as a frontier zone, I contend that the intricate details of these processes are critical to understanding the complexity of raising cattle over time in the tropics, and put into clearer perspective the circumstances of recent years.

Over the course of the late nineteenth century and well into the twentieth, Mato Grosso became a laboratory for tropical and semitropical cattle ranching in Brazil. Recent issues of social and environmental justice in remote cattle regions like the Amazon have precedents in the region, and the state's experience established many of the technical and economic inputs that have been duplicated in most other tropical cattle sectors of the Americas. The results were often paradoxical, whether in forms of economic development, social transformations, or environmental impacts. These factors combined to play a critical role in Mato Grosso and eventually tropical ranching throughout the Americas.[6]

BRAZIL AND MATO GROSSO AS RANCHING HUBS

Usually when we think of the history of ranching in South America, Argentina and Uruguay come to mind. In both countries, cattle and sheep have played prominent roles in their national economies since the late colonial period, and they were settled and populated to a great extent based on the processing of animal products.[7] Yet cattle ranching has a long history in Brazil. Cattle were an integral part of the Brazilian economy virtually from the first permanent Portuguese settlement, and the natural conditions provided opportunities for the growth of ranching and eventual expansion of the nation-state itself. As sugar, tobacco, and coffee plantations occupied the most fertile land, the demand for draft oxen, meat, and hides produced in the more arid interior guaranteed ranching a vital position in the Brazilian economies of the colonial era, continuing into the post-independence period.[8] By examining comparative statistics it is possible to see the tremendous potential for Brazilian ranching by the twentieth century.

On the eve of World War I, Argentina claimed 26 million head, while 30 million animals were pastured in Brazil. At the same time, Uruguay carried 8 million head, not quite a million more than the neighboring Brazilian state of Rio Grande do Sul. By the late 1930s, the Argentine cattle population had grown to 32 million; in Brazil the total was 40 million. After a short-term drain on numbers caused by World War II, by 1952 Brazil counted more than 55 million head, and in Argentina the herd was just over 45 million. Rio Grande do Sul, with more than 10 million head, had surpassed Uruguay's herd of just under 8 million.[9]

These statistics clearly show that at least in terms of numbers Brazil was competitive with its neighbors. Yet there were substantial obstacles. Stephen Bell's fine study of Rio Grande do Sul cattle ranching outlines how this region in the temperate south came to dominate Brazilian production and export from the early nineteenth century; however, geographical and environmental limits to its production had been reached by the 1920s.[10] Production in the rest of the country was constrained for other reasons, as noted by a Uruguayan expert, Agustín Ruano Fournier, who in the 1930s judged Brazil a minor producer of beef but a major future exporter if poor meat quality could be overcome. This competitive disadvantage of Brazilian cattle centered on the demands of the export market (primarily in Europe), breed type, and climatic conditions. Ruano's critique included Rio Grande do Sul but was mostly directed at ranching elsewhere in the country. Cattle raised in the Brazilian tropical or semitropical climate were smaller, carried less flesh, and yielded tougher meat than breeds raised in the temperate climates of Argentina, Uruguay, and North America, and as such ranching practiced in Brazil did not produce meat that normally would pass muster with a fastidious European palate. Not surprisingly, Europeans favored beef with a consistency and tenderness familiar to the average consumer, and that beef primarily came from animals with historical origins in the European countryside. Brazil's struggle to overcome that preference and to penetrate world markets led to the introduction of a completely different animal—the zebu—which initiated a revolution in tropical ranching in which Mato Grosso was at the forefront.[11]

Ruano's analysis was typical of the time and mainly applied the successes experienced in temperate zones as criteria for all ranching. Until relatively recently beef cattle ranching throughout the Americas was open range, with irregular access to reliable markets. Typically ranchers paid limited attention to the controlled productivity of animals, and ranching had to depend on the availability of nutritive grasses and relative climatic consistency. At the same time, regional specificities often determined success or failure. For example, in Mato Grosso raising cattle may have seemed more of a struggle in the semiarid Cerrado than in the humid Pantanal, and in broad terms it was, although this was contingent on favorable climate patterns and forage accessibility. By the same token, the impact of ranching on the environment varied depending on locale and ecosystem. Ranching appears to have left less of an ecological footprint on the Pantanal wetland than in other regions, which prompted some observers to suggest that ranching was instrumental in saving the Pantanal from the uncontrolled development and environmental degradation threatening the rest of southern Mato Grosso.[12] These characteristics reveal the centrality of geography and ecology in determining the path taken by ranching in diverse regions.

As illustrated by Ramos Ferreira's lament, near the end of the nineteenth century opportunities in Mato Grosso were limited. This was destined to change, as over the following decades several recognizable features began to develop, establishing precedents for the modernization of a sparsely populated but potentially rich region of the nation's interior. During World War I, desperate demand in Europe generated a surge in Brazilian beef exports that had profound effects on the Mato Grosso cattle sector. Long before the recent experience of the Amazon, Mato Grosso was penetrated by extensive road and rail construction, agribusiness, and waves of land-hungry peasants (encouraged, if inconsistently, by government developmental interests). Accompanying this boom were problems familiar today: a climate of violence, labor exploitation, aboriginal ethnocide, deforestation, pressure on native species, and spectacular burning of forests and savanna.

Despite the boom, economic growth came gradually and was incomplete even into the 1960s. The natural conditions of the region presented many obstacles: periodic devastating drought and flooding, native grasses of low nutrition, scrawny and feral cattle, unreliable transportation and infrastructure, and even the threat of wildcats. Indeed, how could ranching develop and have a broader influence in an area with so many limitations? Yet it did, which calls attention to the resilience of ranching. The various components of such persistence detailed in this study help to illuminate the contradictory yet decisive role of cattle in leading a remote frontier region into the modern era, and how such change was anything but smooth. The expansion of ranching in Mato Grosso required a measured combination of imposition and negotiation, sometimes at the expense of aspects of the region's character, sometimes in favor of local actors. In the end, this is a complex story of innovation and tradition, perseverance and exploitation, conflict and compromise, that foreshadows and instructs experiences of the later twentieth century in other corners of Brazil and Latin America.

THEMES AND ORGANIZATION

Cattle in the Backlands analyzes ranching in Mato Grosso through three main themes: (1) the economic transformation of a remote region in which a rudimentary productive process eventually absorbed more modern technical inputs; (2) the resulting social changes on the ground, particularly labor structures and land tenure, which shifted as production expanded; and (3) the multifaceted environmental forces that inevitably accompany and often shape the character of livestock farming, including the role of nature in determining the locale and success of cattle raising, as well as the long-term impact of ranching on ecosystems. My contention is that unlike today's understanding of widespread environmental destruction by cattle in the Amazon, the expansion of ranching in Mato Grosso was not as detrimental to local ecology as might be assumed, at least for most of its history. Mitigating factors were multiple, including the number of animals pastured on the land, the developmental level of the industry itself, attitudes of many ranchers, and ultimately the uneven ability of regional and national economies to balance supply and demand.

At the same time, this study outlines the technical, social, ecological, and economic inputs that have shaped the occupation of much of the rest of the Brazilian interior in recent years. I assert that although the regions are not identical, most of the recent approaches to ranching in Amazonia rest on earlier successes and failures in Mato Grosso. Some more-current technologies have been successful; others reflect the contradictions previously experienced on the ranges of southern Mato Grosso decades earlier, as the events unfolding today in Brazil are directly linked to historical activities that have been little examined and even less understood.

Although politics played an important part in many aspects of Mato Grosso's development, for the purposes of this study I have chosen to integrate policy only when it was vital to ranching. I consider it more useful to recount the process from the bottom up instead of the more customary top down. Through an emphasis on the ground level, readers will gain a fuller understanding of the multidimensional role of cattle in transforming the region and nation.

For the most part, chapters are organized thematically. This path tracks how ranching in Mato Grosso evolved over time, in a process that was slow and indeterminate. I maintain that a purely chronological approach detaches from the intricate day-to-day activities of raising cattle, and we cannot understand the interactions of animals, people, and landscape without in-depth examination of the various components that defined the business. Since this approach involves repeated references to historical incidents and details across the text, a timeline of important events throughout the period appears after the table of contents. I hope it will be a useful guide to appreciating the long- and short-term effects of the various on-the-ground details that characterized ranching in the region.

Chapter 1 examines the geographical and ecological structures that provided the foundation on which Mato Grosso grew from a remote region in the eighteenth century into a relatively developed ranching economy of the twentieth. An understanding of the complexities of habitats that determined the opportunities and obstacles of ranching ventures requires an examination of geology, climate, soils, and vegetation. The region's landscapes and ecology are diverse, ranging from annual fluvial floodplain to rolling natural pastures to semiarid scrub savanna. This includes the well-known Pantanal in the west and a sizable percentage of the Brazilian Cerrado, or savanna, which covers almost a third of the nation and half of Mato Grosso. Cattle have grazed here since colonial times. Some environments were more hospitable to ranching than others, and ranching had impacts, although often not as severe as might be assumed.

In chapter 2 I detail the economic growth of ranching in Mato Grosso from the earliest days of nonindigenous settlement to 1914. Mato Grosso was only a peripheral part of the nation until well after the Paraguayan War (1864–1870), but the region was important to both colonial and early imperial authorities in securing the interior from potential Spanish (later Paraguayan and Bolivian) encroachment. Cattle were instrumental in occupying a frontier zone for Brazil and in establishing precedents for subsequent economic expansion. As a result, the development of ranching in Mato Grosso and its role in the early "taming" of the Brazilian west opens a window onto a pattern that has continued into the twenty-first century.

After the Paraguayan War, ranching increased in economic importance in Brazil and contributed to the gradual integration of remote regions into the nation and, if incompletely, to the spread of market capitalism into those areas. Brazil, along with the rest of Latin America and most of the world, went through a major economic transformation from the late nineteenth century as capitalist expansion from Europe (primarily Britain), and later the United States, widened and deepened international trade. It was during this time that agricultural and extractive products like coffee, cacao, cotton, and rubber determined much of Brazilian nineteenth- and twentieth-century economic development,[13] but little has been said about cattle as an important national and international commodity, an omission this study addresses. The expansion of ranching in Mato Grosso through the late nineteenth century and into the twentieth illustrates the important role of livestock in serving economic development in the nation and worldwide.

As shifts in global consumption began to stimulate demand for more cattle products, meat and hides were essential in national consumption and became important export items, especially with the advent of World War I.[14] Mato Grosso became a proving ground for much of the technology used in Brazilian tropical and semitropical cattle ranching, and chapter 3 addresses this development from 1914 to 1950. With the war and an accelerated entry of foreign capital and expertise, live cattle exports to packing plants (*frigoríficos*) in São Paulo and the founding of salt beef factories (*charqueadas*) shaped the conditions for Mato Grosso's integration into the international economy, reflecting or anticipating similar changes in several countries of Latin America. At the same time, access to transportation was essential for all producers, yet transportation structures were chronically limited and frequently inadequate for development, ranching and otherwise. These experiences illustrate that while the more populous areas of the country attempted to establish regular economic and technological relations with their peripheries in order to supply the world and achieve broader economic growth, often inadequate attention was paid to infrastructural development necessary for such success. The result was a protracted struggle to achieve economic growth. Yet by 1950 the region was securely in the forefront of Brazilian tropical ranching; this achievement, however, came with multiple costs, some worthwhile, others not so much.

Land tenure and labor concerns are examined in chapters 4 and 5, to provide a clearer view of the effect of this development on the players—the ranchers, cowboys, Indians, and others. These were fundamental to the history of the industry and determined the everyday performance of tropical ranching over time, above all given the open range nature of ranching and its modification after 1914. Ranching in Mato Grosso displayed features that predate similar processes today, including responses by ranchers to environmental and economic determinants seldom within their control, the contradictory role of governments and their uneven support for ranching, and the impact of these responses on the region's cultural development.

Such changes created space for confrontation and conflict as the human and livestock populations grew in Mato Grosso. Once the region came to national attention after 1870, an inevitable race for land ensued. Immigrants, local ranchers, land speculators, and corporate interests, mostly foreign, competed for access to rangeland, and property values rose. Some well-placed individuals and companies managed to accumulate substantial properties, in some cases measuring in excess of small European countries. Governments tried to mitigate the inevitable conflicts yet were never completely successful, and violence and litigation remained constants in the Mato Grosso experience. Demand for land soared after 1914, causing even more contention, yet the ranching sector was hampered by a chronic lack of reliable credit, a condition that benefited those who had access to investment funds outside the region, such as foreign interests, but which limited the ability of most locals to develop their properties and contributed to the sluggish development commented on by so many outsiders.

Land conflicts had widespread effects on labor relations. In early decades, the terms of employment reflected the remote character of the frontier. Sufficient numbers of workers were not always available, permitting conditions of coercion and the manipulation of labor that included the conscription of employees by some ranchers into their own militias.[15] This could be of mutual benefit, however, at least in an immediate sense, allowing for some flexibility in labor relations. For example, some ranch workers were encouraged by ranchers to become tenants in return for regular or even irregular labor, and several individuals became small ranchers in their own right, though seldom with legal title to property.

With economic expansion, these relationships began to shift, and cowboys and ranch hands gradually became a rural proletariat that no longer depended on the paternalism of the past. At the same time, there was a distinct division of labor on ranches. Many ranch hands lived on properties with their families, but work relations sometimes were unstable. The experiences on ranches tended to extend into other sectors of Mato Grosso society. The influence of ranching on the general culture persisted over time, in some cases to today.

The region also was home to small but significant numbers of aboriginal peoples. Many were forced off their lands by neo-European impositions of landholding and either migrated into other territories or periodically reacted with violence. Although confrontations usually ended in native defeat and, in some cases, elimination, certain groups were able to defend their territory and enter the ranching sector as employees, traders, and in selected cases even as ranchers themselves. They were not numerous, but their presence and activity often meant that ranchers had to negotiate in order to maintain their livelihood with minimal disruption. The impact of neo-Brazilian settlement on native cultures is an important conversation in Mato Grosso historiography and ethnography today, although limited sources on aboriginal relations with ranching preclude a more in-depth analysis of their role here. I hope the information revealed in this study will assist future scholars in pursuing this promising research path.

Chapter 6 explores mundane tasks and inputs, which were just as important as labor relations. Such essential activities as constructing fencing, providing salt, planting pasture, and caring for horses, among others, were the foundations of success or failure for most ranches. Multiple obstacles to successful ranching in tropical and semitropical climates required responses distinct from those of the temperate zones. These included a knowledge of different forage species, an understanding of tropical diseases and pests, an awareness of climatic variations in determining livestock life cycles and selection of suitable breeds, and an appreciation of regional environmental conditions. Faced with these challenges, Mato Grosso ranchers applied an accumulated folk knowledge that facilitated adaptation to often demanding natural conditions, including periodic and sometimes severe drought or widespread flooding.

Advice arrived from foreign and national experts; some of it was accepted, but much was rejected, largely because the experiences of other ranching economies were not necessarily applicable to Mato Grosso. Attempts to modernize the cattle sector included promotion of veterinary posts, pasture improvements, fencing, and the like. For most ranchers it was a hard slog that took decades to provide the promised rewards. The largest obstacles were the challenging environmental conditions and fiscal fallibility, though as more attention was paid to modernizing the business, the inevitable environmental impacts increased, with some becoming quite severe.

As many observers have outlined, the introduction of cattle into tropical zones has caused some spectacular examples of environmental degradation. The effects over decades of what some have called grass mining have caused significant alteration of local and regional ecosystems. In regions like Mato Grosso, where ranching grew slowly, the interaction with local ecosystems was more complex, and often cattle coexisted with local fauna and flora for decades. That is not to say that changes to ecosystems did not occur, but the impact of ranching in many areas depended much more on the number of animals and type of ranching undertaken, and thus for much of the period residents seldom considered pressure on the environment an issue of concern.

That situation, however, was not sustainable, as introductions like planted grasses and modern forms of husbandry became an integral part of Mato Grosso ranching development. As the state became more securely linked into the national and international production and distribution system, more pressure was put on the land, forest, wetlands, and savanna. Environmental disruption increasingly became an issue, though ranchers usually were more aware of these changes than they have been given credit for.[16]

Chapter 7 focuses on the modernization in the choice of cattle breeds suitable to the tropical environment. Well into the twentieth century observers in Mato Grosso regularly commented that livestock "raised themselves." Breeds were regional adaptations of Iberian stock introduced during the colonial period that satisfied national markets, but which were poorly suited to consumer demands overseas. Accordingly, in the twentieth century the technology and experience of ranching in more developed areas of world ranching entered the state. Steadfast belief in the supremacy of European or North American knowledge and methods of production inspired experts to apply them to tropical and semitropical Brazil, and Mato Grosso became a host for such ventures. This included the introduction of European breeds like Hereford and Aberdeen Angus, yet quickly it became clear that temperate-zone animals had little future in the tropical and semitropical ecology of the nation and region. This led to the introduction of humped zebu cattle from India, accompanied by a fierce controversy over the value of "national" breeds versus "Hindu Idols." These remarkable animals soon became critical in transforming ranching in the tropics, first in sectors of the Brazilian interior and eventually throughout tropical Latin America. The zebu led a revolution in tropical ranching in the Americas in which Mato Grosso played a significant role, though the animals had limited effects on the environment, at least for some decades. This paradox is a key feature of the story told here.[17]

Finally, in the conclusion I summarize the main arguments of the study and offer a brief outline of the regional and national cattle sector since 1950. As previously stated, I am interested in relating the experience of Mato Grosso to the ongoing drama unfolding throughout the Brazilian agricultural sector, particularly in the Amazon and Center West, which includes the Mato Grossos. Along with mining and hydroelectricity, ranching and commercial agriculture have come to define development in the region and have received their share of economic promotion and ecological condemnation, nationally and across the globe. It seems abundantly clear, however, that cattle ranching is the main agricultural driver in the occupation of these vulnerable tropical ecosystems. My point is that even though Amazonia and other regions are not environmentally identical to Mato Grosso, there are enough similarities of landscape, ecology, and economic imperatives to warrant bridging more recent developments with those of previous decades. Without trying to force an absolute link between the regions and their histories, I notice that knowledge accumulated in Mato Grosso from the late nineteenth century, especially technologies and their economic successes, was applied in the latter decades of the twentieth century and into the twenty-first, in both the Center West and Amazonia. The circumstances of today should not be read backward into the history of Mato Grosso, but it seems clear that experiences in semitropical regions facilitated expansion into the tropics to the north. Whether any ecological or social lessons have been learned is another matter, though several juries are still deliberating.

Ultimately, it is my expectation that by exploring the history and many relationships of cattle ranching over almost a century, readers will come to appreciate that raising cattle in tropical ecosystems is far more complex than is sometimes assumed, contributing much more to human economy than simply one more productive enterprise. External and internal factors combined to sculpt the character of ranching in Mato Grosso. Ultimately, it provided a model for the development of a significant sector of the rest of the Brazilian interior and similar regions of Latin America, establishing precedents that were carried forward over succeeding decades.

ONE

Mirror of the Land

REGIONAL GEOGRAPHY AND

ENVIRONMENTAL IMPERATIVES

During his tour of southern Mato Grosso in the early 1920s Antonio Simoens da Silva wrote, "Cattle are the mirror of the land."[1] Silva's amazement at the numbers of healthy animals pasturing the region caused him to wax poetic on its natural environment for raising cattle. His observations reflected a prevailing view of the geographical and environmental conditions of southern Mato Grosso, a promised land of ranching opportunity. Yet although sizable herds certainly did indicate successes and potential, conditions were more complex than Silva could possibly imagine.

From its creation as a captaincy in 1748 until the 1940s, Mato Grosso was Brazil's second largest territorial unit. Situated between 8 degrees and 24 degrees south latitude and between 66 degrees and 50 degrees west longitude, and occupying 1,477,041 square kilometers, Mato Grosso included what have become the present states of Rondônia, Mato Grosso, and Mato Grosso do Sul.[2] Within such a vast area, variation in terrain, water sources, climate, soils, and vegetation is considerable. The territory falls within the general climatic classification of tropical and subtropical, exhibiting at least some similarities in climate and vegetation, but it would be an error to ignore the important role played by local and regional biomes. These distinct ecologies are an integral part of this story. They are not always recognized by travelers, scientists, government officials, or even by some local ranchers, yet they dictated different raising practices between regions. As this study proceeds, I hope the reader will begin to appreciate this complexity not only in Mato Grosso but also in the tropics in general.

DIVERSITY OF GEOGRAPHY, CLIMATE, AND VEGETATION

Mato Grosso do Sul covers some 350,500 square kilometers and hosts three major biomes: the western lowland, an inland floodplain or wetland subject to annual pluvial and fluvial flooding called the Pantanal; the Cerrado, a semiarid central and eastern savanna plateau of scrub forest and extensive open areas of coarse grasses; and the southern region of natural grassland dotted with occasional deciduous and semideciduous woodlands, termed campo limpo. (See map 1, p. xii.) All are considered subtropical, though there are a number of distinct subregions, the most significant warranting individual description.[3]

The existence of these three distinct zones in Mato Grosso do Sul was key to the establishment of ranching. Topography, climate, soils, and vegetation differ enough that a rancher familiar with one area would find success difficult in another if these determinants were not taken into consideration. Even within these biomes methods successful in one area were not necessarily applicable to another. Though it could be argued that the Cerrado is somewhat less complex than the Pantanal, it can hardly be categorized as simple. Mato Grosso do Sul offers varied conditions for permanent ranching settlement, many determined by climate.

Historically, climate in the region has been a product of three major seasonal weather systems.[4] The rainy season is from November to March, dominated by a hot and humid continental equatorial air mass, causing seasonal rains and high temperatures. One-half to two-thirds of the annual rainfall occurs in these months, and average temperatures range between 21 and 35 degrees C, with highs often above 40 degrees. From June to September an Atlantic air mass covers the region, bringing drier weather and winds. Mean temperatures range between 15 and 27 degrees C, though during these months occasional fronts sweeping up from Antarctica bring surprisingly cold weather, with temperatures at times dropping to near freezing, while they sometimes cause the formation of uncomfortable fogs as the polar system collides with an equatorial air mass drawn south by the polar mass's approximation. This *friagem* seldom lasts longer than a few days, but it can be a bitter experience for anyone venturing into the region unprepared. In the intermediate months a temperate tropical Atlantic air mass dominates. Temperatures and rainfall are usually moderate, generating a particularly comfortable climate. This is the period when much of cattle reproduction occurs, though conditions manifest themselves differently depending on subregions.[5]

A geological depression of the Paraguay River occupying between 139,000 and 200,000 square kilometers, the Pantanal is one of the best-known inland wetlands in the world. The region is recognized by UNESCO as a World Heritage site, and it recently has become an attraction for ecotourism. Assuming the smaller area, it covers 17 percent of the combined surface area of the modern states of Mato Grosso and Mato Grosso do Sul, 28.5 percent of the latter, and is bordered by hills in Bolivia just to the west of the Paraguay River, by the Maracajú and Bodoquena hills (*serra*) to the east, the Paraguayan Chaco in the south, the cerrado plateau near the city of Cuiabá to the north, and by equatorial forest in the northwest.[6] (See map 2, p. xiii.)

The Pantanal is drained by a number of rivers and streams, and because of its low elevations it is considered the region of widest flooding in South America. By far the most important is the Paraguay River, a strategic and economically important waterway that flows north to south some 1,450 kilometers, bordering Bolivia and then Paraguay to the Apa River, then running a further 1,050 kilometers through Paraguay to its confluence with the Paraná River. It passes 1,260 kilometers through the Pantanal and receives the outflow of a number of other rivers, mainly running east to west, all partially navigable by shallow-draft boats. This hydrological system guarantees an abundance of water to the Pantanal, particularly during flood season. At the same time, the Paraguay River was for over a century the main economic link from the interior of Mato Grosso, including the Pantanal, to the sea through Paraguay and Argentina to Uruguay. Indeed, until the twentieth century both the Paraguay and Paraná rivers were the major arteries of transportation and communication for not only western Brazil but also Paraguay and parts of Argentina. The cattle economies in these regions relied on these highways for decades, as we will see later in the study.[7]

Seasonal rains and river overflow transform the Pantanal into a vast inland sea for six to eight months of the year, prompting the observation that distance means little to residents of the Pantanal because "measurement presupposes the existence of limits, and the Pantanal has no limits."[8] Pantanal topography gradually rises from west to east, away from major rivers and streams with elevations varying between 80 and 150 meters above sea level. The area is characterized by such geomorphological features as fresh- and saltwater lagoons, lagoon outlets and streams, and elevations or hillocks, all highly important in cattle raising.[9]

Baías, the sectors of the Pantanal most compelling to tourists because of their attraction to wildlife, are sweet-water lagoons resulting from land depressions that have been influenced and transformed by animal action. Concentrations of cattle, especially, help to deepen the depressions. Baías are most common in the lower western and central areas of the Pantanal and frequently teem with fish, small caiman (jacaré), and many species of birdlife. *Salinas* have formed in much the same manner, but as the name suggests, they contain high concentrations of mineral salts absorbed from the soil. They are most common in the higher elevations to the east, where evaporation leaves the accumulation of salt deposits along the shores. Should a salina dry up completely, the resulting natural salt lick is given the name *barreiro*. These are a natural salt source for wild and domesticated animals alike. More than one outside observer has been bemused by the scene of a steer with its muzzle thrust deep into a barreiro, calmly eating copious quantities of earth. Interspersed throughout the Pantanal are temporary and permanent watercourses that link baías. These water sources attract great quantities of wildlife, particularly waterbirds and *capivara* (capybara, *Hydrochoerus hydrochaeris*), the largest rodent in the world. Hillocks called cordilheiras, between 3 to 5 meters high, are scattered throughout the Pantanal, with or without woods or forest copses, and normally escape regular annual flooding. Here, ranch houses are built and animals retire when floodwaters cover lower elevations. Cordilheiras not occupied by humans harbor a rich variety of nocturnal wildlife seldom seen in the open countryside, including wildcats, foxes, armadillos, anteaters, and monkeys. As one travels to the east, the number of cordilheiras declines with the gradual rise in elevation.[10]

The Pantanal is subject to extensive annual flooding, but the duration and severity of the flooding differ from year to year and from subregion to subregion. This complexity becomes clear when the observer is presented with studies that divide it into several distinct zones. Though superficially similar, these differ in degree of flooding, soil types, elevation, and native grasses. Antonio Allem and José Valls have divided the Pantanal into nine geographic wetlands, based on principal watercourses and the pattern of historical occupation. The most im-

Cattle grazing by a Pantanal baía, November 2015. Photograph by the author.

portant cattle-raising zones during the period of this study were Nhecolândia (23,574 km²), Aquidauana (10,570 km²), Paiaguás (31,764 km²), Nabileque (15,363 km²), and Miranda (6,399 km²). In the years before and immediately following the Paraguayan War, Cáceres (40,376 km²) and Poconé (15,800 km²) were important ranching regions, but with time they were superseded by their southern neighbors. These divisions underline variability within the Pantanal and counter a common misconception that the region is homogeneous.[11]

Flooding occurs following heavy rains that fall in the tropical northern regions of the Paraguay River beginning in October or November, swelling streams and rivers and taking approximately 15 to 20 days to affect the Pantanal, though this varies depending on downstream location and the effects of local rains. Rainfall varies locally, increasing the amount of floodwater in the west and southwest as it combines with fluvial flooding to cover extensive areas. In the east, which neighbors the drier cerrado, pluvial flooding is almost exclusively local. Significant seasonal and regional variation means that areas in the north flood sooner and dry earlier than those in the extreme south, while the east generally suffers more from drought than elsewhere.

Normally the rains fall from late November to March or April, and widespread flooding extends from January/February to May/June. The driest months are from August to October.[12]

Despite this general description, weather is anything but predictable or constant in the Pantanal. The dry period can extend over several months and even years if air masses are delayed or diminished by global conditions. This has led to such extremes as years of excessive flooding followed by several more years of minimal seasonal flooding and subsequent drought. The effects these variations have had on human settlement in part explain why agriculture is still an insignificant activity in the Pantanal, while ranching has proven more successful, at least within limits.

As might be expected in a region of such size and climatic contrast, vegetation is diverse, determining the character and extent of human occupation. With each year's flooding, fertile clay soils are replenished by mineral deposits carried by river waters from the north, creating conditions for the growth of luxurious natural pastures. If a region is not affected too heavily by local rains and river flooding, it becomes a natural area for cattle raising, such as the western and central zones of Nhecolândia. If, however, the combination of fluvial and pluvial flooding is too intense, ranching becomes much riskier. This is the case in Nabileque and parts of Aquidauana and Miranda. In the elevated eastern regions, where the central Brazilian plateau slopes down into the Pantanal, soils are less fertile, have less humus, and are more permeable, permitting permanent human occupation. But these are also the areas subject to serious drought if local rainfall is too low, forcing the digging of wells when baías and streams dry up. Salinas are more common here, as often soils are salty from constant leaching.[13]

Flooding, Drought, and Ranching Opportunity

Ranchers have both struggled with and flourished in the environmental conditions of the Pantanal. The feeding cycle for cattle, as in every ranching regime where confinement is not employed, has been one of several months of fattening followed by a lean period. Normally in the tropics the dry season deprives animals of easily accessible forage and fattening occurs during the rainy season. In the Pantanal, however, it is the reverse. In some regions, cattle do indeed lose weight during the dry period, but in most they thrive, with only minimal stress occurring in August and September, when reduced moisture causes grasses to lose nutritional value. It is an anomaly of ranching in the Pantanal that the rainy season brings the most stress to cattle, as rising floodwaters restrict access to forage.[14]

Ranchers must know their ranch and local conditions well in order to transfer cattle at the beginning of the season from areas of regular flooding. Otherwise, they would have to devote a good deal of time and expense to swimming stranded animals from isolated dry areas to safer areas nearer ranch headquarters. Though floodwaters seldom rise high enough to drown cattle, frequently they have been extensive enough to force animals to stay put and eventually exhaust local forage. This is directly responsible for overgrazing of available grasses, reducing their ability to regenerate after the waters recede. In addition, the high waters can prevent animals from lying down to rest. Eventually they die of exhaustion, even where sufficient feed is available. The inevitable loss of animals, especially along the rivers, has evoked children's riddles, such as: "Who is it who travels eating his house; when the house sinks, he leaves?" The answer is the turkey vulture (*urubu*), ubiquitous to the region, which finds ready meals in the floating carcasses of dead steers.[15]

Though excessive floods are uncommon, precisely for this reason ranchers frequently have been caught unprepared. According to studies undertaken before the advent of global climate change, unusual flooding came in cycles, and extended over five years. Ranches located closest to major watercourses such as the Paraguay River, and subject to annual flooding over a large area of their territory, were often devastated. The disastrous flood of 1905, for example, claimed some 30,000 head in one region, while in another a single ranch reportedly lost 10,000 animals, leading to a temporary decline in ranching across the Pantanal. Another 160,000 head throughout the Pantanal succumbed to the 1910 flood, and during extensive inundations in 1920–1923 some municipalities lost up to 50 percent of their herds; one rancher alone suffered some 20,000 head drowned. Many small ranchers were ruined. More recently, flooding combined with drought in 1973–1974 took the lives of more than 80,000 head, double the anticipated losses. Similar floods were seen in the 1980s and 1990s, following La Niña cycles.[16]

Flooding determined the character of ranching in other ways. For a period of three to six months each year many ranches were incomunicado, at least before the arrival of radio. During this period it was common for ranchers to send their families to nearby cities, while engaging in building and fencing repairs. In addition, this tends to be breeding season, as animals congregate together, although there was no really distinct season until the arrival of wire fencing. At any rate, between May and November more calves are born, thus they are at most a few months old when waters begin to rise again. As might be expected, the season is a time of high calf mortality, not only due to floods but also because of insect infestation, a problem in some areas. Also, under severe flood conditions, miscarriages rise dramatically.[17]

Periodic drought was also influential in parts of the Pantanal, particularly the higher elevations, where groundwater levels are considerably deeper than in areas near permanent watercourses. Ranchers, depending on resources and their understanding of local conditions, often took advantage of prolonged dry spells. Property improvements were made, pasture expanded, size and productivity of herds increased, and communications extended, prompting the regional saying, "Dry year, rich year." Rancher Renato Ribeiro argued that the droughts of the late 1930s helped to develop the Pantanal since the lack of water forced ranchers to dig wells, providing a regular system of water supply and ensuring that herds would congregate closer to human habitation and care. The same drought period permitted vehicle access, as suggested by the truck journey taken by veteran rancher José de Barros in 1937–1938. An extended drought between 1962 and 1973 created a boom period in the Pantanal cattle industry, coming at a time of widespread economic expansion throughout Brazil. Ranchers extended exploitable territory, outside investment flowed into the region, and a number of ranches were established in areas that in the past had been considered risky for year-round ranching. Beef sales to the voracious market of São Paulo soared.[18]

Drought cycles are less predictable than flood cycles, however, and often have worked in tandem with flooding to wreak devastation on unwary ranchers. While windmills and deep well excavation were introduced by a number of ranchers during the droughts of the 1930s, lack of capital for investment caused the failure of many smaller ranchers in the eastern Pantanal. In addition, prolonged droughts created conditions for ruin among other ranchers, as they expanded herds and pastures into marginal areas, leaving them vulnerable to extensive floods, which inevitably return. This was the case during the dry periods of 1937–1938 and 1962–1973, when expansion and increase in herd size led to greater than normal losses in subsequent years from drought and flood combined, while disrupting the local market because animals could not be rounded up. The 1973–1974 flood would have been even more devastating were it not for early warning from meteorologists.[19]

Vegetation and Pasturing

For ranching, the most important vegetation is what Orlando Valverde called "vegetation of transition," largely mixed forest and pasture, because it dominates in the Pantanal and because ranching was established almost exclusively in this type of vegetation. The bulk of open fields, water holes, and resting areas are located in areas of natural grassland, including less palatable scrub. Human occupation, above all neo-European, altered this landscape somewhat over the years, particularly by using fire to eliminate trees in favor of pasture grasses.[20]

All vegetation in the Pantanal has adapted to the environment by developing extensive and deep root systems in order to survive flooding or periodic drought. Determined by sandy soils of poor fertility, cerrado-like vegetation in the uplands is similar to that found in the cerrado of the central Brazilian plateau. Scrub forest, consisting of woody trees and bushes adapted to semidrought conditions, and tough, unpalatable grass species dominate the area. Cerrado is relatively extensive in Nhecolândia and Paiaguás, as well as in the most easterly area of Aquidauana, comprising yet another third of the entire Pantanal of Nhecolândia.[21]

Vegetation in the rest of the Pantanal is similar to that described above; however, two vegetation types were not fully discussed in Valverde's work. The importance of riverside fields was not emphasized by Valverde because most of his research was done in Nhecolândia, but in areas of clay soils and frequent river overflow, flood-resistant bushes and nutritious tender grasses tend to dominate. At the same time, the most southerly zones of the Pantanal are essentially extensions of the Chaco. Host to the carandá palm (*Copernicia alba*) and with soils of poor permeability, the region is subject to extensive fluvial and pluvial flooding, often making it impassable for vehicles and risky for ranching. The area contains a variety of grasses adapted to semipermanent flooding, as well as to burning, an annual activity with a long history in the carandá stands.[22]

Since soil and vegetation conditions vary from pantanal to pantanal, so do their grasses. To enumerate all the grasses found in the Pantanal would confuse our discussion, since only a few are regularly grazed by cattle. A short description of these will help to clarify once again the ecological diversity of the Pantanal, though it must be emphasized that unlike regions of planted pasture, no one grass dominated in the diet of cattle.

The most common grazing areas have always been by watercourses, after floodwaters recede, and near baías. Grasses found in these areas must be able to survive submersion for a minimum of two to three months. Native species naturally have the highest adaptability, leading to the predominance of two or three species in areas normally grazed by cattle. By far the most common of these grasses are the mimosos, such as *capim mimoso* (*Axonopus purpusii*). Not only are they able to withstand long periods under water and a significant length of time in semidrought conditions, but their growth is also stimulated by cattle trampling, thus ensuring survival and eventual dominance of any area regularly frequented by cattle. For ranching, mimosos are important because they offer high nutritional value to the ruminants, especially when the plants are young and tender after floodwaters have receded.[23]

By far the most frequent of the other grasses is *capim carona* (*Elionurus muticus*), introduced into the Americas during the colonial period. A hardy invader with a tight root system that dies down with flooding, it can return in low flood periods when soils dry out. Carona is often found in association with another introduced grass, *Andropogon selloanus* (*capim pluma-branca*, beard grass), which is most common in the higher areas, covering 10–20 percent of the surface of some ranches. In the late 1980s Allem and Valls speculated that carona's frequency in Nhecolândia was likely due to greater pressure on pasture than in the other wetlands (pantanais). Its ubiquity, coarseness, and low nutritional value for animals made it a plague to ranchers. In the past it was often used to stuff furniture, though today it is being replaced by cultivated brachiaria (*Brachiaria humidicola*).[24]

These grasses survive under various local conditions and, except for carona, are considered good natural forage by ranchers and scientists. Of nine identified species, six are recognized as native to the Pantanal. This is significant, highlighting the unique environment of the Pantanal and the abilities of its species to withstand encroachment, at least as long as human intervention is limited.[25]

The far eastern areas of the Pantanal are less a part of this ecosystem because they border the Cerrado and tend to exhibit grass species common to the cerrado habitat—woody, coarse, of low nutritional value, barely palatable to cattle. Carona is found here, too. It is primarily for this reason that the area was the last to be settled by ranchers and the first to experiment with planted pasture. The imports had limited success, however, as they were incapable of invading the Pantanal, being limited to soils heavily altered by human action, such as agricultural fields and roadsides. Thus the Pantanal as a whole has been overwhelmingly dominated by native pastures, estimated as late as 1974 at 74 percent of total territory, while planted pastures occupied a mere 2 percent.[26]

Nevertheless, where exotic pastures were deliberately introduced, they have had an impact on the local ecosystem. Most cultivated pasture was introduced into the Pantanal only since the 1970s. (The isolated experiments undertaken as early as the 1940s were generally short-lived.) This was in the drier eastern sector of the region and in occasional patches in the cordilheiras. Exotics were not new to Brazil, as they first entered the colony spontaneously from Africa through the slave trade and spread quite widely over succeeding years and decades. Cattle in other parts of Brazil, particularly in the Cerrado, were often raised on invasive *capim gordura* or *melado* (*Melinis minutiflora*, molasses grass), and the species became so widespread that it was considered native.[27] The Pantanal, however, had virtually no experience with this or other exotics until they were deliberately introduced. Into the 1980s the grasses chosen for cultivation were *colonião* (*Panicum maximum*, guinea grass), gordura, jaraguá (*Hyparrhenia rufa*), and in later years *braquiária* (*Brachiaria decumbens*, pará or signal grass).[28] Jaraguá and colonião were common throughout the center-west of Brazil, especially in São Paulo, where most cattle were (and are) fattened before processing in refrigerated slaughterhouses. These grasses were appreciated for their nutritive value, ease of care, rapid growth, and resistance to fire. Planting appears to have been determined by environmental factors. Extensive flooding reduces available pasture, and if this continues for a period of time, or if animal losses to floods are sufficiently high, the natural response is to seek a form of future security elsewhere. This was the case after the floods of 1974, when ranchers began to expand operations into the drier regions.[29]

Pasture was planted generally to fatten steers for market. Although traditional extensive ranching practices continued to predominate in the region, they have been modified by an increase in fattening as improved transportation permits, signaling a new era in cattle ranching for the eastern Pantanal. The ecological result, however, has been a gradual degradation of pasture in certain areas, especially as fire and diking are part of the planting system, and even more serious deforestation in the upper Pantanal regions, with all the attendant residual effects this brings to the rest of the region. One study has concluded that this has caused a form of "disclimax" in some areas of the Pantanal, where woody species have displaced native grasses.[30]

Part of this had to do with one of the most important aspects of ranching: the number of animals pastured in a given area. At first glance it may seem that the Pantanal's extent and the annual renewal of native grasses would ensure that ranchers would face few problems of over- or under-pasturing. This appears to be confirmed by an apparently stable ratio of animals to land area recorded throughout the period. Observers at the beginning of the twentieth century reported one head for between 2.2 and 3 hectares, a proportion that increased to 3.3 hectares per head in the early 1970s. These figures suggest three possible trends: as cattle populations have grown, the amount of land in use as pasture has increased at a slightly higher rate; the average number of head per hectare decreased, possibly due to a gradual decline in native pasture productivity; or, as ranching expands, it moves onto marginal land. It seems all have been at work, but most significantly these figures raise the question of native grass capacity in the Pantanal.[31]

Antonio Allem and José Valls have undertaken the most exhaustive recent study of this subject. Their conclusions are not necessarily definitive, as they admit, but they have accumulated some significant data. The authors argue that in present conditions the Pantanal cannot support more than one head on 2 or 3 hectares, although 2 hectares may be too much pressure, since weed growth has been increasing. But the Pantanal is not uniform in terrain, soils, grasses, and human exploitation; some pantanais were considered overgrazed while others were not, and certain areas within a pantanal have revealed greater evidence of pasture degradation than others. For example, in Poconé in 1974 unpalatable vegetation growth (weeds) was estimated at 15 percent per year, a rate that prompted predictions that some three million hectares would be affected by 1984, nearly 25 percent of total nonplanted pasture in the region. It appears these predictions were accurate. Reasons include recent diking of the region, which has permitted hardy cerrado vegetation to invade areas no longer affected by annual flooding; the excessive number of cattle to available pasture, especially as cattle tend to congregate in areas of the most palatable species; and unstable water levels, forcing animals to move nearer to ranch headquarters for a prolonged period of time.[32]

The regular concentration of cattle in a specific area always creates problems for local pasture survival. This was particularly the case around the baías and ranch headquarters. Near ranch buildings, many years of cattle trampling and excrement accumulation have caused the soil to become impermeable to water, as the undersoil gradually hardens, and the ground level to rise several centimeters. This has restricted the penetration of floodwaters that in the past reached closer to living spaces, a benefit to habitation but not for local vegetation. When there is excessive concentration at the baías for an extended period, these natural watering holes soon become unusable, as the shores become so muddy and dangerous that other drinking sources must be found. On the other hand, in the higher and drier areas regular transit by cattle makes a swampy area passable, as the combined weight over time stabilizes the surface. Excessive compaction of the soil, however, deprives grasses adapted to local conditions of oxygen for survival and helps to introduce species exotic to the region; in the worst cases, some native grasses are eliminated. In part this explains the predominance of the mimosos and carona, though more study is required to determine the long-term ecological effect of mimosos as native invaders.[33]

SCRUB AND DROUGHT: THE CERRADO

The central Brazilian plateau, or Planalto, covers an extensive area of nearly two million square kilometers, 25 percent of Brazilian territory, including large sections of the states of Minas Gerais, Goiás, Bahia, Maranhão, and both Mato Grossos. Here, cerrado and campo limpo are found. Studies of the cerrado as a savanna ecosystem have concluded that it existed long before the entry of humans, not to mention the relatively recent appearance of Europeans. That conclusion is based on a number of factors, such as the presence of many similar herbaceous flora in savannas of Africa and South America, suggesting cerrado existed before the two continents drifted apart some 110 million years ago, and the advance of dry forest (*cerradão*) into the cerrado where no burning or pasturing has taken place. Campo limpo appears to be transitional vegetation between cerrado and cerradão. It seems, in fact, that campo limpo limits the advance of both cerrado and cerradão, thanks to richer soils that support its extensive natural grass cover.[34]

A 1980 census indicated that cerrado made up roughly 51 percent of the state of Mato Grosso do Sul (just over 175,000 km²), a percentage that has not changed significantly over the decades. Cerrado includes the regions north from the municipality of Campo Grande into the modern state of Mato Grosso, east from the Pantanal to Goiás and Minas Gerais, and southeast to the Paraná River. Made up of mostly sandy soils, cerrado in Mato Grosso do Sul is watered by a significant river system—the Paraná River—which forms the division between Mato Grosso do Sul and Minas Gerais, São Paulo, and Paraná. This river flows some 1,500 kilometers from the confluence of the Grande and Paranaíba rivers, where Mato Grosso do Sul, Minas Gerais, and São Paulo meet, to the Paraguay River at Corrientes, Argentina. The Paraná forms the border between Mato Grosso do Sul and the states of São Paulo and Paraná and the nation of Paraguay, a total of some 750 kilometers. Though water sources are less abundant than in the Pantanal, several tributaries of the Paraná in Mato Grosso do Sul run west to east, providing water for ranching. They served as essential communication routes between São Paulo and Cuiabá during the colonial explorations of the seventeenth and eighteenth centuries.[35]

Cerrado topography is uniformly flat and rolling, with the occasional range of hills, particularly in the west. Altitudes range from 200 meters to 700 meters above sea level, the lower elevations occurring in the Paraguay and Paraná valleys. The region is influenced by the same air masses that determine climate in the Pantanal, but extremes are moderated by altitude, particularly during the hot wet season, when temperatures rival those of the Pantanal. Annual mean precipitation is generally quite constant, between 1,200 and 1,500 millimeters, and November to March is the period of highest rainfall. The dry Atlantic air mass dominates from June through September, when average temperatures decline noticeably, ranging between 11 and 31 degrees C, depending on location. Extremes of below freezing are occasionally registered in June or July with the entrance of a polar air mass. The cold can be as penetrating as that of the windswept pampas, requiring the use of heavy blankets and sweaters. As in the Pantanal, the intermediate periods (September–October and April–June) are generally the most moderate. Again, during these periods much of cattle breeding, birthing, and handling occur.[36]

Compared with other sections of the Brazilian cerrado, such as parts of the states of Goiás and Minas Gerais, the cerrado of Mato Grosso receives a greater share of rainfall and retains more moisture in the soil; hence, it does not suffer as much from stress. However, in relation to other regions of the state, the cerrado still is a difficult area in which to raise cattle or engage in agriculture. The threat of drought and soil erosion has been a constant in the cattle industry from its inception. The entire state suffered considerably during the droughts and grasshopper plagues of the 1870s and late 1890s. River levels dropped to record lows, inhibiting transportation and access to drinking water, and thousands of cattle perished, though no attempt was made to determine exact numbers. Severe droughts also occurred in 1916 and 1924–1925, the latter period necessitating import of staples normally grown locally, such as rice, corn, beans, and manioc. Such climatic extremes often have been accompanied by unusual cold, as in 1873 and 1885, when severe frosts covered the region, and snow even fell near Rio Brilhante. The winter of 1893 saw the loss of cacao trees near Corumbá to frost, and the 1902 frosts extended as far north as Cuiabá. In 1917–1918, the notorious frosts of São Paulo that destroyed millions of coffee plants also affected Campo Grande, where temperatures fell to an unheard of −5 C.[37]

The combination of winter aridity and cold has a disastrous effect, as grasses and animals die off. This can be exacerbated by relatively low levels of rainfall during the normally plentiful summer. Ranchers in North America have long emphasized this risk. There is no close comparison between the plains and prairies of North America and the cerrado of Mato Grosso, for the latter region is located in the tropics, with different soils, water retention levels, and vegetation. But it is important to underline how much a ranching industry is dependent on climatic regularity, especially guaranteed water sources. Where natural conditions in the Pantanal (and to a certain extent in campo limpo) preclude the need to worry about necessary elements in cattle survival, cattle raising is considerably easier and cheaper than elsewhere. Both the Pantanal and campo limpo were blessed with extensive natural pastures year round, nutritious to cattle and well adapted to local ecological conditions. This was not the case in the cerrado. Low soil fertility and drought combined to produce generally poor-quality grasses, which regularly limited the growth of cattle during the winter season. Unpalatable invasive species like *barba de bode* (*Aristida pallens*) and carona were common and spread widely in the cerrado, while nutritious native grasses were found only in the humid valleys of rivers and streams. By contrast, campo limpo hosted several native grasses of high nutritional value. With time this changed, as human intervention altered the landscape of campo limpo in profound ways. In fact, the necessity for widespread burning and the introduction of exotics in the cerrado had its influence in campo limpo as well, to the detriment of the local ecosystem.[38]

Soils

Soils of the cerrado are normally quite deep but sandy and of low fertility, and frequently excessively high in aluminum, sometimes causing aluminum toxicity in plants and animals. William Sanford and Elizabeth Wangari argued that despite relatively high annual rainfall, the cerrado hosts mostly coarse grass vegetation of poor nutritional value due to low soil fertility and seasonal drought, at times entering drought stress. Alkali levels are high as well, and the soil porosity determines a low level of water retention, conditions ideal for natural and human-encouraged erosion. Also, many areas of the Mato Grosso do Sul cerrado are covered with termite mounds, especially in the region between Campo Grande and Três Lagoas. The mounds sometimes reach heights of one and a half meters. Their presence indicates soil degradation, usually caused by overgrazing. In the past, most observers considered the cerrado fit only for extensive ranching, since to be economical agriculture requires too much expense in the form of corrective fertilization. Nonetheless, soybean agriculture began to enter the region in the 1970s, and evidence of ecological degradation has grown significantly since.[39]

Vegetation

Tree species in the cerrado are generally deciduous or semideciduous, with trunks often contorted and covered with a thick, cork-like bark. The bark appears to act as protection against fire, a natural phenomenon in the cerrado aggravated by human action over several thousand years, in particular the last two centuries. Deciduous scrub trees have deep root systems, capable of searching out scarce water in the dry season at depths of as much as twenty meters. One tree species useful in ranching is the angico (*Piptadenia* sp.), the bark of which has been used for tanning leather.[40]

By far the most noticeable vegetation in the cerrado, however, is grass. Grass species are numerous, but those of greatest utility as cattle forage are few and subject to invader stress. The most common palatable native species are a mimoso (*Heteropogon villosus*), though mimosos have declined considerably over the past few decades from excessive exposure to fire, insect pests, and disease, a process noticed as early as

Cerrado vegetation, November 2015. Photograph by the author.

the 1910s. The least palatable grasses were those that predominated, including carona and especially the ubiquitous barba de bode. Barba de bode is an incessant invader despised by ranchers and rejected by cattle. Untouched by man or animal, these grasses have been known to grow to heights of two meters.[41]

With such a small selection of palatable grasses, it is not surprising that the cerrado was the first area of Mato Grosso do Sul where ranchers experimented with planted pasture. The 1980 census reveals that 50 percent of all planted pasture in the state was found in the cerrado. Grasses chosen followed period trends; jaraguá and gordura were planted up to the 1970s, when brachiarias took over, dominating since. If managed properly, these grasses are important in increasing weight gain and fertility in cattle. Their expanded use over the years has meant the gradual disappearance of some native species, either due directly to planting or to conditions created by planting that led to the invasion of more hardy, but less palatable, species. This has been the case especially where fire has been employed on a regular and extensive scale.[42]

Pasture planting in the Mato Grosso cerrado became significant only with the entry of the foreign ranching concerns during and after World War I. Contemporary wisdom held that exotics would maintain pasture for longer and that they would permit the pasturing of cattle for fattening. Preferred species were jaraguá and gordura, although alfalfa and rhodes grass were also tested. At the turn of the century gordura was considered a wonder grass, supporting twice as many head as native grasses and surviving up to twenty years before replanting. It did not support burning, however, and since it was more susceptible to frost than other grasses, it was planted in the more elevated areas, while jaraguá was planted in valley bottoms. Reports during World War I noted that gordura produced a "nicer looking" fat on cattle (probably whiter in color), important in the export market of the day, particularly Britain. Jaraguá was resistant to fire, drought, and frost, could produce up to three times a year, and was expected to last at least four years before requiring replanting. Further studies have noted that both gordura and jaraguá are incessant invaders under the right conditions, and gordura will last up to ten years before it is overcome by competition from other plants. Both these species are higher in calcium and phosphorus than other exotics, which was an important reason for their selection. Studies done in the 1890s and in 1913 revealed that of the two, gordura had a higher nutritive value. Surveys in the 1950s found that planted pasture was superior to natural pastures in carrying capacity; in the cerrado, artificial pastures reportedly could carry up to one head to 0.6 hectare, while in areas of native grasses the ratio was considerably less, at one head to 2.4 hectares.[43]

The first rule of planting is to select and fence off an area in order to control pasturing and prevent the invasion of undesirable grass species. However, fencing and planting pasture, without constant attention, damaged the savanna environment and ultimately the rancher. This happened frequently in Mato Grosso, with its tradition of allowing cattle to fend for themselves. The result could be disastrous for future productivity. Andrade noticed as early as the 1930s that in some areas of the cerrado, pastures had declined significantly where there was fencing. He attributed this to a lack of attention on the part of ranchers. Primavesi explains that uncontrolled trampling, combined with the limited aboveground biomass of the planted species, causes the soil to lose its fertility as it is exposed to the elements and to the incursions of weeds. In addition, monoculture leads not only to soil degradation but also to an increase in grass and animal parasites, with subsequent productivity loss. Thanks to low stocking rates, the impact in Mato Grosso was minimal until the 1970s, but it seems to have accelerated recently.[44]

Finally, the presence of legumes in the cerrado should not be ignored, for here they are more important in cattle diet than in the Pantanal. Under the conditions of a long dry period such as experienced in the cerrado, cattle prefer to eat grasses during the wet season, turning to legumes in the dry season, and native legumes are quite abundant in the region. Until recently, legumes were ignored by ranchers (though not necessarily by cattle). Today recognition of the importance of these native legumes has led to some tentative steps toward consortium planting.[45]

A WORLD BETWEEN: CAMPO LIMPO

Though campo limpo is the smallest geographical region in this study, it has had a powerful influence on the development of the state's cattle sector. A more fertile extension of the cerrado, campo limpo is primarily made up of natural pasture and scattered woodlands, the latter particularly in the river valleys and hilltops. Because it is so attractive to ranching, its historical name, Vacaria (cattle country), has endured. The area covers roughly 10 percent of the state, about 35,000 square kilometers. Campo Grande and Ponta Porã are its northern and southern limits, respectively. In the west, it is bordered by the Serra de Maracajú, and in the east it extends to the Paraná River flatlands, today occupied by soybean farms, forest, and cerrado.[46]

The region is crossed by several streams and rivers, flowing west to east from the Serra de Maracajú to the Paraná River, which have guaranteed regular access to water for ranching and agriculture. The most southerly section of this region along the Paraguay-Paraná border historically was covered in semitropical forest, the most westerly extension of the Mata Atlântica (Atlantic Forest), which included abundant natural *erva mate* stands (Paraguayan tea, *Ilex paraguariensis*). This changed considerably from the 1970s, as forest and erva have been devastated and farming and ranching have taken over. Today, these areas are considered more cerrado than anything else.[47]

The Vacaria is affected in much the same way by air masses as the Cerrado and Pantanal. There are some important differences, however. The 1970s Brazilian government *Radambrasil* study observed campo limpo as free of seasonal drought. Annual precipitation is between 1,500 and 1,700 millimeters, distributed throughout the year; thus, there are fewer instances of drought, and seasons are slightly different from those of the Cerrado and Pantanal. In the dry season temperatures range between 12 and 28 degrees C. Occasionally, the polar front brings subfreezing temperatures to higher locations like Ponta Porã. In the warmer rainy season periodic squalls cause short-term local flooding, an extreme nuisance in the countryside, where dirt roads are made impassable for days at a time and become as treacherous as driving on snow or ice, especially in the higher elevations. Temperatures fluctuate between 18 and 35 degrees C, though as elsewhere in Mato Grosso do Sul extremes of 40 C have been registered.[48]

As in the other regions, the prominent vegetation in campo limpo is grass. Considered steppe grassland, the region is interspersed with shrubs and trees, comparable to cerrado in many respects. Principal nongrass vegetation is virtually the same as in the Cerrado. It is unclear whether this is the result of cerrado encroachment on campo limpo or the disappearance of cerrado vegetation within the campo limpo area due to the action of fire. Historically, the lush southern forests, especially along river courses, attracted considerable attention for extraction of valuable hardwoods and erva, but today predatory logging has stripped most of the region well into Paraguay, leaving room for the expansion of ranching and more recently sugarcane as biofuel.[49]

As in the Cerrado, the most common native grass species is the mimoso. Indeed, in the past the Vacaria also was known as Campos Mimosos, but mimosos have suffered over the years, victims of fire and drought. Edgar Kuhlmann noted as early as 1954 that most pastures had been taken over by *capim branco felpudo* (*Andropogon* sp.), which is fire resistant (E. V. Komarek called the andropogons "fire grasses"). Barba de bode is also present, particularly on the most degraded pastures.[50] To the extent that campo limpo has been an area of small ranches, with properties seldom exceeding two square leagues,[51] it is reasonable to assume that the degree of degradation by fire and overgrazing has been significant, though the introduction of planted pasture has tended to offset the damage or at least delay it. As in the Cerrado, exotics like jaraguá, colonião, gordura, elephant grass, and brachiaria have been planted in extensive areas, particularly since the 1970s.[52]

One of the distinctive features of the south of Mato Grosso is a broad diversity of geography and ecosystems, a great deal of which are favorable to raising cattle. Yet, conditions vary between subregions, such that an understanding of local ecology has been essential for any productive economic endeavor, though this has not always guaranteed success. Experience with ranching in the Cerrado, for example, provided limited ecological knowledge for raising cattle in the Pantanal. Nonetheless, the natural conditions were favorable for a rudimentary cattle industry. As more attention was paid to local conditions and the periodic introduction of exotics developed, cattle numbers increased and a nascent ranching sector became crucial to the state economy.

TWO

Establishing Roots

THE RANCHING ECONOMY TO 1914

The history of ranching in Mato Grosso after Brazilian independence is the economic history of the region overall. Certainly there was other activity, often considered more important than cattle, but it was cattle that sustained the region from at least the 1820s. Several obstacles existed, above all the vast distances between the Brazilian coast and the ranching regions and even between habitations within the province. And there was the issue of jurisdiction, since both Brazil and Paraguay claimed the same territories. This was a major stimulus for the devastating Paraguayan War (1864–1870), which almost eliminated Paraguay as a nation and severely disrupted life in Mato Grosso. It was only after the war that the region truly became integrated in the Brazilian empire.[1]

PRE-1870 RANCHING

Livestock entered Mato Grosso–Paraguay with the first explorers. The conquistadores brought their horses upriver upon founding Asunción in 1537, and as more Spaniards arrived at the settlement, livestock increased. Over time, expanding settlements, forts, and missions north of Asunción also imported cattle overland, often becoming easy prey to the raids of native nations. On the Portuguese side, *bandeiras* and later *monções* (gold-seeking and slave-raiding expeditions) tried with difficulty to carry domestic animals with them, though the rigors of travel usually precluded large livestock. Nonetheless, the region soon hosted feral cattle, horses, and pigs, which aboriginal peoples either hunted or eventually captured and raised for their own use and to trade.

Missionaries also taught their native charges how to raise cattle. Until their expulsion in 1767, Jesuits were especially active in Paraguay, in the Chiquitos and Moxos regions of present-day Bolivia, and in the short-lived mission of Itatins or Xerez located in the Vacaria of Mato Grosso do Sul. From the mid-1600s, cattle and horses were raised successfully in these missions, and they were often traded to the Portuguese gold miners and settlers of Cuiabá in the 1740s, before the Treaty of Madrid (1750) consolidated frontiers and suppressed trade between the two colonies. The missions also were raided by native peoples, especially the Guaikuru, known as the "Indian cavaliers" thanks to their equestrian skills on horses lost by or stolen from the Europeans. With the eviction of the Jesuits, the missions foundered, leaving many feral cattle to roam the region before permanent neo-European settlement developed in the early 1800s.[2]

While settlement came early in the Spanish domains, and with it ranching, the Portuguese in the region largely were slave raiders and gold prospectors, hardly activities conducive to pastoralism. The discovery of gold and the founding of Cuiabá in 1718 set the stage for the eventual sporadic entry of cattle into Mato Grosso. As the region grew demographically, the need for local food production to supply the mines stimulated the establishment of several ranches near Cuiabá.

The first record of cattle in Cuiabá was in 1727, when four animals made the journey overland from Goiás. Regular influx of cattle into Cuiabá occurred from the 1730s, when the Cuiabá-Goiás road was opened and protected from aboriginal attack. Large land grants called *sesmarias*, typically of 13,068 hectares each, were granted by the crown to selected settlers and were easily expanded, with or without official permission, into a series of properties. It was expected they would be populated by cattle and planted with food crops. This was the case for the supply post of Camapuã southeast of Cuiabá, established in 1726, which raised cattle, along with foodstuffs like corn, beans, sugar, and manioc. However, it took much longer to build up ranches elsewhere.[3]

By the 1750s ranches were established along the Cuiabá River near the mining center. Others were created as a direct result of the 1777 Treaty of San Ildefonso, when the Portuguese erected military garrisons along the Paraguay River. At Albuquerque, near the future Corumbá, the intention was to eliminate the reliance of the troops on cattle bartered with the Guaikuru, who raided ranches in Paraguay in order to trade with the Brazilians. Ranching expanded gradually along the rivers during the last decades of the 1700s, and by the early nineteenth century several landholdings existed, most in the northern Pantanal. Some holdings were quite large. It was reported in the 1820s that the ranch called Jacobina, located near São Luis de Cáceres on the Paraguay River, was made up of 16 sesmarias totaling 240 square leagues (more than one million hectares!), pasturing in excess of 60,000 head, and growing commercial crops such as corn, sugar, beans, manioc, and even cereals; it considered itself completely independent of local and imperial government. According to Virgílio Corrêa Filho, owner João Pereira Leite "dispensed with government assistance when he didn't reject it."[4] By 1828, in all of Mato Grosso there were estimated to be nearly 175,000 head, 160,000 of them in Cuiabá district, which encompassed the Pantanal, north and south (including Jacobina). Of 363 sesmarias, 92 were dedicated to raising cattle. Southern Mato Grosso between Camapuã and Amambai also was recognized as an ideal region for raising cattle, though no one was sure of the number of private ranches, especially considering the custom of squatting on public land.[5]

The Portuguese crown itself was not inactive, establishing several ranches (fazendas reais) in Mato Grosso during the latter years of the eighteenth century and first decade of the nineteenth.[6] These included three in the north of the Pantanal and three more in the south, as well as plans for others in the Paraguay border region. The intention appears to have been to establish official Portuguese sovereignty over border regions still considered in dispute, despite earlier treaties. Caiçara, near present-day Cáceres, was acquired in 1779 and by 1784 allegedly had some 3,600 head. By 1822, the year of Brazilian independence, it was said to hold 12,000 cattle, though in 1845 this had diminished drastically, to 2,000, plus 300 horses for army use. Miranda, a fort established in 1778, initially called Mondego, became a national ranch called Betione around 1810, said to have "many cattle." A Brazilian deserter to Paraguay reported in 1825–1826 that 3,000 head were pastured at Miranda and another 1,000 at Albuquerque. He also revealed that some 4,000 head of cattle and horses had been sent from Miranda to Cuiabá in previous years. Luiz D'Alincourt, traveling from Santos to Cuiabá at the same time, was told the ranch held 9,500 head of cattle, not counting feral stock, and 750 horses. A provincial presidential report indicated that in 1872 Caiçara had only 100 head and was used almost exclusively to raise horses for the military, while nearby Casalvasco held 5,000 beeves and Betione 3,000.[7]

Effective settlement in southern Mato Grosso away from the Paraguay River did not begin until the 1830s–1840s. Frequent conflicts with

Spanish authorities and settlers in Paraguay and then with the Paraguayan military after independence caused the government in Rio de Janeiro to encourage occupation of the area. Claimed by both nations, the region was a focus of mutual distrust that contributed to the eruption of the Paraguayan War, yet occupation occurred only sporadically before the war, almost exclusively by Brazilians.

As settlers trickled in, political authorities in both newly independent countries sought to strengthen their territorial claims. The Paraguayans periodically sent military patrols in a show of jurisdiction as Asunción was concerned that the gradual influx of Brazilians threatened Paraguay's claim to the region. In some cases, Brazilian ranchers were harassed or even imprisoned on the grounds of occupying Paraguayan lands without permission. The Paraguayan government recognized that the nation lacked population and resources to settle the area; hence, population pressure and the unequal land tenure system in more settled regions of Brazil virtually guaranteed southern Mato Grosso a predominantly Portuguese-speaking settler population.[8]

Aside from the establishment of national ranches, the government was sluggish in securing the interior for the nation. Military agricultural colonies, of which the national ranch structure was a part, were set up at several locations around Brazil through the 1850s and 1860s. The intention was to establish colonies of soldiers and their families in remote regions with the incentive of free land grants after military service, with the hope that civilians would be attracted to the area, eventually opening it up for general settlement and definitive Brazilian occupation and claim. Authorities anticipated that the settlements might also draw foreign immigrants and thus offer a partial solution to a potential national agricultural labor crisis caused by the closing of the slave trade in 1850. This policy of Brazilian human and economic expansion into the nation's frontier regions lasted for several decades, but the initial attempts were desultory, since they were not given the infrastructural support they required.[9]

Several colonies were set up on or near the Brilhante, Dourados, and Miranda rivers between 1853 and 1861. Nioác showed the most promise, as it was part of an ambitious project of the Baron of Antonina (João da Silva Machado), senator for Paraná, to link the southern coast of Paraná with the Paraguay River and eventually with Cuiabá. The project was undercut in 1857 by the opening of the Paraguay River to Brazilian shipping via Asunción. Losing its initial economic rationale and subsequent governmental interest, Nioác stagnated. The same was true for the other colonies. Little fiscal investment was directed to the settlements, soldiers were only too glad to leave when they completed their service, and few civilians dared risk a venture so ignored by the government itself. While the colonies established an official Brazilian presence in the region, they did nothing to help develop it, with few animals, little agriculture, no medical facilities, and insufficient military personnel to protect the few settlers from aboriginal and Paraguayan harassment.[10]

The Paraguayans reacted to the colonies with exasperation. After hearing of the establishment of Dourados the year before, in 1862 the Paraguayan government sent a patrol to Miranda and Dourados that allegedly ordered the colonies dismantled. This was denied by the Paraguayan officer in command of the patrol, but the surprise Paraguayan invasion of December 1864 quickly saw the colonies overrun and razed. They would be reoccupied after the war, but they served no better purpose than they had before.[11]

Southern Mato Grosso before the war experienced more effective occupation by Brazilian civilians who were not tied to the colonies and were almost exclusively engaged in cattle ranching. For the most part, these settlers were small ranchers from the Triângulo Mineiro region of western Minas Gerais bordering Mato Grosso, who sought an escape from the growing human, economic, and environmental pressures on land in their home areas. They originally settled on unoccupied territory near Santana do Paranaíba, a cerrado similar to that in Minas. The first recorded ranch was that of the Garcia Leal family, established in 1829. Many more ranches sprang up in succeeding years. Alfredo D'Escragnolle Taunay, an officer in the Brazilian military expedition sent in 1865 to challenge the Paraguayan invasion, reported several large ranches in the area between Baús, on the Goiás border, and Coxim, including the Garcia Leal ranch. The family soon became the potentate of the region and encouraged other settlers from the Triângulo to seek their fortunes in Mato Grosso. The Lopes family, and later the Barbosas from Franca in São Paulo, among others, headed to Paranaíba in the 1830s, but learning of luxurious natural pastures near the Paraguayan border, they soon moved south in search of an easier area in which to settle. Encounters with feral stock, probably descendants of mission animals released after the expulsion of the Jesuits in the eighteenth century, sealed their resolve to remain in the Vacaria. These families became the prominent ranchers in the region before the war.[12]

The area continued to prosper, with marriage alliances celebrated among families, including with ranchers from Paraná expanding westward. As early as 1838, a reported 72 ranchers thrived in southern Mato Grosso. They were not rooted to the land, however, as wealth was measured only in the number of feral cattle that could be culled for personal use or local sale, while land titles were rare. As a result, these ranchers moved from pasture to pasture in a form of transhumance, taking advantage of climate and forage conditions in diverse locales. Land tenure was so precarious that in the 1840s or 1850s a ranch was bartered for an old shotgun and a tame horse. Virtually all the ranchers were squatters on public property, which left them vulnerable after the passage of the national land law of 1850.[13]

The law was one of the most important of the imperial period. Land had largely been squatted after independence, and in the interior, property holders with resources and force tended to dominate. The law legalized all landholdings after survey and registration, a circumstance that favored the stronger claimants, who had the resources to claim more than they had occupied. For example, in the 1840s the Baron of Antonina in Paraná sent an expedition into the area under the leadership of a member of the Lopes family. It appears the Baron intended to extend personal influence over the region through the concession of a road or railway from Curitiba to the Paraguay River and, after promulgation of the law, by buying up land from local ranchers. His "purchases," all dated 1849, were based on declarations of prior "continuous occupation," as required by the 1850 law. Local ranchers knew the region very well, but few had ever spent as many as five years in occupation of a specific locale, and registry was seldom considered until the Baron's land agents entered the area after 1850. This became a major judicial case in the twentieth century when descendants of the Baron claimed vast amounts of territory based on these dubious documents. They were unsuccessful, and it is ironic that the Baron himself never set foot in the region, dying in São Paulo in 1875. Meanwhile, legal or illegal possession of lands in southern Mato Grosso had little effect on many ranchers, as the authorities were incapable of regulating land use or even registering many holdings, considering their weak presence in the region. Besides, there was still the question of national jurisdiction, since Asunción claimed much of the area as Paraguayan.[14]

Most reports of the prewar period do not mention cattle populations, probably as much from ignorance of numbers among the ranchers as from neglect on the part of the chroniclers. Despite sporadic branding, the nature of the stock precluded the gathering of statistics with any accuracy. Residents had traditionally relied on common law and local cooperation to round up (or "hunt") semiferal animals, a system that worked only as long as there were no outside claims. However, estimates by Paraguayan occupying troops and others in the region offer some idea of the extent of ranching. In 1858, an expedition from Jataí in Paraná to Miranda encountered a ranch on the Nioác River downriver from the military colony, allegedly with 10,000 head and 200 horses. Two other ranches were reported to have 2,000 and 5,000 head each. Yet another ranch at the time of the Paraguayan invasion was said to hold 10,000 cattle. Paraguayan troops reported that in January and February 1865 they had been able to round up nearly 3,000 head of cattle and 300 horses. Alfredo Taunay estimated that there were some 30,000 head in the hands of the Paraguayans in the Miranda region in mid-1866, rounded up from ranches along the army's land invasion routes, between Bela Vista on the Apa River and Miranda, a distance of roughly 200 kilometers. Undoubtedly they also included cattle from ranches farther afield, but the numbers do not take into account how many had already been sent to Asunción, either live or in the form of jerky and hides. These reports indicate a sizable cattle population in southern Mato Grosso before the war, though other parts of the province, including the Pantanal, also hosted many head.[15]

The decline of gold mining in the Cuiabá region in the mid-eighteenth century saw the establishment of rudimentary cattle ranches on the lush fields of Poconé and Cáceres in the northern Pantanal. By the 1830s, a number of adventurous budding ranchers turned to the south of the Pantanal to establish holdings. Among them were the future Baron of Villa Maria (1862), José Gomes da Silva, son-in-law of Pereira Leite at Jacobina and founder of the ranch Firme; as well as the Mascarenhas, Paes de Barros, Barros, and Rondon families. Other ranchers entered the region after strong nativist movements developed throughout Brazil following the sudden abdication of Emperor Pedro I in 1831. Riots in Cuiabá in 1834, known as the Rusga, left scores of Portuguese dead, and several of the perpetrators escaped punishment by moving south into the Pantanal. These included João and Generoso Alves Ribeiro, who established their families on the Taboco River in the eastern Pantanal. Linked by marriage, the bulk of these ranchers brought between 300 and 500 head apiece and within decades possessed herds of several thousands.[16] (See map 3, p. xiv.)

It appears many ranchers took advantage of the 1850 land law and legalized their squatted holdings, using the right of "pacific occupation" provided in the law. There were more than ten major ranchers in the region before the war. In 1864, the Paraguayan army was told there were some twenty ranches in the Pantanal between Miranda and the São Lourenço River, and it was estimated that farther south between 12,000 and 15,000 head could easily be rounded up for consumption. Many more were observed near Albuquerque. According to Paraguayan reports, throughout the two-year occupation the army had no difficulty supplying cattle to its forces, to local civilians, and to Asunción in the form of jerky. The Brazilian expeditionary force also found many cattle wherever they camped. Near Taboco, a ranch owned by José Alves de Arruda was estimated to hold 13,000 head, though the force was prevented from taking full advantage of this food source by a virulent attack of *mal das cadeiras* (surra), which decimated their horses and mules, making roundup impossible.[17]

The presence of so many ranches in the decades before the war reinforces the observation that southern Mato Grosso possessed favorable conditions for the raising of cattle. It also raises the likelihood that despite the region's remoteness, there must have been some sale to the rest of Brazil. In the 1820s the market throughout Mato Grosso was almost purely local. Cuiabá, Corumbá, and the military garrisons absorbed most production, but favorable environmental conditions led to a natural overproduction that kept prices low. Ranchers, though encamped on vast expanses of land and often claiming large herds, maintained little more than a subsistence existence that was spartan in contrast to its potential. With the exception of properties located near Cuiabá, ranchers were poor. Cattle drives to markets in the east began to be organized only in the mid-1850s.[18]

Until then, Mato Grosso was at a disadvantage. The most promising markets were the expanding city of Rio de Janeiro and the growing coffee plantations in the provinces of Rio de Janeiro and São Paulo, but competition came from ranchers closer to consumers, in Minas Gerais, Goiás, and São Paulo. Clearly, it was cheaper to trail steers from these areas than from Mato Grosso. Also, the export of salted beef via the Paraguay River, allowed by treaty with Paraguay only in 1857, was closed off in 1864 when the Paraguayan War isolated the province once again.

Yet a boom in demand in Rio de Janeiro and elsewhere in the 1850s had an important impact on ranching in Mato Grosso. A report from 1858 revealed that Mato Grosso contributed some 17,000 head annually to the Rio market, sent through Minas Gerais, though this had been diminishing at the time of publication. Prices had risen significantly in the country, with quality steers in Rio sometimes reaching 100 milreis. Mato Grosso ranchers apparently sold their steers off the ranch at 13 to 15 milreis to Mineiro drovers, who herded the animals to winter pasture in Minas. After weight gain, the animals were driven to Rio and sold to the slaughterhouses there, though often the uncontrolled market resulted in supply gluts, a situation exploited by the slaughterhouses in the capital. Adding drover expenses and costs of preslaughter pasturing, the purchase of corn in the dry season, slave acquisition and maintenance (which apparently increased after the African slave trade ceased in 1850), salt, and other necessities, it is understandable why the final price of a steer in Rio could be so high, though speculation by the slaughterhouses also was common. The resulting cost of beef in the city often was prohibitive to the consumer and a contentious issue for residents. There was concern over the future of the industry, as many Mineiro ranchers had switched to buying and selling, cows were slaughtered indiscriminately, and stocks declined. Nonetheless, the 1850s proved to be the start of a regular cattle industry in south-central Brazil, including Mato Grosso, when rising urban demand stimulated the sector there and helped to double or even triple steer prices of only a decade earlier.[19]

For a number of reasons Mato Grosso's participation in this boom was short-lived. Not least was the opening of the Paraguay River to Brazilian shipping. Now there was a route for shipment of salt beef and jerky to Rio de Janeiro that did not depend on the precarious transport of cattle on the hoof. Almost immediately ranchers, mostly those in the Pantanal, turned to producing jerky for sale to shippers in Cuiabá and Corumbá, who were pioneers in a business that would become highly important after the Paraguayan War. Another reason for the decline in cattle sales through Minas was a very specific ecological imperative that limited the number of horses in the Pantanal, mal das cadeiras, which coincidentally arrived in the 1850s. Little understood, the disease ravaged the Pantanal for decades, and the regular loss of horses elevated the cost of a steed to as much as 100 to 200 milreis apiece. Saddle steers, a makeshift substitute used for transport, were ill-suited for roundup, especially since cattle were semiferal and frequently could be "captured" only by the bullet. Heavy losses on the drives to Minas Gerais caused the industry, in the words of Joaquim Moutinho, president of Mato Grosso, to "disappear completely" by the 1860s. In addition, the quality of Mato Grosso beef was considered poor by all observers, even those in Cuiabá, where animals herded to market were left for days in corrals without food and water before slaughter. Similar conditions reduced the quality of salted beef as well.[20]

At the same time, periodic smuggling of cattle from Mato Grosso to Paraguay was common, particularly to the poorly supervised Chaco. In 1859, reports of cattle being driven across the Paraguay River south of the Paraguayan fort of Fuerte Borbón (Olimpo) spurred the Paraguayan government to send a gunboat to investigate. Corrals, chutes, canoes, implements, horses, a small jerky operation, and "a considerable number of cattle" were found, though the perpetrators escaped into the nearby Chaco and Pantanàl. It appeared the smugglers were Guaikuru and at least four nonnative Brazilians, identified by their dress. In view of the considerable investment in the operation, it appears local Brazilian entrepreneurs were involved. Indeed, a Brazilian deserter interrogated in Concepción accused the residents of the municipality of Miranda of organizing smuggling operations to the natives of the Chaco, supposedly with the consent of the president of Mato Grosso. The Brazilian Minister in Asunción quite naturally denied involvement at higher echelons. It is not clear how widespread such operations were, or for how long they had been under way, but it is not difficult to believe that such activity was relatively common, with or without the approval of the provincial president. Considering the expansion of the industry in Mato Grosso, it is reasonable to expect that any outlet, especially one as relatively simple and close by as the Chaco, would be exploited by local ranchers. On the other hand, reports of trade with Paraguay across disputed land borders are rare. This seems to indicate that the centuries-old dispute over jurisdiction in the area limited the amount of contact permitted between Paraguayans and Brazilians, probably because of actions by the Paraguayan government. The prompt Paraguayan official response to the smuggling operation reinforces this conclusion.[21]

Notwithstanding these deterrents to a prosperous ranching industry and President Moutinho's pessimistic assessment, cattle were found throughout Mato Grosso, a market had been established, and many ranchers were becoming relatively prosperous. By the 1850s the remote province was gradually being drawn into the national economic structure, impressing upon the national government the potential of this vast, little-known western region of the empire. Just as Mato Grosso's economic star appeared to be on the ascendant, however, war brought devastation to the south of the state, reduced herds, and scattered ranchers and other settlers.

WAR AND CATTLE

Traditional Brazilian historiography has maintained that the Paraguayan invasion completely destroyed the limited degree of prosperity of southern Mato Grosso, as human and animal populations were dispersed, captured, or consumed by the voracious appetite of a murderous Paraguayan war machine. Typical is the assessment that the Paraguayans took more than 300,000 head from the region through which they passed, leaving virtually no cattle behind. There is no question that local cattle were captured and slaughtered for troop and prisoner consumption or to make jerked beef for shipment to Asunción, but it appears that the number of cattle in the hands of the Paraguayans was limited and, even more important, that the invaders apparently conserved a resource essential for the support of a nation engaged in war. Joaquim Moutinho admitted as early as 1869, when the war was still raging in Paraguay, that many thousands of head had been preserved by the Paraguayans, who had not had time to drive or slaughter all before their withdrawal from Mato Grosso. He reported that the animals, most of them feral, carried the Paraguayan government brand "LP" (*la patria*), and thanks to the Paraguayans the two prewar Brazilian government ranches in the Miranda district, Betione and Poeira, were well-stocked with healthy cattle.[22]

Part of the reason the Paraguayan army was not able to round up more cattle in the region, despite having two years to complete such an operation, may have been the outbreak of mal das cadeiras at precisely the moment troops entered Mato Grosso. Several Paraguayan military reports in 1865 mentioned the devastating effects of the disease on their horses. Not only were troops immobilized by floodwaters (the invasion came in December–January 1864–1865, as Pantanal waters were beginning to rise), which impeded any further prosecution of operations to Cuiabá, but the high equine mortality also left them unable to support a land assault on the Mato Grosso capital. Incidences of the disease continued for a year or more, restricting the Paraguayans' ability to round up the semiferal cattle in the south of the province. Initially, they were forced to divert horses from the south of their country, a policy that soon became impossible as allied forces invaded Paraguay itself. As strategic considerations required the reinforcement of the southern army, in 1867 Paraguayan troops were withdrawn from Mato Grosso. The reduced number of horses then at their disposal explains in part why they did not slaughter or take more cattle with them.[23]

Brazilian troops faced the same problems, severely restricting their ability to respond to the Paraguayan invasion. Alfredo Taunay reported that while the expeditionary force sent from São Paulo was camped in Coxim in early 1866, it lost almost all its horses and mules to mal das cadeiras. The few remaining mounts were soon affected during the subsequent march from Coxim to the Taboco River through the Pantanal. Though many cattle were seen, the number of available horses was insufficient to round up more than a few head. This situation persisted into 1870, when the commander of the Brazilian force in Miranda, Caetano da Silva Albuquerque, reported that he could not follow orders to drive cattle from the Bela Vista area because the disease had decimated horses in the vicinity. He had to rely on drovers from Minas Gerais, who were bringing steers and mounts overland and selling the animals at "fabulous prices," though apparently too late for the needs of the force. His alternative was to try and obtain animals from the overall Brazilian command.[24]

Meanwhile, rancher reaction to the invasion was mixed. For the most part ranchers, natives, settlers, and some military fled. Some stayed, however, either because they could not escape or because they decided to put their fate into the hands of providence. Most of those who remained ended up as prisoners of war in Asunción, and some did not survive. The majority returned to their homes only in the early 1870s, but several managed to stay on their ranches throughout the war, particularly those in areas where Paraguayan patrols only passed through. One region was near the Taboco River in the eastern Pantanal. According to Brazilian commander André da Costa Leite d'Almeida, who was captured while trying to escape the invasion, the owner of Fazenda Taboco, Joaquim Alves Corrêa, had misdirected retreating Brazilian troops by falsely stating that Paraguayan troops had taken Coxim. This forced the soldiers to attempt their escape via Camapuã to Santana do Paranaíba, during which they were captured. Almeida accused Alves Corrêa of deliberately misleading them in order to avoid having his cattle and horses requisitioned by the military. It appears that the rancher told most of the truth after all, however, as Paraguayan troops passed through the ranch in mid-January 1865 and sacked part of Coxim at the end of the month. By land, this was as far north as the invaders advanced.[25]

On the other hand, Renato Ribeiro argues that Taunay's contention that the Paraguayans took 5,000 head from his ancestor's ranch was highly exaggerated. In fact, the Alves Corrêa family felt they had been badly treated by *Brazilian* forces, who camped on the ranch for three months in 1866, consuming many cattle in the process. Multiple appeals over the years to the Brazilian government for compensation fell on deaf ears. Regardless, it seems clear that the Paraguayans, no doubt against their ultimate intentions, did not eliminate southern Mato Grosso's cattle population, despite extensive exploitation of resources at hand. By war's end in 1870, there were still sufficient numbers available for the reconstruction of many ranches. This explains to a great extent why ranchers eventually were able to return and renew and expand the local cattle industry.[26]

THE MATO GROSSO ECONOMY, 1870–1914

After the war, southern Mato Grosso became a modest arena for outside investment and immigration. Private investment was directed into urban expansion (particularly in Corumbá and Cuiabá), the extraction of erva mate along the now definitive Mato Grosso–Paraguay boundary, a brief rubber boom in the Amazon, and cattle. Though the imperial government was criticized for not paying sufficient attention to its more remote territories, there were some efforts to incorporate Mato Grosso into the nation in the years after the hostilities. Brazilian government investment included a shipyard near Corumbá, military outposts, attempts to revive some prewar national military colonies, and public works, particularly river dredging and construction in Cuiabá. The Mato Grosso economy expanded gradually, primarily through the export of erva mate and cattle products from the south and of rubber from its remote northern jungle region; though, with the exception of a brief rubber boom at the beginning of the twentieth century, it was the two southern products that sustained the government and bureaucracy in Cuiabá.[27]

According to the national census, the Mato Grosso human population in 1872 was near 60,000, although the count did not include native nations, estimated at a further 24,000. Perhaps as many as half of the nonnative population lived in the south. Prewar estimates are little better than guesses, but if we accept the overall figure of 67,500 inhabitants in 1862, including native peoples, and a generally assumed 2 percent average annual growth rate, without the war, the total Mato Grosso population in 1872 would have been roughly 82,500. Depopulation caused by the invasion and a devastating smallpox epidemic indicates the province experienced little demographic growth between 1862 and 1872[28] (see table 2.1). Presidential reports for the decade of the 1870s reflect concern over the state's weak demographic and economic condition. The almost total lack of agriculture and industry was attributed to the small population, vast distances from other regions of Brazil, and poor transportation. Presidents proposed solutions through incentives for agricultural colonies, promotion of immigration, and land grants.

With limited finances the provincial government could do little on its own and deferred to national authorities for subsidies. A law promulgated in 1869 opening the port of Corumbá to free import and export of goods for two years was one step in that direction, leading to a modest expansion, but even as late as 1872 Rio de Janeiro had to provide emergency food relief. Only the decision to build a naval shipyard (*arsenal*) at Corumbá in 1873 and the withdrawal of Brazilian troops from Paraguay in 1876 gave Mato Grosso the jump start it required. In an attempt to stimulate opportunities in several remote frontiers, Rio began a program of transfer payments to provincial governments. Imperial investment through the naval and war ministries then trickled into the province. Mato Grosso presidential reports indicate that between 1873 and 1887 Cuiabá received a total of 15,000 milreis in transfer payments, an average over 1,000 milreis annually. Initially the region faced a problem of insufficient labor. The imperial government thus saw an opportunity in the abysmal postwar conditions in Paraguay to relieve the labor shortage in Mato Grosso. In 1874 it passed a law permitting the Brazilian consulate in Asunción to issue free one-way passages to Corumbá for Paraguayans. Response in Paraguay, still shattered by the war, was enthusiastic.[29]

Despite a temporary downturn in the Corumbá economy by mid-1878, Brazilian government investment served its own interests and the Mato Grosso economy well.[30] It had acted as a stimulus to business activity, not only around Corumbá but in Cuiabá as well. Import-export houses were established, mostly by entrepreneurs from Buenos Aires and Montevideo; investors showed interest in expanding cattle ranching; and a number of extractors, Brazilian and Paraguayan, had already exploited the erva lands along the Paraguayan border. The Rio government also made some attempts to stimulate the agricultural sector in remote regions, particularly along the international boundary. The provincial government was given permission to grant parcels of up to one square league (3,600 hectares) for colonization within the federally restricted 10-league (66-kilometer) strip along borders. National agricultural military colonies, which had functioned desultorily before the war, were reestablished with the familiar intention of attracting settlers, reinforcing sovereignty, and "pacifying" local native peoples.[31]

In 1879 the provincial president gave glowing descriptions of ideal natural conditions for farming—fertile soil, benign climate, plentiful water—and predicted great prosperity for the colonies. But despite good intentions, federal financial support was unreliable. Reports from colony directors throughout the 1870s and 1880s constantly pleaded for the provision of tools, seeds, and building materials, to little avail. They were not even useful as police and fiscal posts, as they rapidly lost their initial civilian populations and soldiers who had deserted or died were seldom replaced. History soon repeated itself, as the financially strapped imperial government found little incentive to support the colonies during the uncertain 1880s, and in 1896 the new republican regime withdrew all military personnel and support, leaving the colonies in civilian hands. Deprived of even minimal guaranteed financing, the colonies disintegrated shortly thereafter.[32]

The experience of the colonies indicates that federal attention to the development of Mato Grosso was short-lived. Though provincial government income rose between the early 1870s and 1889, it did not increase at the rates desired by the province's leaders. Provincial tax income from exports was 18,000 milreis (18 contos, US$8,800) in 1871, and by 1885 this had risen to 51 contos (roughly US$19,400). Property and import taxes and irregular transfer payments from Rio de Janeiro added some income, but the provincial treasury habitually operated at near-deficit levels. These figures do not indicate the entire state of the economy, however, for there was a greater proportional increase in the elaboration of cattle products and the exploitation of erva mate than in other sectors.[33]

Export statistics collected at Corumbá reveal considerable growth in cattle production between 1875 and 1885. In this ten-year period, exports of hides increased modestly, but dried beef exports skyrocketed by a factor of 10. In addition, 1883 figures reveal the export of more

TABLE 2.1. STATE AND SELECTED MUNICIPAL POPULATIONS

Location	1862	1872	1890	1900	1910	1920	1940	1950
Aquidauana[a]						9,826	20,949	21,258
Bela Vista[a]						9,735	13,775	16,436
Cáceres						11,316	17,603	19,262
Campo Grande[a]						21,360	49,629	57,033
Corumbá[a]	3,000	9,800	—	15,000		19,547	29,521	38,734
Cuiabá						33,678	54,394	56,204
Coxim						6,899	11,203	8,508
Dourados[a,b]						6,238	14,985	22,834
Entre Rios[c]						—	8,375	8,838
Maracajú[a]						—	5,160	5,799
Miranda[a,d]	3,700	11,300	—	91,000		6,819	10,622	7,419
Nioác[a]						7,907	4,757	6,742
Paranaíba[a]	2,800	5,000	—	35,000		10,143	14,105	22,482
Poconé						7,088	16,313	13,438
Ponta Porã[a]						19,280	32,996	19,997

Porto Murtinho[a]					3,586	7,185	8,436
Três Lagoas[a]					9,044	15,378	18,803
Mato Grosso	67,500	84,400	92,800	118,000	185,800 (est.)	246,612	432,265	522,044

Sources: Data from Joaquim Ferreira Moutinho, *Notícia sobre a província de Matto Grosso seguida d'um roteiro da viagem da sua capital á S. Paulo* (São Paulo: Typographia de Henrique Schroeder), 1869, 113; Mato Grosso, "Recenseamento feito no ano de 1872," Documentos avulsos lata 1872-F, APMT; Brazil, Ministério da Indústria, Viação e Obras Publicas, Directoria Geral de Estatística, *Synopse do recenseamento de 31 de dezembro de 1890* (Rio de Janeiro: Oficina da Estatística, 1898); Brazil, Instituto Brasileiro de Geografia e Estatística (IBGE). Séries estatísticas retrospectivas, vol. 1, *Reportório Estatístico do Brasil Separata do Anuário Estatístico do Brasil, Ano V, 1939–1940* (Rio de Janeiro: IBGE, 1986), 14; Brazil, Directoria Geral de Estatística, *Anuário estatística do Brasil, 1908–1912* (Rio de Janeiro: Typ. Nacional, 1912), 251; S. Cardoso Ayala and Feliciano Simon, *Album Graphico do Estado de Matto Grosso*, EEUU do Brasil (Corumbá/Hamburg: n.p., 1914), 414–420; Brazil, Ministério da Agricultura, Commercio e Obras Públicas, Directoria Geral de Estatística, *Recenseamento realizado em 1 de Setembro de 1920, vol. 4, parte 1 "População"* (Rio de Janeiro: Typ. da Estatística, 1926), 408–409; Brazil, IBGE, Recenseamento Geral do Brasil (1° de Setembro de 1940), Série Regional, Parte XXII, *Mato Grosso: Censo Demográfico, Censos Económicos* (Rio de Janeiro: Serviço Gráfico do IBGE, 1952), 51; Brazil, IBGE, Conselho Nacional de Estatística, Serviço Nacional de Recenseamento, Série Regional, Volume XXIX, *Estado de Mato Grosso: Censos Demográficos e Económicos* (Rio de Janeiro: IBGE, 1956), 64.

Note: All of these municipalities were primary cattle districts.

[a]Southern municipalities.

[b]Dourados was a district in Ponta Porã municipality in 1920, normally counted as part of the total. In this case, the totals are separated.

[c]Renamed Rio Brilhante in 1948.

[d]In 1900, the municipalities of Nioác and Campo Grande are included in the Miranda total.

TABLE 2.2. MATO GROSSO CATTLE-DERIVED EXPORTS, 1875–1910

Year	Hides (units)	Jerky (kg)	Live cattle	Beef bouillon (kg)
1875	32,100	8,000	—	—
1879–1880	25,000	80,700	—	—
1883	40,000	—	—	111,600
1884	41,400	—	—	75,000
1885	47,000	—	5,000	—
1890	52,800	—	—	96,500
1891	—	—	—	115,000
1895	53,200	—	—	157,400
1900	48,000	130,000	—	141,000
	13,000 (Descalvados)			
1905	59,000	133,000	30,000 (1906, Paranaíba)	29,400
	2,500 (Descalvados)			
1910	77,200	482,000	55,000	38,000

Sources: Data from presidential reports ("Mensagem do Presidente do Estado de Matto Grosso") found on microfilm in the Núcleo de Documentação e Informação Histórica Regional, Universidade Federal de Mato Grosso, Cuiabá (NDIHR); various "Colletorias: Impostos de importação e exportação," 1883–1915, Arquivo Público do Estado de Mato Grosso, Cuiabá (APMT); Delegacia Fiscal do Tesouro Nacional em Mato Grosso, Alfândega de Corumbá, Capatazia, Guias de Exportação, Importação, e Reexportação, 1884–1905, NDIHR.

than 111,600 kilos of beef bouillon (*caldo de carne*) (see table 2.2). Erva exports through the poorly controlled customs posts along the Paraguayan border also showed significant increase. They generated tax income that rose from 16.8 contos in 1885 to over 40 contos the next year, reflecting the growing importance of erva in the region.[34]

The declaration of the republic brought many institutional changes to Brazil, the most important for this study being the designation of greater fiscal, jurisdictional, and legal powers to the states. This new freedom and responsibility meant that government income was generated locally, while public lands came under the control of the state government. Access to land theoretically became easier, stimulating migration into the state, while land sales and rental became both a source of state income and chronic corruption. By 1890 the population had risen to 92,827, including perhaps 24,000 native peoples, and an estimated 118,025 persons inhabited the state by 1900.[35]

Between 1890 and 1900, state income catapulted from over 272 contos to 1,433 contos. Yet expenses almost always exceeded income by a small margin, leaving little leeway for economic planners. Mato Grosso's single most important export and tax earner, at least at the beginning, was unquestionably erva mate, which in addition to taxes provided the state a guaranteed annual income of over 50 contos through the lease by a private extractor of more than five million hectares along the Paraguayan border. Import taxes also brought in some income, although authorities were besieged by requests for exemptions on articles for the nascent cattle and sugar industries, and there was always the nagging problem of contraband. Smuggling, particularly of erva and cattle through Paraguay, deprived the state of income while draining funds into maintaining fiscal posts and border patrols.[36]

This was exacerbated in the 1890s and into the next century by political upheaval throughout the state. Armed conflict first flared up in 1891 as competing political forces jockeyed for power. Local potentates, largely ranchers, then squared off, arming hundreds of retainers and initiating a long period of intermittent civil disorder. In a typical response to real and perceived neglect from Cuiabá, the south was affected by an element of separatism. Many southerners believed that the politicians and bureaucrats in Cuiabá extracted a sizable income from the south and returned virtually nothing to the region, spending it on themselves and Cuiabá. The opposition often financed itself by trafficking in contraband with Paraguay. Mercenaries were then contracted in Paraguay, though many of these became bandits whose main intent was to take advantage of the upheaval to plunder the countryside, particularly of cattle, which were then sold in Paraguay. To make matters worse, fighting occasionally broke out among groups over land allotments illegally handed out by local land agents.[37]

Between 1899 and 1906, rebels and bandits crisscrossed Mato Grosso, disrupted commerce in the south, forced the exile of many participants, and ended up fighting pitched battles in various southern localities and for Cuiabá. Political fighting ceased with the death of the state president in 1906, giving victory to elements sympathetic to southern interests. The upheaval naturally restricted the collection of taxes by the state and in 1901 forced the Cuiabá government to contract loans totaling 400 contos, backed by erva exports, to pay for the military effort. This led the government to estimate in 1903 that the value of contraband with Paraguay was more than double that of legal commerce. Such statistics and the ongoing state of semianarchy in the region motivated an agreement with the federal government to establish a federal frontier police force in 1902 and a federally run customs post at Bela Vista in 1904. The post also would collect state export duties between Mato Grosso and Paraguay, simplifying Cuiabá's position in the politically charged climate of the day.[38]

Despite these upheavals, Mato Grosso grew and some sectors prospered. State income rose to a high of 5,200 contos in 1910, almost a fourfold increase from 1900, and was maintained at over 4,000 contos to 1915. Much of the increase was driven by the short-term rubber boom in the Amazon region, which in 1913 accounted for over half of all revenues. Erva and cattle products contributed between 250 and 300 contos each. Remaining income came from taxes on land, land sales and rental, and import duties.[39] Still, the budget was seldom balanced, as income consistently lagged behind expenses.

At the same time, populations increased significantly in various districts. Corumbá grew from 3,000 in 1862 to 9,800 in 1872 and more than 15,000 in 1900, while Paranaíba saw its population catapult from fewer than 3,000 in 1862 to 35,000 by 1900. In the entire south, the population rose from roughly 4,000 in 1872 to 12,000 in 1890 and over 90,000 by 1900. This demographic expansion helped to establish the framework for future prosperity.[40]

RANCHING REESTABLISHED: PROMISE AND REALITY

In the early 1860s, the future of cattle raising was anything but bright and assured in southern Mato Grosso, but war brought attention to the region and led to greater official interest than before. Though the attention was still minimal compared with that paid to other areas of the empire, the presence of large herds of cattle and vast "unopened" spaces attracted more ranchers and settlers from other regions of Brazil and abroad. Many of the troops decommissioned after the hostilities either stayed or returned home to encourage family and friends to risk setting up ranches in Mato Grosso. Many of the cattle were feral, but they were found in sufficient numbers to warrant the effort to round them up. Estimates in 1885 suggested that there were more than 800,000 head in the province. Projecting back, with limited import of animals at the time and a generally accepted annual growth rate of 20 percent, minus export and consumption (10 percent), probably as many as 200,000 head roamed Mato Grosso in 1870, though numbers varied from region to region.[41]

A small percentage of this total was in government hands. It is amazing that the prewar national ranches were still operating in the 1870s and 1880s; they were more of a headache than a benefit. Their combined cattle populations in 1879 were said to be more than 8,000 head. Because of mal das cadeiras, these ranches lacked horses, which hampered the taking of inventory (among other tasks), and I have found no reports on personnel or stock. The ranches were a drain on the treasury, and in 1872 the government put them up for sale, but by 1883 most were still unsold. At this point the archival trail becomes obscure; anecdotal evidence suggests the ranches simply reverted to public land, some of it no doubt occupied by squatters over the years, and the land was sold piecemeal by state governments after the demise of the empire in 1889.[42]

The numerous cattle roaming the province were a strong magnet to settlers, even if they were in remote areas. In 1870 there were no local businessmen with sufficient capital in Mato Grosso, but opportunities eventually attracted several merchants upriver from Buenos Aires and Montevideo. Many had well-placed connections in these centers and access to credit not available to their Matogrossense counterparts. There also was a need in postwar Argentina and Uruguay to find alternative regions to raise cattle and process jerky, because most suitable land was already occupied and sheep ranching had inflated property prices. Taxes were also an important factor; on the establishment of factories and the export of cattle products, taxes were very low or nonexistent.[43]

At the time, many cattle were found in the upper Paraguay River valley, where Paraguayan troops had not ventured. The major ranch was Jacobina. According to Francisco Antônio Pimenta Bueno, who investigated the area for the Ministry of Agriculture and later became provincial president, in 1879 Jacobina encompassed over one million hectares and pastured some 600,000 head of cattle. This number, although undoubtedly exaggerated, indicates that substantial herds in the

The Descalvados jerky factory, in the 1910s, viewed from the Paraguay River. Reprinted from S. Cardoso Ayala and Feliciano Simon, Album Graphico do Estado de Matto-Grosso, EEUU do Brazil *(Corumbá/Hamburg: n.p., 1914), 293.*

northern Pantanal, most semi- or fully feral, presented opportunities to entrepreneurs seeking new business ventures and encouraged a government eager for regional development.[44]

The most important of these early entrepreneurs came from Buenos Aires by way of Paraguay. Rafael del Sar, who initially financed himself by trading manufactured items in Corumbá for hides, began a jerky factory called Descalvados near Jacobina in 1873–1874, on land bought from Pereira Leite. In 1873, with capital of 5,000 ounces of gold and tax exemptions on the export of jerky for two years, del Sar began processing cattle from Jacobina, slaughtering as many as 5,000 head a year by the late 1870s. According to del Sar, he could have processed more except for mal das cadeiras, though he did import some horses from Paraguay. Del Sar also exported hides, horns, hair, tallow, and shoe leather to Buenos Aires, which he sold at a handsome 300-percent profit. By 1881, he apparently had enough, whether exhausted by personal problems or satiated with his profits; he sold the property to an immigrant from Montevideo, Jaime Cibils y Buxareo.[45]

Cibils represented a new era in the Mato Grosso cattle products industry, one that was more dynamic and attracted greater interest in the region from abroad. From a prosperous merchant family in the Uruguayan capital, he arrived in Mato Grosso in 1881 with sufficient capital to purchase not only the del Sar operation but also the entire ranch of Jacobina, which was put up for auction after the death of Pereira Leite in 1880. The acquisition was named Cambará-Descalvados, later known simply as Descalvados, and Cibils set up a *caldo de carne* (beef bouillon) factory, drawing on a more believable cattle population of 150,000 to 175,000 head.[46]

The Cibils operation was in direct competition with the pioneer Liebig plant in Uruguay, supplying hospitals and poorhouses in Europe and colonial armies in Asia. It showed surprising promise from the beginning, with exports of 111,600 kilos in 1883, and won prizes at major world exhibitions each year between 1882 and 1885. At its height, the quality of the bouillon was judged superior to that of Liebig. But the operation soon encountered problems, as feverish exports led to the rapid decline of animal stocks to an estimated 100,000 head by 1886. The still-precarious nature of the industry, which had to overcome long distances to market (requiring at least twice as much travel time as the Uruguayan competition), with increased risk of spoilage, prompted provincial presidents in 1886 and 1887 to urge the elimination of export taxes in return for the company's obligation to transport the province's mails. Cibils also offered to settle an indeterminate number of immigrants.[47]

The provincial president also suggested the elimination of duties on the export of beef jerky and the tiny leather industry, leaving only hides and tallow to continue to be taxed. The obvious intent was to stimulate what was considered to be the best potential source of future wealth for the province. In a prescient appeal, President Galdino Pimentel went so far as to argue that the bouillon industry "should be protected by the public powers in the interest of developing one of the branches of the [cattle] industry that everything indicates in the future will be the principal wealth and source of income for the province."[48] It appears such lobbying was successful, as subsequent records register the export of all cattle products tax-free, including even hides, at least until after 1895.[49]

Other factors, however, made Cibils's efforts difficult, if not impossible. The frequency of cattle rustling by border bandits was a serious concern, so much so that in the early years, when Cibils filed legal action against two brothers who herded cattle from Brazil to Bolivia claiming that the cattle were Bolivian government property, he himself ended up in jail in Bolivia. Only a promise to drop charges brought his release. The plague of rustling and banditry in the region continued for many decades, exacerbated by the fact that after the definitive establishment of boundaries between Mato Grosso and Bolivia in the early 1880s, 20 square leagues (87,000 hectares) of Descalvados ended up in Bolivia. The Bolivian government apparently made no effort to expel the company, but it was difficult to protect cattle and pursue rustlers since the company had to apply for permission each time it wished to send guards into its remote Bolivian properties.[50]

A more literal plague in these years was mal das cadeiras. The disease was endemic to the Pantanal and struck regularly through the 1880s, reducing horse stocks and forcing their regular import from the Río de la Plata, a draining expense for the company. One traveler reported that as early as 1880 horses at Descalvados were selling for 180 milreis apiece (US$80), twice their value elsewhere. Although the export of bouillon rose to 115,000 kilos in 1891, other issues, largely difficulties in competing with Uruguayan production and Cibils's indirect participation in an ill-fated coup attempt in 1892, plus a dispute over his actual legal holdings, contributed to his decision to sell the operation to a Belgian syndicate in 1895.[51]

The Belgians, a group with links to King Leopold II, brought an element of modern capitalism not previously seen in Mato Grosso. Descalvados and other investments, particularly rubber, exhibited many of the classic characteristics of an enclave economy: management contracted in Europe, imported machinery, workers largely hired from outside the region (Argentines, Uruguayans, and Paraguayans), foodstuffs produced on the ranch, and most exports sent in company ships to market in Europe. Mato Grosso received only some export tax income, 600 reis per kilo. The ranch and bouillon factory were initially capitalized at 3.5 million Belgian francs (US$675,000), which increased to 5 million francs (US$964,000) in 1899 when they passed into the hands of the Banque de l'Outre-Mer, the principal financial arm of Leopold's colonial ventures. Operations expanded with the purchase of the 500,000-hectare Metello ranch of São José. Its 40,000 head of cattle, added to the estimated 100,000 head at Descalvados, provided the company with a sizable stock of animals.[52]

The factory processed between 20,000 and 30,000 head a year, not on a par with Liebig's 140,000 head but sufficient to export more than 157,000 kilos of bouillon in 1895 and another 141,000 kilos (including dry meat extract) in 1900. A combination of factors led to dramatic declines in bouillon exports after that date. Smallpox in 1900, drought and flooding between 1902 and 1906, mal das cadeiras, endemic rustling from Bolivia, and excessive exploitation of the basic resource reduced animal stocks and production. Exports in 1905, for example, amounted to fewer than 30,000 kilos of bouillon and a decline to 2,500 hides from a high of 13,000 in 1900. The company was forced to seek further capital from the Belgian bank in 1905, but its management was primarily to blame for falling production. Massive overculling of stocks, including cows and heifers, revealed an absolute lack of consideration for future production, which caused a Brazilian commission to investigate the operation. The commission condemned Descalvados for virtually destroying a local resource, for acting arbitrarily in the running of the factory, and in meting out its own justice to rustlers and unmanageable workers. The report, released in 1907, led to a state homestead law that attempted to limit the size of latifundios, though this proved no easy task.[53]

The supervisor of Descalvados was singled out. Captain Franz Van Dionant, a close friend of the heir to the Belgian throne and a Congo veteran, was appointed Belgian consul at Descalvados when he arrived in 1899, and Asunción awarded him the post of Paraguayan consul. This latter position was considered useful for the hiring and control of laborers. Rio de Janeiro refused to recognize the positions located in so remote a region, though he was accepted as vice-consul in Corumbá in 1901. Apparently Van Dionant and his many arbitrary actions had so antagonized the local population and power brokers that when he left Mato Grosso in 1904, allegedly for reasons of health, he had eroded much of the political support for the company and reportedly had suffered threats on his life. These events, plus fading Belgian interest in all its operations by 1904–1905 and especially after the death of Leopold in 1909, set the stage for the sale of Descalvados and its other properties in 1911 for one million Belgian francs (US$192,000) to yet another foreign entrepreneur, American Percival Farquhar.[54]

In addition to Descalvados, Farquhar's operations (Brazil Land, Cattle and Packing Company) included 480,000 hectares of ranch land purchased from various private owners in southern Mato Grosso in anticipation of a railroad, then under construction (see map 3, p. xiv). The land purchase cost over 1,000 contos (US$320,000), high for the time, leading to a rapid rise in land and cattle values throughout the south. Brazil Land represented expansion and innovation in Brazilian ranching, as Farquhar contracted with an experienced Scot, Murdo Mackenzie, to run his ranches. Mackenzie, the former manager of the Matador Land and Cattle Company operations in Texas and ex-president of the US Cattlemen's Association, arrived in early 1912 and immediately set about organizing. The company imported breeding stock from the United States and began to string fences and plant pasture grasses, techniques extremely rare among local ranchers.[55]

The purchase came on the eve of financial trouble for Farquhar, caused by overextension of investments. The syndicate was forced to disinvest some of its many South American holdings after the global stock market retraction of 1913–1914, and Descalvados and the other ranching operations were placed under the control of receivers from 1914 until 1919. Descalvados produced mostly beef jerky in the years leading up to World War I, in quantities of about 480,000 kilos (1910). Mackenzie returned to the United States after the expiration of his contract in 1919, but he kept in contact through his interest in the Boston Cattle Company's operations near Brazil Land. His technical expertise and hiring of US cattlemen guaranteed success to Farquhar's cattle operations, which continued until they were nationalized in the 1940s.[56]

Meanwhile, ranching in other areas of the Pantanal grew slowly but steadily between 1870 and 1914, despite isolation and the threat of rebellion and banditry. Although presidential reports regularly bemoaned the "primitive" character of the cattle industry, settlers entered the region in increasing numbers. Some of the ranches registered under the state land law of 1892 were enormous. In the northern Pantanal, two fazendas owned by the Costa Marques family totaled 350,000 hectares and carried an estimated 50,000 head. Firme, which was to become the center of a ranching empire controlled by Joaquim Eugênio Gomes da Silva (Nheco) in the central Pantanal, today called Nhecolândia, covered 176,000 hectares. It was said to hold 15,000 head in 1907. Some other large ranches in the vicinity were Palmeiras (106,000 hectares), Rio Negro (119,000 hectares), and Taboco (345,000 hectares), the latter owned by José Alves Ribeiro, who wielded considerable influence in the eastern Pantanal.[57]

In the most southerly part of the Pantanal, along the lower Paraguay River north of the Apa River, Antônio Joaquim Malheiros claimed more than 800,000 hectares, estimated in 1887 to support 18,000 to 20,000 head. By 1898 his fazenda, Barranco Branco, held 70,000 animals, though subdivision in 1907, which involved the entry of Argentine capital, reduced the ranch to 435,000 hectares and its easily retrievable cattle population to 5,000. Argentine interests entered Mato Grosso in 1905 through an unidentified intermediary, who had received a concession of four million hectares. In 1908, this was transferred to Fomento Argentino Sul Americano, and two years later the company applied for the purchase of one million hectares, in return relinquishing the remaining concession. The land was within the 10-league (66 square kilometers) international frontier zone, requiring federal approval. It took

The town of Aquidauana in the 1910s. Reprinted from Cardoso Ayala and Simon, Album Graphico do Estado de Matto-Grosso, *405.*

over ten years to receive a final response, but in 1921 the application was rejected by Rio, even though state authorities had granted conditional title to some of the land. It appears no definitive title was ever awarded the company.[58]

As large as these ranches were, they were somewhat atypical in two respects: first, they were located in the Pantanal, where environmental conditions encouraged extensive territories, and second, in that region there was less pressure for land than in the drier campo limpo and cerrado. In these latter areas, which became the most active ranching zones, especially since they were closer to markets in São Paulo, there were two major waves of rancher settlement. As early as 1872 or 1873, ranchers from Minas Gerais and Goiás began to establish themselves in the future Campo Grande and Vacaria regions with cattle driven from their home areas. For the first years the market was Paraguay, which was profitable, although ranchers often found themselves dependent on itinerant merchants (*mascates*), who brought manufactured items and salt from Concepción in return for payment in live cattle. Immigration was sufficiently extensive that only thirteen years later, in 1886, a prospective settler reported that the best land had already been taken. He observed that ranches of between 35,000 and 45,000 hectares were not uncommon, and some ranchers possessed as many as 1,000 head. Though this observation of the scarcity of good land was probably related to accessibility to cattle markets, it illustrates the influx of a large number of settlers within a relatively short period. Growth led to the creation of a number of towns, including Campo Grande in 1872 and

Ponta Porã as a customs post in 1873. Aquidauana was established in 1892, followed by Entre Rios (Rio Brilhante) in 1900 and, as a direct product of the railroad, Três Lagoas in 1915.[59]

At the same time, the border region was being settled, mostly through erva collecting and smuggling to and from Paraguay. Cattle often fit into this latter category. With the granting of the Mate Larangeira erva concession in 1882, cattle became important not only for traffic to Paraguay but also for the company, which maintained two ranches for draft animals and meat. Santa Virginia totaled 106,000 hectares in 1896, and Santa Margarida covered 135,000 hectares. Together, they pastured 11,000 head in 1897 and 10,000 in 1904. Another ranch, Tres Barras, which in 1898 became the site of the company town of Porto Murtinho on the Paraguay River, held 8,000 head in 1895. Mate Larangeira, which became Larangeira Mendes e Cia. in 1903, with headquarters in Buenos Aires, supported a large workforce, more than 2,700 workers in 1897, and required animals for food and to pull more than 500 carts, each requiring six to eight oxen. In 1911, a total of 30,000 head pastured on the ranches, with another 4,500 on a ranch in Paraguay. The company also owned 4,500 horses, including 1,000 across the border, and it operated 80 mule carts. The number of oxen, no doubt part of the total cattle count, was said to approach 20,000, although this probably refers to all animals raised for traction.[60]

The border region also saw increased settlement of small ranches, sometimes complementing, sometimes in conflict with the Larangeira operation. Despite possessing the ranches, Larangeira periodically purchased cattle and foodstuffs from local settlers, mostly for consumption by its workers. A report in 1882 counted 28 small ranches in a 5-league (33 kilometer) radius of Colonia Dourados alone. These ranchers were soon dwarfed by immigration from Rio Grande do Sul after the bloody Federalist civil war in that state (1893–1895). Encouraged by compatriots who had settled in Mato Grosso in the 1880s, *Gaúcho* rancher families flowed into southern Mato Grosso between 1895 and 1910, making the torturous journey across Argentina and Paraguay in trains of up to 30 wagons pulled by teams of six to eight oxen. Some trains took as long as three months to reach Mato Grosso, and an estimated 10,000 Gaúchos entered Mato Grosso in this fifteen-year period.[61]

Conflict between the settler-ranchers and Mate Larangeira soon ensued. The settlers did not respect the Larangeira concession and often established ranches on or abutting erva lands. Pastures were burned to promote fresh forage for cattle, and fires regularly spread into erva stands. Larangeira lodged complaints to the government in Cuiabá as early as 1897 and continued to oppose the growing immigration over the following decade. As Larangeira was usually well connected, the separatist movement that sprang up in southern Mato Grosso later included a large number of Gaúcho supporters. The company was defended by state police forces, although it did maintain its own private security force, which often was used against its enemies and employee runaways. Opposition victory in the civil war of 1906 led to a project for division of the Larangeira concession into small plots once the concession expired in 1916, but by then the company was no longer a player, having managed to exhaust most of the state's erva. Mirroring the coffee regions of Rio de Janeiro some decades earlier, ranching was already occupying some of the abandoned erva regions.[62]

Ranches throughout the border region remained relatively small. Miguel Lisboa estimated in 1907 that most ranches in the Vacaria were no larger than 7,000 hectares and pastured between 1,000 and 2,000 head. In the cerrado region near Campo Grande, properties were larger, as much as 70,000 to 100,000 hectares, with up to 15,000 head on the largest ranches. Comparing these figures with the estimates from 1886, it is clear that the southern zone had become very attractive to those small ranchers forced off their land in other states like Minas Gerais and Rio Grande do Sul. To those with some minimal capital resources, southern Mato Grosso offered opportunities not found elsewhere.[63]

To the east, the municipality of Santana do Paranaíba on the Minas Gerais, São Paulo, and Goiás borders was certainly the least dynamic area of the state. Yet, thanks to its location, it served a vital function in the development of cattle raising throughout Mato Grosso. Its proximity made it an extension of the Minas Gerais ranching industry in that state's western hub of cattle raising, the Triângulo Mineiro. Cattle were imported through Paranaíba, new ideas and technologies in cattle raising arrived there first, and, above all, the most important exit port for cattle driven to winter pastures in Minas or São Paulo was in the municipality. In the latter years of the nineteenth century, animals passed through Paranaíba from as far away as the Vacaria and the Pantanal. The Paraná River crossing at Taboada, where export taxes and river passage tolls were collected, was an important money earner for the Mato Grosso treasury.

Yet the cattle industry here never achieved the prosperity it did in the more benign climate of the deep south of the state. Part of the reason was the location of the municipality in the heart of some of the most unforgiving country of central Brazil. This is cerrado, as in the Campo Grande area, but here frequent drought, poor soil conditions, and limited reliable water sources combined to make scratching a living off the land arduous at best. Part of the cause may have been the gross mistreatment of the land by the earliest ranchers, who had burned and overpastured the most fertile sectors. Whatever the reason, throughout the postwar decades Mato Grosso's presidents received a litany of complaints from the region lamenting the lack of official attention.

Another factor was the unequal tax burden. For example, a letter from the Paranaíba municipal council in 1881 reported that cattle drovers from Minas, discouraged by provincial import and export tariffs, had chosen to seek animals in Goiás, where taxes were half those of Mato Grosso. In response, many ranchers had already left the municipality, and indeed, many of the subsequent settlers in the Vacaria and Campo Grande regions came from Paranaíba. This seems to have been a constant during the last years of the empire.[64]

Despite these restrictions there were significant numbers of cattle in the region. A census taken in 1895 of ranches with herds of 600 head or more counted a total of more than 88,000 head on 76 ranches. The report stressed that the population was certainly higher, since many ranchers had fewer than 600 head and some animals on the larger ranches were probably missed. This seems reasonable, especially since the municipality was huge, extending over 75,000 square kilometers, and settlers were isolated not only from the rest of the state but from one another. Even in the 1920s access to Paranaíba from the rest of Mato Grosso was only along cattle trails, an issue that became an integral part of the debate over development in the state.[65]

CATTLE TRAILS

Before the Paraguayan War, the inconsistent market for Mato Grosso cattle in eastern Brazil drew on the trailing of animals to Rio de Janeiro through Minas, most crossing into the Triângulo Mineiro near Paranaíba, then on to Uberaba in Minas Gerais where they joined with cattle from Goiás. From there they were driven through Minas to Rio, although an irregular meat market in the national capital often caused delays that resulted in inadequate care of the animals, starvation, disease, and significant losses for the drovers. By the 1880s and 1890s, when Rio de Janeiro state began to produce more beef for its urban market, especially as coffee production declined and plantations were converted into pasture, the market shifted to São Paulo, fast becoming the economic dynamo of the nation. As that state converted virgin forest and plains into coffee plantations, and as the city of São Paulo began to grow, cattle were needed to feed the expanding population. The principal sources were Minas, Goiás, and Mato Grosso.[66]

Until the early twentieth century, there were only two ways to trail cattle out of Mato Grosso to Minas. One was from the Coxim area into Goiás and to Uberaba. More common was one that roughly followed the traditional route taken decades earlier: through the port of Taboada near Paranaíba and, depending on the river level, across the Paraná River either in rafts or by swimming, then on to Uberaba. The drive could extend over 850 kilometers if cattle came from as far as Nioác municipality. In Uberaba, the animals usually were mixed with others from Goiás, depending on their physical condition at the time, and then driven to the Rio Grande, which separates the states of Minas and São Paulo. Until the 1910s there was no bridge, and animals swam across the river before being driven to Franca or Barretos in São Paulo, where they grazed on a mixture of natural and artificial pasture until ready for slaughter. The distance from Uberaba to Barretos or Franca was another 100 kilometers.[67]

This traditional route was long and arduous, there was irregular access to water en route, and animals arrived in Minas and São Paulo so thin that they required several months of fattening before slaughter. Such conditions stimulated the proposal of a formal cattle trail from Campo Grande to the Paraná River some 450 kilometers downriver from Taboada. Proponents argued that this route would reduce the distance from the Vacaria by 600 kilometers and provide southern Mato Grosso a direct link to the expanding São Paulo market. In approving the initial petition for construction of the trail in 1896, the president of Mato Grosso observed that São Paulo was the natural market for Mato Grosso cattle. Cuiabá, however, offered no financial aid in its construction, in contrast to the government in São Paulo. It took a local entrepreneur, Manoel da Costa Lima (Major Cecílio), to undertake the challenge, and with capital earned from the sale of cattle in São Paulo he began construction in 1902. He finished the trail in 1904, established Porto XV de Novembro on the Paraná, and, shades of Fitzcarraldo, had a tug for moving cattle rafts across the river brought overland from Paraguay (taking almost two years). In 1906 the route was inaugurated.[68]

Meanwhile, Francisco Tibiriçá received a similar concession from the Cuiabá government in 1904, for a road from Campo Grande to Porto XV. Tibiriçá formed an association with Arthur Diederichsen, who took charge of the São Paulo section of the road, through a company named Companhia de Viação São Paulo-Mato Grosso, and in 1907 Tibiriçá bought Major Cecílio's concession. By 1908 the road from Campo Grande across the Paraná River and well into São Paulo was complete and functioning. The company became a power on both sides of the river, owned steam launches not only on the Paraná but also on smaller rivers, and held more than 160,000 hectares of land in São Paulo and 300,000 hectares in Mato Grosso. It also acquired cattle and was able to sell off much of its land to small holders, in São Paulo for coffee and in Mato Grosso for ranching.[69]

The road was a blessing to Mato Grosso, since it reduced the distance by as much as 400 kilometers, and thus the time traveled, as well as transportation costs in the export of cattle, while providing an opportunity for yet another customs post for the state. Statistics indicate that the trail and the port stimulated even greater export of southern Mato Grosso cattle directly to São Paulo and was a further incentive to the settlement of the region. In 1911, the first year that exports were recorded at Porto XV, more than 9,000 head left the state, a figure that increased manyfold over the following two decades. Despite the completion of the Estrada de Ferro Noroeste do Brasil railroad in 1914, the trail and port were utilized by Mato Grosso ranchers well into the 1960s, especially as the company provided ferry service between Mato Grosso and São Paulo, linking the railroad at the Paulista railhead of Jupiá. Yet part of the reason the trail lasted for so long was the inability of the railroad to provide sufficient service to the Mato Grosso cattle sector, a subject explored at length in chapter 3.[70]

The early years of ranching in Mato Grosso reveal the economic potential, but also the obstacles, faced by settlers in the region. Isolation, poor communications, inconsistent official interest, and threatened and real international and local conflict limited opportunities. The Paraguayan War ended up stimulating the occupation of Mato Grosso territory, though the Mato Grosso cattle industry between 1870 and 1914 was in some ways a reflection of economic development in the region, a foil to that development in others. Most settlers, ranchers, and government officials simply did their best to get by with whatever resources were available. For most, this meant raising cattle as cheaply as possible and seeking the best prices for their animals. This was not easy, and sometimes required reliance on contraband, but the markets were there and they did expand, if gradually. So did cattle populations and production. Mato Grosso was on the periphery of a larger national and ultimately international system of trade and finance. Why the region did not develop faster than it did had much to do with the competitive edge held by other regions that were closer to markets. Still, there was sufficient profit to be made to entice investors to the state, and as the costs of land and production in other areas increased, Mato Grosso became even more attractive for investment and settlement. This was its role: greater opportunities for access to cheap land than most other areas of the country, and proximity to expanding neighboring markets, specifically São Paulo. Admittedly, the region did not exactly develop on its own terms, but this would not have been realistic in Brazil before 1914.

Government, whether in Cuiabá or in Rio de Janeiro, had few resources and lacked incentive to invest seriously in regional development. There was significant federal intervention in the coffee economy, but there was no national policy for other products, including cattle. Far-flung frontier regions naturally did not share the same attention given those nearer the coast. Even when governments made efforts to stimulate the sector, they were usually ad hoc, too limited, and often hijacked by politics. Mato Grosso and its cattle industry were destined to remain a peripheral economy until either other cattle-producing regions reached their productive limits or some extraneous force intervened. World War I was that force, though as we will see, it proved to be somewhat less dynamic for Mato Grosso ranching than anticipated.

THREE

A Boom of Sorts

THE RANCHING ECONOMY, 1914–1950

In 1914, two events transformed the economy of Mato Grosso, especially in the cattle sector—the completion of the São Paulo–Corumbá railway and the outbreak of World War I. By that time, hides and jerky had begun to overtake erva mate as the most lucrative products and exports of the south. Soon the lure of cheap land, an already established cattle population, and promised access to growing markets caused many to realize that despite distant markets and small human populations, in the great expanses of the Brazilian interior, ranching was the most secure investment one could make.

This was reinforced by the explosive demand for meat and cattle products brought on by World War I, which stimulated the construction of refrigerated packing plants in São Paulo and beef jerky factories (*charqueadas*) in Mato Grosso. The war was a watershed in global economic history, and in Brazil the conflagration stimulated an economic expansion across the country, including in cattle.[1] For Mato Grosso, ranching came into its own, and over the following decades, southern Mato Grosso became the economic hub of the state, a position it held into the 1970s, though for many the path was frustratingly arduous, with considerable trials and tribulations along the way.

THE MATO GROSSO ECONOMY, 1914–1950

Apart from a rubber boom between 1900 and 1912, by 1914 Mato Grosso's economic success was based primarily on the production of erva mate for export to Argentina and cattle and cattle products destined for São Paulo and the Río de la Plata. Demand from the global conflict contributed to a significant rise in total state income by 1919. Land taxes and sales brought in over 1,500 contos, and all exports earned a total of 3,200 contos. Erva provided a steady 350 contos, but cattle contributed a substantial and growing 1,500 contos.[2]

The state economy underwent significant transformation. The gradual pace of prewar land acquisition escalated into a torrent of petitions, sales, and squatting. Land became the focus of much more speculation than before, and the population increased, thanks to the railroad, which became a key factor in attracting settlers into the region. The new municipality of Campo Grande, for example, grew from roughly 5,000 inhabitants in 1912 to about 22,000 by 1920, an increase of almost 500 percent, while the state population increased by more than 55,000. Corumbá, by contrast, saw a population increase for the same time period of only 4,500, reflecting its decline as the state's commercial center. Before 1914, Mato Grosso was gradually entering into the complexities of international capitalism, but the conjuncture of rail and world war thrust the state abruptly into the system, which thereafter profoundly affected the character of the regional economy.[3]

The postwar Brazilian economy experienced several fluctuations and readjustments, including a recession as export demand declined and imports became more expensive. A modest recovery began in the mid-1920s, but along with the rest of the world, Brazil suffered from the Wall Street crash of 1929. Exports and imports declined to one-third of their mid-1920s values, though increased exports of some goods, including animal products, in the mid-1930s helped to soften the downturn. After 1937, all Brazilian products, with the exception of cotton and refrigerated meat, experienced an export decline that did not improve until World War II.[4]

At the same time, the 1930s saw an increase in industrial production in Brazil, mostly through import substitution. Compared with the rest of the industrial economy, however, production of animal products (meats, hides, and wool) grew slowly, indicating the continued importance of local consumption of hides and meat products (fresh and jerky). As a principal supplier of the raw material, Mato Grosso played an important role in Brazilian attempts at market expansion. Despite the effects of the Great Depression, the human population of the state nearly doubled between 1920 and 1940, to more than 400,000 inhabitants; by 1950, the state's population had grown to 522,000. This growth reflected Mato Grosso's attraction in relation to other regions of the country, a trend that continued over subsequent decades.[5]

TABLE 3.1. MATO GROSSO EXPORTS, 1914–1935

Year	Live Cattle (head)	Hides (units)	Jerky (kg)	Erva (kg)	Rubber (kg)
1914	51,469	61,515	1,733,973	5,370,041	2,633,000[a]
1915	54,798	65,250	2,703,267	4,584,786	n/a
1916	51,134	108,918	3,755,310	n/a	n/a
1917	66,689	71,122	4,052,811	n/a	n/a
1918	62,545	166,925	4,144,736	n/a	n/a
1919	128,091	127,173	2,983,848	7,100204	4,606,000
1920	88,152	126,796	2,535,662	6,798,580	n/a
1921	67,752	71,588	2,175,126	7,954,650	2,824,000
1922	82,122	111,361	4,775,320	9,395,490	n/a
1923	110,134	145,015	5,969,067	11,374,150	n/a
1924	106,222	141,555	7,297,427	n/a	n/a
1925	152,561	160,870	7,366,388	n/a	n/a
1926	119,946	82,810	3,293,698	11,281,322	3,160,419
1927	176,621	102,978	3,752,549	10,290,249	3,640,917
1928	183,263	129,341	5,848,176	13,626,265	3,042,269
1929	129,732	76,719	3,704,582	16,386,924	2,437,336
1930	104,203	96,659	4,328,738	14,319,515	1,892,235
1931	142,525	90,489	3,794,443	12,236,168	1,131,146
1932	124,711	80,356	4,236,339	11,117,449	959,312
1933	183,878	116,438	4,443,615	8,541,154	930,517
1934	209,336	113,982	4,700,572	7,673,029	841,754
1935	227,470	124,386	4,171,819	9,569,480	806,994

Sources: Data from Virgílio Corrêa Filho, *A propósito do Boi pantaneiro*, Monografias Cuiabanas, vol. 6 (Rio de Janeiro: Pongetti e Cia., 1926), 55–58; Virgílio Alves Corrêa Filho, *História de Mato Grosso* (Rio de Janeiro: Ministério da Educação e Cultura, 1969), 691–692; Mato Grosso, *Mensagem apresentado à Assembléa Legislativa pelo Presidente de Mato Grosso, Dr. Annibal Toledo, 13 de maio de 1930* (Cuiabá: Imprensa Oficial, 1930), annexo; Mato Grosso, "Exportação de Matto-Grosso, Decennio de 1926–1935," *Relatório do Interventor Federal de Matto-Grosso, Manoel Ary Pires, 13 de junho de 1937* (Cuiabá: Imprensa Oficial, 1937), annexo. [a]Estimate for 1913.

Reflecting the same trends as the nation, Mato Grosso state government revenue declined after World War I, from 5,600 contos in 1919 to 3,900 contos in 1922. Income then fluctuated throughout the 1920s, but fiscal outlays continued to drain the state treasury, and only in 1923 did revenues exceed expenses. Despite some income growth compared with preceding decades, the drain became even more acute in the 1930s, when expenditures periodically exceeded revenues by as much as two-thirds.[6]

Depending on the product, exports grew after the postwar depression, but this was not constant or necessarily maintained. Tax income from erva mate exports rose after the war, but exports began a steady decline after 1929 due to the depression of the 1930s and expanded Argentine production. The Mato Grosso erva industry never recovered and had to direct its production to the tiny local market. Rubber exports also declined through the 1920s. With the exception of a boom stimulated by the federal government during World War II, latex never contributed a significant proportion to state income again.[7] As in the prewar period, postwar state export income predominantly derived from live cattle and cattle products (see table 3.1).

RANCHING: CONVOLUTED OPPORTUNITIES

With the exception of beef jerky and bouillon, Brazilian meat products had virtually no overseas buyers before 1914. For decades South American meat exports came primarily from Argentina and Uruguay, although Brazil had great potential as a cheaper supplier to the ever-expanding European market. After all, in 1912–1913 Brazil claimed over 30 million head of cattle, mostly beef, while Argentina counted 25 million and Uruguay between 8 and 9 million. War demand rapidly changed this dynamic. Brazilian exports of hides rose appreciably during the war years, to a high of 61,000 metric tons in 1919, while frozen and chilled beef exports in particular showed a dramatic increase, from a negligible 1.5 tons in 1914 to more than 65,000 tons by 1917. In the same period, exports of canned meats climbed from less than 230 tons in 1913 to 6,500 tons, and jerky exports from 20 tons to 8,700 tons. The war alone created the Brazilian frozen beef industry, which previously had been handicapped by Argentine and Uruguayan domination of the market and by the heavy reliance of Brazilian producers on jerky for domestic consumption. There was an abrupt slowdown of frozen beef production after the war, but by 1919 the industry was firmly established and became a significant economic player over the following decades.[8]

Apart from beef bouillon and jerky, before World War I the Mato Grosso cattle industry did not benefit from overseas customers. Yet despite its remoteness, Mato Grosso contributed more than 2.5 million head to Brazil's total bovine population, the fourth largest herd in the nation. Until the war the state's markets were limited, but with urgent demand generated by the fighting Mato Grosso became an important source of cattle for the São Paulo refrigerated slaughterhouses (*frigoríficos*). The first operating frozen meat plant was the result of the efforts of São Paulo entrepreneur Antonio Prado, who constructed a frigorífico at Barretos in São Paulo state in 1913, though it did not begin production until the following year. He was matched by Percival Farquhar, who had invested in the construction of a refrigerated packing house at Osasco in the suburbs of São Paulo in 1912. The factory was conveniently completed in 1914 and opened in 1915. These operations were soon joined by four others during the war, financed by foreign capital, in São Paulo and Santos and in Rio de Janeiro and Santana do Livramento, Rio Grande do Sul.[9]

The majority of Brazilian frozen beef exports were processed at Osasco (Wilson, later Continental Products Co.) and Armour in São Paulo, which at the height of their operations processed 800 and 1,500 head per day, respectively. With war's end, the postwar recession restricted the market and some operations temporarily closed their doors or sold out. Prado, for example, sold his plant to the British Vestey firm in 1923. By the 1930s the market had picked up again, and several more frigoríficos began operations. In 1931 there were ten such packing houses—five in the state of São Paulo, one in Rio de Janeiro, one in Paraná, and three in Rio Grande do Sul. All but two were in foreign hands, including Vestey (four) and Armour (two). Frigoríficos were well established, serving the ranching community throughout the south of the country, although they relied primarily on the transport of animals from more remote regions. It was many years before such production decentralized and operations were established in the centers of cattle raising, such as Mato Grosso.[10]

According to presidential reports, between 1914 and 1919 Mato Grosso state income from the export of all cattle products increased from 478 contos to 1,492 contos, the latter figure making up 46 percent of all state export income. Live cattle exports rose dramatically during the war years, from 30,000 head before the war to more than 128,000 head in 1919. In 1915, the Osasco slaughterhouse alone processed over 36,000 animals from Mato Grosso. If the totals of slaughter for hides and jerky are included, in 1919 Mato Grosso contributed 265,000 head of cattle, roughly 10 percent of the total herd. But this war-driven marketing fever led to serious complications in the cattle sector.[11]

Brazilian and state authorities complained about the indiscriminate slaughter of cows and heifers during the period. Between 1912 and 1916, Brazil's total cattle population declined from over 30 million to under 29 million. In Mato Grosso, although the number of animals grew to 2.7 million, an increase of only 200,000 in four years indicates overconsumption. Between 1914 and 1920, official figures counted more than 500,000 live animals exported from Mato Grosso, while an additional 720,000 were slaughtered for hides and jerky. Contraband may have made up as much as another 20 percent. This was an average of over 180,000 head per year. These figures caused great alarm because, as many officials argued, the reproductive base of the herd was under threat and if excessive production was not halted it would be disastrous for the industry. The state government responded by passing legislation in 1918 to prohibit the slaughter and export of cows of calf-bearing age. Similar measures were undertaken throughout Brazil, with apparent success, as the national herd had increased to more than 34 million by 1920. Given the average age of steers at slaughter of three years, the minimal rate of growth for both Brazil and Mato Grosso between 1916 and 1920 indicates disaster may well have been narrowly averted[12] (see tables 3.2 and 3.3).

World War I also stimulated other cattle products that beforehand had enjoyed only a modest market. The European nations not only required meat for their soldiers and civilian populations but also leather for boots, holsters, cartridge holders, belts, slings, and other necessities in a foot soldier's kit. As noted, Brazilian hide exports increased from 37,000 metric tons in 1914 to more than 61,000 tons in 1919, while Mato Grosso contributions grew from 810 tons to 1,800 tons, representing almost 140,000 head. Beef jerky production more than doubled, from 1,700 tons in 1914 to a high of 4,200 tons in 1918. The market for jerky was primarily domestic, particularly in the northeast and Rio de Janeiro. Ordinarily, it was supplied by Rio Grande do Sul and Uruguay, but since cattle from those sources were diverted to supply Río de la Plata frigoríficos and canned meat plants, Mato Grosso found a ready market for its normally unappealing jerky. This was really the beginning of the Mato Grosso jerky industry, which, though small, we will see was able to fill part of the postwar void left by a declining Uruguayan industry while complementing an uncertain one in Rio Grande do Sul.[13]

With the postwar recession, however, cattle prices that had reached as much as 110 to 120 milreis per head (US$28 to $31, enough to buy

TABLE 3.2. BRAZIL CATTLE POPULATION BY STATE, 1912–1950 (IN MILLIONS)

State	1912	1920	1938	1950
Brazil	30.7	34.3	40.0	47.1
Mato Grosso	2.5	2.8	2.7	3.4
Goiás	1.9	3.0	3.0	3.5
Minas Gerais	6.8	7.3	11.1	10.5
Rio Grande do Sul	7.2	8.5	7.9	9.2
São Paulo	1.3	2.4	3.5	5.9
Bahia	2.6	2.7	3.2	4.0

TABLE 3.3. MATO GROSSO CATTLE POPULATION BY MUNICIPALITY, 1912–1950

Municipality	1912	1920	1940	1950
Aquidauana	130,000	187,510	209,454	339,539
Bela Vista	200,000	212,736	90,656	136,632
Cáceres	110,000	123,779	60,002	175,410
Campo Grande	650,000	372,919	276,907	215,604
Corumbá	250,000	202,042	358,729	568,576
Coxim	220,000	195,746	n/a	110,746
Maracajú	—	—	109,121	105,532
Miranda	92,000	177,198	39,292	91,184
Nioác	87,000	158,474	35,039	74,309
Paranaíba	260,000	146,083	54,812	108,599
Poconé	170,000	159,959	132,495	142,396
Ponta Porã	160,000	239,089	104,974	146,840
Porto Murtinho	98,000	87,748	60,351	104,109
Três Lagoas	—	164,153	204,606	142,985
State total	2,550,450	2,831,667	2,136,278	3,442,599

Sources: Data from Brazil, Fundação Instituto Brasileiro de Geografia e Estatística, Séries Estatísticas Retrospectivas, vol. 1, Repertório estatística do Brasil: Quadros retrospectivos, Separata do Anuário Estatístico do Brasil, Ano V, 1939–1940 (Rio de Janeiro: IBGE, 1986), 25; Brazil, Ministério da Agricultura, Indústria e Commercio, Directoria do Serviço de Estatística, Synopse do censo pecuário da república pelo processo indirecto das avaliações em 1912–1913 (Rio de Janeiro: Typ. do Ministério da Agricultura, Indústria e Commercio, 1914), 36, 62; Brazil, Ministério da Agricultura, Commercio e Obras Públicas, Directoria Geral de Estatística, Recenseamento realizado em 1 de Setembro de 1920, vol. 3, parte 1 "Agricultura" (Rio de Janeiro: Typ. da Estatística, 1926), 401; Brazil, IBGE, Recenseamento Geral do Brasil (1º de

TABLES 3.2 AND 3.3. (*CONTINUED*)

Setembro de 1940), Série Regional, Parte XXII, *Mato Grosso: Censo Demográfico, Censos Eco-nômicos* (Rio de Janeiro: Serviço Gráfico do IBGE, 1952), 210; Brazil, IBGE, Conselho Nacional de Estatística, Serviço Nacional de Recenseamento, Série Nacional, Vol. 2: *Censo Agrícola do Brasil, 1950* (Rio de Janeiro: IBGE, 1956), 151; Brazil, Fundação Instituto Brasileiro de Geografia e Estatística, Recenseamento Geral do Brasil, Série Nacional, Vol. 2, 2ª Parte: *Censo Agrícola de 1960* (Rio de Janeiro: IBGE, 1970), 114.

Note: The cattle census for 1940 was acknowledged as incomplete. It was estimated there were at least half a million more head in the state than the census reported.

a square league of land in the Pantanal) plummeted to an average of 50 milreis (US$6.50). Ranchers sold their cows, despite low prices and official proscription, in order to pay off debts contracted during the boom. The market was the local jerky plants, which took advantage of the collapse of the frozen beef market to export jerky to Rio de Janeiro and Uruguay.[14]

The postwar crisis combined with extensive flooding in 1921 to push some ranchers to the brink of bankruptcy. Desperate pleas were made for tax and transportation relief. State president Pedro Celestino Corrêa da Costa responded by revoking municipal taxes on the purchase of livestock, extending the period for payment of state taxes, and canceling fines for late payment. Meanwhile, a commission of federal deputies from cattle-producing states formed to investigate and seek solutions to the national cattle crisis. In his 1992 annual address, President Celestino offered a number of suggestions to the commission: (a) reduce the salt import tax, as the jerky industry nationwide used imported rather than national salt; (b) decrease freight charges for salt and beef, including the federal transportation tax; (c) cut São Paulo state taxes on frigorífico production and fattening pastures and eliminate the tax per head of cattle originating in other states; and (d) extend the time period for charqueadas to upgrade their facilities, as required by federal law, since most of the small operations found it virtually impossible to finance the improvements.[15]

In support of these recommendations, an article appearing a month after the presidential address revealed that taxes on an animal sent from Mato Grosso to the São Paulo slaughterhouses totaled over 40 milreis per head, with more than 25 milreis going to the federal and São Paulo state governments. Though federal response was predictably slow, in mid-1924 the inflated cost of meat in Rio de Janeiro and São Paulo prodded Rio into declaring tax and internal tariff relief on certain meat products, particularly jerked and dried beef. The measure, initially for sixty days, was extended to year's end. It had a salutary effect on the cattle industry and on the cost of living for urban residents, particularly the poor, who were the main consumers of dried and jerked beef.[16]

By the mid-1920s, Brazilian cattle and cattle product exports had benefitted from the global economic recovery. The national herd had increased to more than 40 million head by 1925, and Mato Grosso's share to 3.5 million, while ranching expanded in neighboring states like Goiás and São Paulo. Still, by the mid-1920s the national market had become largely domestic, with only some 27 percent by value of beef production exported in 1928. In fact, Earl Downes argues that the production of fresh beef saved the São Paulo frigoríficos. He goes on to note that the combined markets of Santos, Campinas, and São Paulo city consumed as much as 50,000 head a month at the time. Certainly this did guarantee a continued market for the packing plants during the postwar recession, while the Mato Grosso economy became fully dependent on the cattle industry. But the structure of the industry throughout Brazil began to change, as Brazilian meat gradually reentered the world market, particularly in Europe and the Middle East. Between 1922 and 1930, Brazil's exports averaged more than 50,000 metric tons a year. At the same time, Mato Grosso contributed one million live animals to neighboring São Paulo frigoríficos and their fattening pastures, an average of more than 130,000 per year. The expanding charqueadas shipped 36,000 tons of beef jerky and 21,000 tons of hides out of the state, representing 700,000 to one million head (between 90,000 and 125,000 head per year). The market was largely Europe for hides and Brazil (through Uruguay) for jerky.[17]

Despite the advent of world depression and the expanded domestic market, the 1930s saw an increase in cattle product exports (see table 3.1). Average annual Brazilian meat exports rose to a respectable 72,000 tons, thanks largely to expanded sales to Italy after 1935. Between 1930 and 1935, Mato Grosso exported over one million head, and in 1935 more animals were sent to São Paulo than ever before (227,470). Jerky and hide production also benefited. In the mid-1930s, Mato Grosso probably processed between 300,000 and 350,000 head per year, for on-the-hoof and jerky exports, and local consumption (50,000 head a year).[18]

For local and national consumption, evidence is spotty, but with a total population of just under 375,000 in 1936, Mato Grosso per capita beef consumption was reported at between 13 and 18 kilos annually.

This was a meager 35 to 50 grams per person per day, in a region dedicated to cattle ranching. By comparison, in 1940 average consumption in Rio de Janeiro was allegedly a substantial 50 kilos a year, 137 grams per day. Only the city of Campo Grande seemed to have greater access to beef. In 1927, it slaughtered enough cattle for roughly 250 grams of beef per person per day, although undoubtedly not all went to local residents. Still, by today's lean dietary standards, this is considerable. One can imagine that meat availability was hardly spread equally over the entire population, however. Given the ubiquity of cattle, it may well have been greater than these figures suggest, as a great deal of beef consumption in rural areas went unreported. More research is needed on domestic consumption.[19]

During World War II Brazilian cattle production remained relatively stable in relation to the late 1930s, despite the demands of warring armies for beef and leather. Yet, after peace, postwar demand expanded and the number of cattle slaughtered increased significantly, from 4 million in 1945 to just over 6 million in 1949. The value of these animals skyrocketed, doubling in only five years. By 1950 the Mato Grosso cattle population had increased to 3.5 million head, while jerky production had risen to exceed 7 million kilograms. The cattle economy saw a major boost with the construction of a small frigorífico in Campo Grande the same year. The creation of this establishment finally allowed ranchers to sell fattened cattle for slaughter locally, gradually reducing the number sent to São Paulo and contributing to the demise of the charqueadas over the following years.[20]

The market price of Mato Grosso cattle differed, however, depending on the animals' destination. Local charqueadas paid on average half what drovers paid per head, and drovers were able to sell their purchases in São Paulo for up to double what they paid in Mato Grosso. For example, the costs of the drives in the late 1920s, including taxes, feed, right of way, and other expenses, were estimated at 30 to 35 milreis per head from the Vacaria region to São Paulo, leaving a profit of between 30 and 60 milreis per animal, depending on market conditions. Even with estimated losses of 5 to 10 percent per drive, the drover still managed a reasonable return. For most ranchers, then, the wise choice was to sell to the more lucrative São Paulo market. On the other hand, biological constraints meant drovers would buy only the heaviest animals that could withstand long drives of up to three months. As a result, small, scrawny, sickly, and old animals, as well as cows, tended to end their days in the charqueadas.[21]

CHARQUE AND CHARQUEADAS:
A PARTICULAR PROSPERITY

Salt beef has a long history in the Río de la Plata region, with production in Brazil centered in Rio Grande do Sul. Through most of the nineteenth century, the principal markets were the slave populations of Brazil and Cuba; after emancipation, poverty and acquired preference caused continued consumption of the product.[22] As noted, Mato Grosso was a minor producer until the stimulus of World War I. For decades ranchers in the Pantanal had processed salt and dried beef on their ranches and sold it in Corumbá, from where some ended up at market in Montevideo. The dried beef, inadequately processed with little or no salt, was of poor quality, but its preservation time was considerably shorter than that of jerky. Most jerky exports were from Descalvados, although there the making of jerky was secondary to the production of bouillon. This changed with the stimulus of the war, although by that time other producers were beginning to enter the picture.[23]

New operations included a charqueada established at Miranda in 1907 and, around the same time, another at Barranco Branco on the lower Paraguay River near Porto Murtinho, both by Uruguayan entrepreneurs. Markets ranged from Rio de Janeiro to northeastern Brazil and Cuba. Traditionally, they relied on the river for transportation, but there was some hope that the railroad under construction would offer a faster and cheaper route. The Miranda factory could process up to 20,000 head a year; Barranco Branco, with facilities to process 250 head a day, was established with the intention of using the river as its transportation outlet, drawing on cattle from nearby ranches transported to the factory by barge.[24]

By the outbreak of the war, another factory had been established near Porto Murtinho, also owned by a Uruguayan firm. With the exception of Descalvados, all charqueadas were owned by Uruguayans, indicating the amount of capital available in Montevideo for such investments. For most of the previous decades, South American jerky production had been dominated by Uruguay, Rio Grande do Sul, and, to a lesser extent, Argentina. The war expanded alternatives, however, and since Uruguay had been moving to produce frozen beef for the hungry overseas market, a temporary opportunity arose in Mato Grosso to fill the gap. By 1919, a total of 12 charqueadas were operating in the state, 8 on or near the Paraguay River, especially near the Pantanal, where ranchers organized to take advantage of the boom. In

Working jerky at Charqueada Pedra Branca, Miranda, in the 1910s. Reprinted from Cardoso Ayala and Simon, Album Graphico do Estado de Matto-Grosso, *291.*

1922, the number of jerky plants had risen to 14, and another 11 were added by 1925. This was the height of jerky production, when twenty firms ran 25 charqueadas in nine municipalities. In 1927, Mato Grosso slaughtered more than 45,000 animals for jerky, second in the nation. Through the 1920s and the later 1930s, jerky enjoyed a steady market, though the number of establishments declined as markets and consumer tastes gradually changed. World War II provided a short boom, at which time producers began selling directly to the Brazilian Northeast instead of utilizing middlemen in Rio de Janeiro. Most of the animals were cows, since steers were sent directly to the frigoríficos in São Paulo. As late as the 1950s, near the end of the jerky era, there were 9 charqueadas operating in Mato Grosso. The last closed in 1962, in response to government authorization for construction of a frigorífico after years of lobbying by ranchers and local merchants.[25]

Production in these many operations varied, but most were tooled for between 5,000 and 20,000 head a year. The zenith of jerky exports was 1924–1925, when they topped 7 million kilos. Subsequent years saw a decline to an average of about 4 million kilos annually, figures that remained stable into the 1940s (see table 3.1). The most important charqueadas were along the Paraguay River and its tributaries in the municipalities of Corumbá, Porto Murtinho, Miranda, and Aquidauana. There were also charqueadas in Campo Grande, Três Lagoas, Poconé, and Cuiabá. The biggest producers relied on the river system for export, while the ten inland plants were dependent on the railroad. In the end, the riverside factories survived the longest, since transportation by way of the railway proved chronically inadequate.[26]

Jerky Technology

The technology for producing jerky had been developed in the previous century, primarily in Uruguay, and was modified little since then. Herbert Smith, a US naturalist traveling in Rio Grande do Sul in the 1880s, described a process that remained in use in Mato Grosso into the 1940s. Smith noted that there were several stages to the operation. First, a steer was lassoed and secured and a knife driven into its neck, allegedly stunning but not killing the animal; it was then dragged out of the slaughter area on a small cart and taken to where it was slain with a knife to the heart, bled, and stripped of its hide. The meat was sliced from the bones into eight cuts and passed on to specialized workers, who carved it into uniform widths of roughly 15 centimeters (called *xarquear*); the meat strips were then piled on top of one another, interspersed with salt. The weight of the meat caused its liquids to mix with the salt. Approximately 8 to 10 kilos of salt were used per animal. This was referred to as dry salting. The piles of meat were left to season for two days. If there was no rain, the meat was hung out in the sun on horizontal wooden beams to dry, covered by leather sheets at night. If the weather was not ideal, the meat could be kept stacked until dry weather arrived, up to several months if necessary.[27]

In Mato Grosso, the slaughter seasons (*safra*) tended to be October to November and April to May. Expert meat cutters were imported from Uruguay, and along the Paraguay River slaughtering was done at night in shifts, to avoid the stifling heat and humidity of the day. Meat was prepared during the day. Cattle were primarily local breeds (pantaneiro or tucura), small animals with tougher meat and less fat than other breeds. Their hides were also said to be more uniform. Approximately 95 kilos of meat could be rendered from a thin steer of roughly 200 kilos. Bone accounted for 22 percent of the animal, and 30 to 40 percent of the weight was lost in drying.[28]

Enormous amounts of salt were needed. In 1924, for example, Wilhelm Pinsdorf, who owned a charqueada in Aquidauana, reportedly bought 50,000 kilos of salt in Buenos Aires, which he calculated would last his operation only five days. The salt was imported almost exclusively from Spain (sal de Cádiz), considered to be better than the Brazilian product from Rio Grande do Norte, although there appears to have been no reason to reject the national product as long as it was aged properly before use. Brazilian salt had been offered on the Mato Grosso market from at least 1892, and the origin of salt became a national debate throughout the period and into the 1940s. Concerns were cost, since imported salt was sometimes cheaper because of production and transportation efficiency, bulk import, and market exchange (salt merchants often purchased hides for export). This price difference was exacerbated in 1924, when the federal government acceded to demands of jerky producers in Rio Grande do Sul by reducing the tariff on imported salt. Northeastern salt quality also was inconsistent, since often it was marketed before sufficiently aged, thus leading to bacterial contamination of the meat called *vermelho*, or in crystals too large to be effective (the main arguments of the Gaúcho producers). Allegedly there was also an element of prejudice against the Brazilian product among Spanish technicians in charge of the charqueadas. By the late 1930s, Brazilian salt began to replace the Spanish product as the Spanish Civil War intervened and northeastern producers marketed their salt more aggressively. Sampaio Fernandes reported that by 1937 imports into Brazil had fallen from an average 100,000 tons to between 20,000 and 30,000 tons. The onset of World War II ended regular importation.[29]

By the 1940s, processing required between 45 and 50 kilos of salt per animal, a significant increase from Herbert Smith's time. Salt also was needed for treating hides. Hides were left in the brine runoff from the meat piles for 24 hours, then covered in salt, folded over to expose the underside, and left to dry. A pure saline solution could also be used, usually in combination with salt layering. The hides were then left to dry by hanging, much like jerky. This brining process was termed wet salting; it later became the most common method of preserving both hides and jerky. Fat was either stripped off by hand or steamed away. The process could also be energy- and cost-efficient, depending on the plant and local needs, as the bones could be burned to fuel the steaming procedure and the resulting ash then sent overseas for fertilizer. Possibly because there was so little organized modern agriculture in the vicinity, ash from the charqueadas was seldom used for fertilizer in Brazil, including Mato Grosso, though in later years bones were processed

Preparation of hides, 1910s. Reprinted from Cardoso Ayala and Simon, Album Graphico do Estado de Matto-Grosso, *293.*

into bone meal (*farinha de ossos*) for national agriculture. Other products like tongues and horns were normally exported whole to be processed elsewhere, but the entrails were discarded, unlike in Argentina, where they were used for fertilizer and pig feed. In later years meat was injected with a saline and chemical solution, a cheaper and more efficient method of preservation. Hides underwent chemical treatment using arsenic, magnesium salts, or chromium (called bluing or wet blue, in reference to a temporary blue coloring caused by the chemicals). The reader can well imagine the pollution created by these processes, but although it has been commented on in other jurisdictions, I have encountered no reference for Mato Grosso.[30]

Responses to waste were almost certainly parallel throughout the sector. In Rio Grande do Sul, the traditional approach to nonmarketable production of the charqueadas was simple disposal, with a predictable environmental impact. Water was essential to processing, thus all charqueadas were located near water sources. The drawing of water did not occasion concerns over erosion or other impacts, but the streams and rivers were the recipients of liquids from the production line, which included salted water and blood, processed bone matter, and lost fats and solids. Although as much of the animals as possible was used, unusable or lost waste (*guano das charqueadas*) was simply dumped into streams or into the open. The noxious smells associated with the factories have been attributed to this refuse. Blood, as much as 15 to 20 kilos per animal, was not used in early charqueadas. It ended up in the streams or in open pits; in Rio Grande do Sul it became a form of mortar to hold back the erosion of coastal sand dunes. The stink for seaside residents can only be imagined. At certain times of the year, streams ran red with the detritus. It is said that in the nineteenth century the Arroio Pelotas in Pelotas, Rio Grande do Sul, was referred to by locals as Rio Vermelho (Red), because of the waste of more than forty jerky factories along its banks.[31]

Beginning in the 1930s, the Brazilian government sought to eliminate or at least reduce this pollution, but because of the precarious economic situation of the establishments, especially in Rio Grande do Sul, the measures were modest. Only in the mid-1940s were improvements implemented that utilized waste more efficiently and reduced pollution. By that time, the industry had declined in the state as frigoríficos came to dominate the beef processing sector, but in Mato Grosso the regulations led to cleaner, more sanitary, and economically efficient operations.[32]

Jerky and hides were either shipped down the Paraguay River to Montevideo, where they were transshipped to Rio, or sent by rail to São Paulo and Rio. Neither form of transportation was ideal, but the river route was marginally more reliable. The route was monopolized by a single exporter, Joaquim Vivo & Cia., which operated out of Montevideo and transshipped Mato Grosso jerky destined for the Brazilian market. It was this arrangement, however, that annoyed many Mato Grosso producers and provoked considerable consternation in Rio Grande do Sul, Brazil's major jerky producer, as it created conditions for smuggling.

Jerky and Industrial Contraband

Smuggling was endemic in Mato Grosso, as is usual in remote frontier regions. After all, chronic transportation and taxation issues, plus periodic political instability, caused producers to attempt to overcome limits to their incomes and realize profits however possible. Contraband was the natural outcome. Most traffic was with Paraguay and, in later years, across the Paraná River into Minas and São Paulo. It was obvious to state authorities that insufficient attention to the traffic across the international border was costing them a sizable fiscal income. Erva smuggling alone was reputed to cost the state tens of contos in export taxes, but other items, particularly cattle commonly destined for Paraguayan jerky producers, also crossed the frontier duty-free, not to mention the majority of imports consumed in southern Mato Grosso outside Corumbá. Part of the problem was the lack of resources and attention in Cuiabá to outfit and maintain an effective police and border force, which contributed to chronic corruption on the part of underpaid customs agents.[33]

Officials and local residents simultaneously lamented and engaged in smuggling, of cattle, cattle products, salt, erva, and necessary imported items. In the face of Cuiabá's inability or unwillingness to control the traffic, locals addressed the issues on their own terms, leading to constant official correspondence regarding the loss of revenue and the "criminality" of participants.[34] As Walther Bernecker argues for nineteenth-century Mexico and Latin America overall, given the limited presence of State investment in a relatively distant frontier region, contraband and the corruption that accompanied it reflected different developmental levels of the state and was not a question of morality at all.[35] In effect, contraband could be considered a positive in Mato Grosso, since without smugglers' ingenuity and flexible morals, the region may not have been able to develop even as much as it did. That conclusion may be debatable, but by far the most prominent example of smuggling might be described as "industrial" contraband, which involved the well-organized switching of products in the warehouses of Montevideo.

As explained by Stephen Bell, Rio Grande do Sul had led Brazil in cattle population and hide and jerky production for nearly a century. Much of its wealth had been based on the illicit trade of cattle and jerky across the Uruguayan frontier, with local cattle barons and jerky producers in occasional conflict. By World War I, however, Rio Grande do Sul had developed a modern jerky industry that processed its own herds. After a brief flirtation with chilled and frozen beef during the war, in the 1920s the state economy returned to the production of jerky for the Brazilian market. This upsurge was temporary, and by 1940 the number of animals slaughtered for jerky declined significantly. In the decades after World War I, jerky exports, which were largely to Cuba and Mexico, slowly sank to insignificance.[36]

Rio Grande do Sul's greatest competitor was neighboring Uruguay, which by this time was already following Argentina's lead by converting from charqueadas to frozen meat plants. Slaughter for jerky in Uruguay diminished dramatically, from an average of 500,000 head in 1900–1910 to a minuscule 7,000 head in 1931–1934. Uruguayan export statistics reveal corresponding declines. Until 1915 most of the exports were to Brazil, in competition with the gaúcho product, but after that time Cuba became the main destination.[37]

But Rio Grande do Sul suffered a crisis in the jerky business during the 1920s, prompting calls for control of contraband jerky entering from Uruguay. This was where Mato Grosso entered the picture. Since it was a national product, jerky from the state transshipped through Montevideo was exempt from import tariffs into Brazil. Yet, the passage through the territory of a foreign competitor created a golden opportunity for fraud. If Mato Grosso jerky could be substituted for Uruguayan jerky, which was highly appreciated in Brazil because it was marbled with fat, tariff barriers of as much as 60 to 100 percent ad valorem could be avoided. The upriver jerky then could be exported as Uruguayan to Cuba, a smaller and distant market, where lean jerky was prized because it could survive the long sea journey without becoming rancid (as was allegedly the case with Uruguayan and Argentine jerky). This was exactly what happened, causing protest in Rio Grande do Sul, whose product suffered competition and as such commanded lower prices in the internal market than it would have done otherwise. Ironically, the Uruguayans also imported jerky produced in Rio Grande do Sul for resale abroad. This jerky, similar in quality to that of Mato Grosso, underwent the same transition, replaced by Uruguayan meat that was then marketed in Brazil as native.[38]

Smuggling was encouraged by liberal Uruguayan import laws, which permitted the free entry of Brazilian products in transit to the Brazilian market, plus warehousing for up to a year tax-free. This was part of a 1906 Uruguayan government initiative to stimulate the economy through port improvement, railway expansion, and increased trade, in partial response to the construction of a port at Rio Grande in the neighboring Brazilian state. The result was that Mato Grosso jerky enjoyed special treatment en route to markets in Rio de Janeiro and the Brazilian Northeast. Licenses for the use of Uruguay as a transit point were issued by the government in Rio, and many licensees often resold them to a Uruguayan jerky factory, which would switch products and reexport tax-free. The practice was common, with license resale prices even quoted publicly in the Río de la Plata financial market. One of the companies most deeply involved in the activity was the sole exporter of

Mato Grosso jerky, Joaquim Vivo & Cia. Vivo's contacts in Montevideo included well-known Uruguayan jerky producers J. Schroeder and Pedro Ferrés & Cia.[39]

Estimates alleged that out of a total of 16,000 tons exported by Uruguay to Cuba in 1927, more than 10,000 was Brazilian in origin, 6,000 to 7,000 tons of it from Mato Grosso. The Brazilian consul general in Montevideo investigated, eliciting denials of any wrongdoing from Ferrés and Schroeder. Nonetheless, consular officials concluded that the only way to avoid substitution was to require a customs document from the Uruguayan authorities noting the date of arrival of the shipment and its location while in Montevideo, and to legalize transit licenses only after the ships carrying the jerky had left for their ultimate destination. This they proceeded to do.[40]

Pressure from gaúcho politicians, particularly Getúlio Vargas, at the time the Brazilian Treasury Minister (Fazenda), spurred Rio de Janeiro to more solid action on their behalf. In 1928, the Rio Grande do Sul government proposed that all Brazilian goods transiting foreign territory be "denationalized" and forced to pay normal import duties on reentry into Brazil. The reaction from Mato Grosso to the proposal was predictable. Jerky producers argued that as long as there was no improvement in the quality and cost of transport out of the state, particularly by rail, Mato Grosso had no choice but to export through Montevideo. The federal government, however, operating on logical political expediency, since Rio Grande do Sul wielded much greater political power than did Mato Grosso, accepted the gaúcho proposal without modification, passing a law to that effect in late 1928. In order to placate the Matogrossenses, however, the Rio government allocated funds for the resumption of government shipping by Lloyd Brasileiro to Corumbá and a reduction of rail rates on the railroad to São Paulo.[41]

There was little satisfaction with the announcement in Mato Grosso, however, since it did nothing to improve opportunities for the region in the short run. Nonetheless, charqueadas continued to be important in the state up to the 1960s; eventually they faced competition from newly established frigoríficos (some set up in former charqueadas) and changing consumer preferences, internally and abroad. Still, the conflict with Rio Grande do Sul interests was illustrative of both the promise and the disappointment of ranching in Mato Grosso. Certainly opportunities developed, and contraband diminished (though never disappeared), perhaps in part because of such high-profile disputes. The result was that attention continued to be given to the quality of cattle and the need to diversify beef production. Transportation, as indicated already, remained an obstacle, especially in the promise of a railway.

TRANSPORTATION: DILEMMAS AND CONTRADICTIONS

The Paraguay River was the traditional highway for Mato Grosso exports and imports, along with the few rugged cattle trails to Minas Gerais and São Paulo. The completion of the railway in 1914 offered a crucial alternative because, among other things, relying on the river meant dependence on environmental conditions. In the dry season, August and September, the river was sometimes too shallow for larger vessels. At the height of the rainy season, January and February, smaller craft and barges were frequently lost to the powerful current. There was also the danger of submerged trees, the result of riverbank clearing, which caused innumerable delays and regular shipwrecks. For the most part, however, these problems were surmountable. What most affected river commerce was the irregularity of service and the high freight rates charged by the few companies that plied the river, obstacles it was hoped a railway would overcome.[42]

Shipping

With the opening of regular river communications between Rio and Mato Grosso after the Paraguayan War, it was hoped that the region's isolation was at an end. But the vicissitudes of the economy and irregular attention from Rio meant the province/state continued to depend on river transportation that left a great deal to be desired. As early as the 1870s, provincial presidents expressed a desire for a railroad to open up the region, since the fluvial route was long, had to pass through three foreign nations, was dependent on river levels, and was dominated by two companies. By the end of the 1880s, however, there were many boats plying the river from Montevideo to Corumbá, including a federally subsidized service, and others that serviced smaller ports. There was little service to the more remote regions of Mato Grosso, such as Cáceres, Miranda, and even Cuiabá, as boats of more than one-meter draft could not sail beyond Corumbá. This caused Corumbá to become an entrepôt to the rest of the region and the most important commercial center in the state.[43]

Venus, *a vessel in the government shipping line Lloyd Brasileiro, at Corumbá in the 1910s.* Reprinted from Cardoso Ayala and Simon, Album Graphico do Estado de Matto-Grosso, 67.

Until World War I, shipping on the Paraguay River was dominated by three companies, the Brazilian Lloyd Brasileiro, the Argentine Cía. Mihanovich, and the Paraguayan Vierci Hermanos. Lloyd Brasileiro, owned by the federal government, was formed in 1890 to regularize the erratic shipping operations that connected Brazilian ports and plied inland waters. The Lloyd operations were plagued by financial problems, almost from their inception, and were passed back and forth from public to private hands until 1913, when they were directly assumed by the government and remained in federal hands for the duration of the war. These problems hindered local economies that relied on the service and that were often considered of secondary importance by the directors in Rio de Janeiro. Government financial support was sporadic, sailings were irregular and often delayed, freight charges were onerous, and goods were often damaged or stolen.[44]

Part of the slack was taken up by the other companies, but their priorities were not in Mato Grosso and so freight rates were often prohibitive. In 1912 a ton of freight was charged US$4.10 between Montevideo and Europe or North America, a journey of three weeks to a month, but for the two-week upriver trip between Montevideo and Corumbá, the rate was nearly double. From the interior town of Nioác to Corumbá, the same freight paid over $60. These were rates for imported items, generally cheaper than for exports. In 1907, charges for the export of a ton of beef jerky from Miranda to Rio de Janeiro exceeded US$40. Rates changed little in later years. Indeed, they appear to have increased, as revealed in 1928, when Mihanovich charged 40 Argentine gold pesos (US$38.60) per ton (1.2 cubic meters) between Buenos Aires and Corumbá, decried as exorbitant by local exporters. Admittedly, these price differentials were partly due to the inequities inherent in volume shipping, but they also reveal Mato Grosso's geographical vulnerability and limited insertion into the broader national and international market.[45]

Despite these obstacles, in the early twentieth century the volume of shipping via Corumbá increased. Between 1906 and 1912, the number of ships entering Corumbá grew from 48 to 128, while vessel tonnage expanded accordingly. The increased volume was almost exclusively in cattle products, particularly beef jerky. Exports of jerky from Corumbá jumped from 52,000 kilos in 1905 to 1.6 million kilos by 1912, and between 1919 and 1925, exports from the charqueadas along the lower Paraguay River doubled from 800,000 kilos to 1.5 million kilos.[46]

The fundamental problem of shipping was not resolved by this flurry of activity. Lloyd Brasileiro suspended service between Montevideo and Corumbá in 1914 because of financial and administrative

Port of Corumbá, in the 1910s. Reprinted from Cardoso Ayala and Simon, Album Graphico do Estado de Matto-Grosso, *110.*

problems, renting out its boats to a local import-export firm, Companhia Minas e Viação de Matto Grosso. The merchants of Corumbá were not pleased with that company's service, however, characterizing it as abusive. Lloyd had been operated by a negligent administration and employees, but Minas e Viação went even further, detouring ships to serve Paraguayan and Argentine commerce and virtually ignoring Mato Grosso. As a result, Mato Grosso ports "waited months without notice of national ships." Mihanovich continued its twice-monthly service, but it was insufficient for the goods waiting to be transported. In March 1923, over 800 tons of beef jerky warehoused in Corumbá awaited shipment to Montevideo. A sailors' strike in Asunción prevented Mihanovich from carrying it, but even so this meant the profits would go into "the pocket[s] of Argentine capitalists" instead of those of Brazilians. Besides, the Argentine company, despite running large modern vessels, offered irregular cargo service since it was governed by strict schedules that prioritized ports in Paraguay and Argentina, seldom in keeping with the needs of Mato Grosso commerce. Meanwhile, service from Corumbá to Cuiabá was also uncertain, with only one sternwheeler plying the route twice a month. The trip took almost six days upriver during the rainy season, longer in the dry, when transfer to smaller boats was common. The negative impact on local commerce of irregular service at exorbitant rates can easily be imagined.[47]

Lloyd eventually took over service from Minas e Viação, but little improved, at least at first. A newspaper article in 1924 observed that while Mihanovich covered the round-trip Montevideo-Corumbá run in 25 days, it took Lloyd up to three months. Indeed, by 1928, scarcely one year after three new ships had been inaugurated on the Montevideo-Corumbá run, only one was operational. The other two had suffered damage en route from Rio de Janeiro to Montevideo. In addition, the boats could carry only 750 to 800 tons per voyage. By contrast, the Mihanovich ships had a capacity between 5,500 and 8,000 tons.

By 1930 it appears service by the Brazilian carrier had improved. The three new ships had finally come into service, and total annual shipping in tonnage entering Corumbá from foreign ports almost tripled between 1928 and 1931. Also, state-subsidized service between Corumbá and Cuiabá was instituted in 1925, utilizing four paddle wheelers, and there was regular service between the railhead of Porto Esperança and Corumbá to coincide with the rail schedule, and even a scheduled service to Cáceres. Mihanovich continued its passenger service with weekly runs between Corumbá and Asunción.[48]

The disorganization that swept over shipping on the upper Paraguay River after 1914 seemed to have been overcome by 1930. Corumbá, which still was not served directly by the railway, maintained its position as a commercial entrepôt for Cuiabá and points north. Nevertheless, shipping was unable to compete with the railway in the long run, and between 1930 and the 1950s service gradually declined. Yet the railroad, which certainly offered a challenge to inadequate shipping, did not deliver quite as expected.

The Noroeste do Brasil

The concept of a railway across Mato Grosso had been envisioned before the Paraguayan War, and pressure for links between the Brazilian coast and Cuiabá began to mount after the cessation of hostilities. Despite numerous proposed projects and awarded concessions, nothing concrete was undertaken until 1903, when geopolitical considerations, more than economic concerns, convinced Rio de Janeiro that a railroad linking its remote territories with the coast was essential.[49] A railway was proposed from São Paulo to Cuiabá in 1905, but unofficial promises to Bolivia and the strategic concept of a transcontinental railway linking the Atlantic and Pacific caused it to be redirected to Corumbá in 1907. Fernando de Azevedo later described this new line as Brazil's "first political highway."[50] (See map 3, p. xiv, for railway routes.)

An added incentive for the railway was that it could act as a stimulus to economic expansion in the state of São Paulo, particularly the opening up of previously remote public lands to coffee cultivation in the state's west. Belying its claim to strategic necessity, the federal government awarded the railroad concession to a consortium backed by a mix of Brazilian, French, and Belgian capital. The group then divided construction into São Paulo and Mato Grosso sections. The São Paulo route was called the Estrada de Ferro Noroeste do Brasil (EFNOB); the line across Mato Grosso was the EF Itapura-Corumbá. Construction began at Bauru, São Paulo, in 1905 and was completed to the Paraná River in 1910. In Mato Grosso, work began in 1908 at Itapura on the Paraná River and at Porto Esperança on the Paraguay, and the railroad started partial operation just as World War I broke out. In the interim, work on the Mato Grosso segment had gone too slowly for the government's interests, and in mid-1913 that part was taken over by Rio. Undoubtedly, this was a relief for the consortium, since the Mato Grosso

A train on the Noroeste do Brasil railroad, in the 1910s. Reprinted from Cardoso Ayala and Simon, Album Graphico do Estado de Matto-Grosso, *152.*

section was thoroughly unprofitable. Apparently only the attraction of healthy profits carrying coffee in São Paulo had persuaded the entrepreneurs to accept the Mato Grosso obligation as well. As it turned out, the São Paulo operation was poorly constructed and maintained as well, causing Rio to take it over in late 1917. The entire route then became the EFNOB, more commonly referred to as the Noroeste, and extended a total of 1,273 kilometers, 837 across Mato Grosso.[51]

Although the inauguration of service in Mato Grosso in 1914 was timely, the difficulties of building across parts of the Pantanal delayed completion of the link to Porto Esperança until 1917. For most of the route, construction was across relatively flat, open terrain with few serious obstacles. The route chosen ran from the Paraná River to Campo Grande and on through Aquidauana and Miranda to the Paraguay River. This was largely open cerrado, crossed by only minor streams until just west of Miranda, where the lower Pantanal extends south. Seasonal flooding in this section made construction arduous and maintenance expensive, but compared with the headaches of running lines along the wooded ridges of São Paulo, Mato Grosso offered generally unproblematic conditions for the building of a railroad.[52]

Still, service was insecure west of Miranda, especially during the rainy season, a situation frequently commented on by travelers passing through the region. Two other major physical obstacles prevented secure regular service between São Paulo and Corumbá: the Paraná and

Paraguay rivers. Until a bridge was completed linking the two states in 1928, rail cars had to be rafted across the Paraná, and service from Porto Esperança to Corumbá was continued by small river steamers until another bridge was finally completed across the Paraguay in 1947 and service extended to Corumbá in 1953. The railroad had provided mostly aggravation to the residents of Corumbá and Cuiabá until then.[53]

Expectations were high among Matogrossenses as the railroad neared completion. This was especially so among cattle ranchers, who envisioned raising and fattening animals at home for direct export to the São Paulo slaughterhouses; a journey of 36 hours by rail would replace three months on the trail. Time would be saved and costs reduced, ensuring greater profits for ranchers, large and small. Ranchers also hoped to gain more influence over a market still in the hands of the cattle buyers and slaughterhouses with their winter pastures. In addition, greater access would reduce import costs and attract settlers into a region that experienced a chronic population shortage. Both human and financial capital were seen as natural extensions of efficient transportation and would contribute to freeing the state from trade controlled by a small number of economic agents in São Paulo and the Río de la Plata. It was hoped that Mato Grosso could then determine its own destiny.[54]

Rail construction in the Pantanal, 1910s. Reprinted from Cardoso Ayala and Simon, Album Graphico do Estado de Matto-Grosso, *155.*

The railroad's impact was slower in coming than expected, but by 1920 it was clear that it had opened up the state to development possibilities unimaginable before. The influx of people, many of them speculators, land grabbers, adventurers, but also some settlers, transformed areas along the railway line and, eventually, all of southern Mato Grosso. Dusty, torpid small towns expanded overnight, others were created out of the wilderness, and modern facilities and ideas flowed into the state. From only 5,000 inhabitants in 1912, Campo Grande municipality catapulted to more than 21,000 by 1920, while the township increased from 1,200 to 6,000 inhabitants. By 1940, the municipality counted 50,000 inhabitants, at least half in the city. At the same time, Aquidauana increased from roughly 6,000 inhabitants in 1914 to 10,000 in 1920 and 21,000 by 1940. Três Lagoas, which began life with the railroad encampments in 1909, had a population of 9,000 in 1920 and more than 15,000 by 1940. (See table 2.1.) Along with the larger population came regular postal service and telegraph communications, luxuries residents of Campo Grande and other towns had only dreamed of before.[55]

Economic activity increased proportionately as exports of live cattle and cattle products took off. Part of this dynamism was the product of foreign investment, which had been attracted to the region with the completion of the railway. Brazil Land had already established its operations before the line was fully operational, but other investors, like the British Brazilian Meat Company and the French Fazendas Francesas, entered during and after the wartime boom. The railroad facilitated an inflow of essential materials such as salt, wire, grass seeds, breeder bulls, and machinery. It also permitted cattle buyers to travel from town to town much more easily than on horseback, transforming the marketing of cattle, at least as far as access to ranches was concerned. Several charqueadas sprang up in the towns along the railroad; twelve of the twenty-five plants in existence in 1925 were located on or near the rail line, an increase from only one in 1914.[56]

Yet by their very ambition, the expectations of the ranching industry raised by the arrival of the railroad were exaggerated. Mato Grosso, though important to the São Paulo meat processing industry, was not a priority in the global plans of the Brazilian government. Even during the export boom of World War I, cattle and cattle products did not exit the state by rail to the degree anticipated. As early as 1916 the state president complained about service. He was especially concerned about freight rates, arguing that unless they were reduced, Mato Grosso production had no future. He then compared rates for similar products in

Rio Grande do Sul. The EF Itapura line charged over three times the rate levied by the EF Rio Grandense for a ton of hides over the same distance. Cargo volume was probably much greater in Rio Grande do Sul than in Mato Grosso, but these were still exorbitant rates. The result inhibited the expansion of hide production in Mato Grosso and forced those producers already in business to rely on river transport, the very dependency they had hoped to escape through the railroad.[57]

Freight rates became a constant issue over subsequent years. As beef jerky production increased considerably in the years after World War I, cattle prices dropped and a large stock of animals grew. Rates charged by the EFNOB, however, were as high as $70 a ton during the 1920s, 80 percent more than those of the exorbitant Mihanovich fluvial service. This obstacle remained, despite frequent calls for cheaper freight rates and lower taxes, until 1928, when the federal government denationalized all Brazilian jerky transshipped through other nations. The response from charqueadas and the Mato Grosso government was immediate.[58]

Once the industry was denied transshipment at Montevideo, Mato Grosso producers requested that rail freight rates be reduced by 50 percent and that a direct port-to-port fluvial service be inaugurated by the government. The Ministry of Transport agreed to reduce rates for charqueadas that filled the rail cars with a minimum of 20 tons of jerky, but transport was guaranteed only to Bauru, with re-dispatch to market from there. This revealed the other major stumbling block in Noroeste service, rolling stock. As a response from the Aquidauana town council pointed out, aside from the chronic shortage of rail cars, the cars that *were* provided were inadequate to satisfy the ministry offer, since their maximum capacity was only between 16 and 18 tons. Transferring freight in Bauru would have threatened cargos, since the railroad did not have sufficient warehousing space. Any delay in transport (and delays were likely, given the EFNOB's concentration on the transport of coffee in São Paulo) would have left the jerky exposed to the elements, increasing the risk of spoilage. Indirect dispatch also would have dried up credit, since banks were unwilling to finance goods in transit. Intensive lobbying by Mato Grosso federal deputies in Rio won the government over, and the 50-percent reduction was awarded on any full cars, and shipment was to be dispatched directly to the destination. The decision probably temporarily saved the industry in Mato Grosso, as export figures reveal steady activity from 1929 into the mid-1930s.[59]

Ironically, although the railroad provided momentary relief for the jerky industry, it also contributed to the ultimate demise of jerky production in the state. The jerky industry served as a fallback market for the sale of animals when the international frozen meat market slowed down in the immediate postwar period, softening the blow to Mato Grosso ranchers. Meanwhile, cattle buyers began to enter the region by rail, which facilitated direct contact with ranchers. By the mid-1930s, when the export of Brazilian frozen meat began to rise again and national consumption also increased, ranchers began to divert their stocks from the charqueadas to the long but lucrative cattle drives to the São Paulo frigoríficos and fresh meat markets. The jerky plants soon felt the pinch, and several were forced to shut down, particularly those along the rail line. Jerky production became concentrated in a few plants, particularly those near Corumbá, and most cattle continued to be driven to São Paulo.[60]

The railroad was supposed to relieve rancher reliance on long drives by facilitating speedy export of fattened live cattle directly to the São Paulo frigoríficos. Instead, ranchers faced the same problems that plagued the charqueadas—insufficient rolling stock, small rail cars, and inconsistent freight rates. Though rates by the mid-1930s were comparable between cattle drives and the railroad (US$4.50–$5 per head), the problem lay largely in the company's inadequate provision of rolling stock. Most cars could accommodate at most 18 to 20 head of live cattle, which was not cost-effective for the railroad, since the return from cattle did not cover the expenses of running the car. Service was also infuriatingly slow. In 1924, the journey from Campo Grande to Rio took a total of 52 hours, with ideal connections and barring no delays, which were common. It was another 26 hours between Campo Grande and Corumbá, including 12 hours by boat from Corumbá to the rail head at Porto Esperança. There were no facilities along the route for feeding cattle or even providing them with water; thus, cattle transported from Campo Grande to the São Paulo stockyards endured a minimum of 40 hours without food or water—even under ideal conditions. In fact, most cattle suffered seven or eight days in the cattle cars, with limited if any attention. It takes little imagination to visualize the deplorable state of the animals at the end of their journey. Considering the costs in purely economic terms, this required a period of recuperation that was simply prohibitive for rancher and processor alike.[61]

Above all, though, the number of cars available for transporting live cattle was insufficient. It was simple financial logic to favor the São Paulo sector over that of Mato Grosso. By transporting coffee, the railroad operated at a profit in São Paulo, while it ran a constant deficit on its Mato Grosso run, where only cattle products were available for transport. While most of Mato Grosso complained about the freight rates, the company in turn lamented that if it were to make its operation profitable in the state, it would have to raise rates so high no one would be able to pay. Certainly the railroad had some serious budgetary problems throughout its existence. These were aggravated between 1930 and the mid-1940s, when the proportion of coffee cargos declined as the export market shrank during the world depression, and exhausted coffee lands in São Paulo were converted to ranching. Service did improve temporarily during the early 1930s, when the number of head transported by rail in Mato Grosso and São Paulo combined jumped from 73,600 in 1932 to 108,600 in 1933, but the shortage of cars remained a problem. In May 1934, for example, ranchers in certain areas of Mato Grosso had been waiting over a month for transportation. Many turned to the drives, since they were operating on short-term credit and could not afford delay. Some improvements were reported near the end of the year, but they were too few and were not continued. By the 1940s, rail service once again had deteriorated in Mato Grosso.[62]

Neither was the company blessed with a visionary administration. Fernando Azevedo suggested that to save itself the EFNOB should promote agricultural diversification and the construction of frigoríficos in Mato Grosso, in order to create future sources of income. This meant introducing refrigerated cars, not part of the Noroeste system even as late as the 1950s. Economic diversification eventually did occur in the state when a frigorífico was established in Campo Grande in the early 1960s, but the railroad could not claim to have played much of a part. The isolation of Matogrossense ranchers extended into the 1960s, and cattle drives to São Paulo continued.[63]

Another consequence of the EFNOB was the decline of Corumbá as an important commercial center. Unlike the rest of the state, during World War I Corumbá's population actually declined. The biggest and best ships were diverted for war service, while the railway drained off a good deal of the rest of commerce. Campo Grande displaced Corumbá as the commercial hub of the state and became the economic center of the south. Conditions improved marginally during the 1920s, but the depression of the 1930s again slowed the economy of the city. Had the railroad been completed to Corumbá as planned, it may have been able to limit the city's decline, but it did not reach Corumbá until 1953. Corumbá never recovered its previous status, although extension of the railroad into Bolivia during the 1950s for a time ensured a minor importance as an international terminus.[64]

The railway also produced a temporary depression on some cattle ranches, particularly in the Pantanal, as the Noroeste competed for labor. Many workers sought better opportunities as rail maintenance crews and as construction laborers in the growing towns. This contributed to the stagnation of some of the more traditional holdings and the subdivision of many ranches, especially as outside buyers entered the region willing to pay top money for land and cattle. Some ranchers eventually enjoyed modest benefits from the railway, as it stimulated the opening of cattle trails and roads from remote regions to the railheads. Several trails even followed the railway right-of-way between regional markets, and the far south near the Paraguayan border experienced significant population growth as ranching and agriculture flourished. This was stimulated by the creation of a national agricultural colony at Dourados in 1943, during the Estado Novo dictatorship of Getúlio Vargas, and the completion of a spur line from Campo Grande to Ponta Porã in 1953.[65]

World War I and the postwar years saw consolidation of the cattle industry and a time of modernization, or at least reorganization, of ranching and related production. Charqueadas played an integral part, but not all of Mato Grosso's problems were solved during this time; in fact, precious few found resolution. Other problems were created as the state became more integrated into the international economic structure dominating in much of the rest of Brazil. These outside pressures imposed new contradictions on the state. The very liberation of the region through expanded export production and a direct link to the economic center of the country also permitted a form of dependence on outside forces that was more subtle but no less limiting than the previous experience. Traditional sectors that had held sway in the state's economic world were challenged by more dynamic and cosmopolitan forces. Outside capital and entrepreneurs contributed to the development of Mato Grosso but also acted, by the very nature of ranching in the region, to keep the state dependent on one set of commodities for trade. Mato Grosso was at once integrated into the expanding national economic structure and denied much of the benefit of that growth.

Such development did not necessarily relieve the isolation of the frontier. Although links with the past largely had been loosened and the mystery of the frontier revealed, most of the ranching economy continued in its traditional form. As we will see, some activities were indistinguishable from their conduct of decades earlier. To 1950, the process was slow, uneven, and without obvious guarantees, a condition that required adaptation and innovation on the ground. Issues involving land tenure, transportation, labor relations, indigenous populations, and the routine tasks of raising and marketing cattle depended on a flexibility that is characteristic of frontier zones, but which sometimes motivated a certain recalcitrance grounded in tradition. This is not a contradiction, as we will see in subsequent chapters.

As for transportation, in the end the railroad did not bring the immediate expansion eagerly anticipated by Matogrossenses. It was limited in stimulating exports, often acting more to introduce people and goods from outside the state than the reverse. The railway, then, was no panacea for the drawbacks of isolation. The failure of shipping and rail to live up to expectations only underlines the limits of transportation facilities; they cannot deliver quick solutions to frontier underdevelopment when other economic influences, including needs in competing regions, are not favorable.

FOUR

Land Access

OPPORTUNITIES AND OBSTACLES

The availability of grazing land was the prime factor attracting the influx of settlers into Mato Grosso after 1870, but the sticky questions of legalization remained. As time went on, it became increasingly difficult to settle on land claimed by others or to ignore other claims to one's own holdings. All lands were considered public (*terras devolutas*) until registered, and legalization required registration and fees if there was to be peace in the ranching areas. Despite numerous laws and attempts at statewide surveying, government seldom delivered on its obligations, which contributed to conflict, especially in the years leading up to World War I. Cattle were a major factor in this drama.

Access to land in the period before World War I was quite chaotic, and the postwar era saw nothing short of a feverish rush. There were many obstacles—political, judicial, and financial—in the way of legalization, but as in so many frontier regions, residents found ways to acquire and to control properties, often at the expense of other settlers and the native population. As population increased markedly between the war and 1950 (see table 2.1), land was occupied and developed, and by the 1920s ownership was generally legalized. It was hardly a smooth process, however, and the issue of land tenure generated many struggles, contributing to political and social instability well beyond that time frame.

Part of the problem also was access to credit, which offered an added set of concerns that certainly aggravated that instability. In reality, reliable credit hardly existed through most of the period under study, a situation that limited the potential growth of the cattle sector and, by extension, the entire state. Many observers argued for establishment of dependable banks, but it wasn't until after World War II that Mato Grosso benefited from consistent access to credit.

Even more important, an underlying context was key to understanding accessibility to land in remote regions—the presence of native peoples. Without question, the first inhabitants of Mato Grosso were soon overwhelmed by the influx of settlement, but they were not passive victims in the usurpation of their traditional lands. Their reactions, however, were mixed and sometimes contradictory, while governments proved inadequate to address predictable conflict, even when honest efforts were made. The inevitable result, as in other ranching regions of the Americas, was that indigenous groups struggled for but usually lost land to encroaching settlement.

AMBIGUOUS TENURE: THE LAND QUESTION TO 1914

The destruction of land records in Corumbá during the Paraguayan occupation created a major headache for the imperial government and prewar settlers. Incomplete prewar documents suggested a total of 52 registered holdings in the Corumbá area (Albuquerque). Some of these ranchers never returned, but those who did had to re-register their holdings. Proof of previous occupation, as stipulated in the 1850 land law, depended on testimony by neighbors and prewar officials, who were not easy to find. A sizable proportion of ranchers in the lower Pantanal, then, simply squatted until legislation could be enacted to regularize holdings. This took time, for the imperial government insisted on the law of 1850 and the qualifying decree of 1854, ultimately forcing the development of a system of provisional property licenses dependent on final adjudication in Rio. Based on the recommendations of the municipal property judge (*juiz comissário*), provincial presidents had to request permission from the imperial agriculture ministry in order to grant land to petitioners. Since the judges were appointed by the provincial presidents, normally the process was a formality, but it still meant extra time and expense for petitioners, many of whom simply dispensed with registration. Considering the remoteness of most of the lands in question, officials seldom were able to survey the properties, a situation that continued until the state land law of 1892.[1]

Other problems caused by this bureaucratic vulnerability were difficulties in assigning property judges and controversies over appointments. The appointees received a great deal of power since they determined the structure of landholding in their districts, and their positions were political plums or time bombs, depending on the recipients' relationships to local power brokers. In some regions the position was not only politically sensitive but also physically dangerous. Naturally it could be and was exploited for personal gain. The problem plagued all of Mato Grosso and was yet another lament by provincial presidents up to the declaration of the republic.[2]

Even when a judge was available and inclined to honesty, there was still the problem of resources. Few had access to sufficient funds for surveyors, tools, even transportation, severely limiting their effectiveness and contributing to conditions that bred corruption. Nonetheless, claims were made during the last decade of the empire and efforts begun to legitimate some holdings, above all those occupied by petitioners with greater access to financial resources. The area most sought after appears to have been along the Paraguayan border. As the region was within the 10-league (66-kilometer) zone under imperial jurisdiction, permission had to be received not only from the Ministry of Agriculture but also the Foreign Ministry. Besides the Mate Larangeira concession, a number of petitions were filed from small and large ranchers who already occupied the lands they requested along the frontier.[3]

Virtually all requests were for sesmarias. Officially these colonial land grants were no longer awarded after 1822, but they continued to be handed out in remote regions of the country in an attempt to stimulate settlement. Petitioners requested grants of varying sizes; the standard sesmaria was one league by three leagues, a total of just over 13,000 hectares. In theory only one sesmaria could be awarded per applicant. In fact many petitioners were granted several. On paper this permitted the occupation of large expanses of remote areas and extended Brazil's claim to its vast interior, though often the land was not occupied as envisioned by governments. The ultimate effect of granting so much land to so few was to aid in the creation of enormous latifundios and to limit opportunities for future small agriculturalists and ranchers. This policy (or lack of one) paved the way for a similar preference for large properties after 1889.[4]

An illustrative example is the case of Antônio Joaquim Malheiros, who claimed more than 800,000 hectares near Paraguay. Some of this land was not within the border zone, but very little of it actually had been demarcated. In 1886 and 1887 Malheiros, who also was the local government agent for the Kadiwéu natives who lived on land he claimed, sought to legalize some of his holdings, more than 25,000 hectares. He was awarded provisional title by the provincial government on 1,000 hectares in 1888, but apparently the land was not surveyed until the 1890s. Other requests by gaúcho immigrants in the early 1880s indicate that the government favored ranch settlement in the area in order to guarantee Brazilian occupation of the border region. This is reinforced by the refusal of Rio to award land in these areas to foreigners; whenever a non-Brazilian filed a petition, it was refused and the suggestion made that land be bought from private holders. This likely was the main reason the claims of Portuguese-born Malheiros were not legitimated by Rio. Nevertheless, illegal occupation of most of Mato Grosso continued through the declaration of the republic and the transfer of land control by Rio to the states, and in that period Malheiros managed to create a local fiefdom that lasted to the end of the century. With the benefit of direct revenues from land sale and taxes after 1892, Cuiabá authorities finally made an effort, if not always successful, to regularize landholding in Mato Grosso.[5]

Still, in the final decade of the nineteenth century only a few large ranches were scattered throughout the Pantanal. (See map 3, p. xiv.) Nhecolândia was settled by no more than a dozen families, raising perhaps as many as 100,000 head. By 1920, however, there were some 80 ranches in the area with nearly 200,000 animals. The Brazilian census of that year counted more than one million head throughout the Pantanal. Figures from 1941 put the total Pantanal herd at 1.7 million, and by the mid-1950s the number of ranches in Nhecolândia totaled about 100, with an average of 3,000 to 5,000 animals per ranch and a total human population of 5,000 to 6,000. In the early 1970s, some 3,500 ranches were counted throughout the Pantanal, carrying more than five million head. Nearly half were found in Nhecolândia.[6] (See table 4.1 for cattle populations by environmental region.)

Pantanal ranches varied in size, but on the whole were large. In the 1890s, the largest holdings ranged between 106,000 and 380,000 hectares each. By the 1920s, some ranches had been subdivided among family members, but others had expanded, more than half extending over 100,000 hectares. Between 1920 and 1940, the rate of ranch occupation of the total Pantanal had risen marginally from 46 percent (6.4 million hectares) to 48 percent (6.8 million hectares). More than 4.6 million hectares were natural pasture, and 1.2 million hectares were covered in forest. From the 1950s to the 1970s, subdivision increased the number of medium-sized ranches but neither eliminated large ranches nor permitted any significant expansion of small properties that could

TABLE 4.1. CATTLE POPULATION BY ENVIRONMENTAL REGION

Year	Location	Number	% of total
1887	Mato Grosso	800,000	100
1890s	Nhecolândia	100,000	—
1907	cerrado	250,000	33
	campo limpo	250,000	33
	Pantanal	200,000–250,000	25–33
	Mato Grosso	750,000 (?)	—
1912–1913	cerrado	1,000,000	40
	campo limpo	500,000	20
	Pantanal	800,000	32
	Mato Grosso	2,500,000	100
1920	cerrado	800,000	30
	campo limpo	500,000	18
	Pantanal	1,100,000	40
	(Nhecolândia)	200,000	7.5
	Mato Grosso	2,700,000	100
1941	cerrado	1,500,000	38
	campo limpo	700,000	18
	Pantanal	1,700,000	43
	Mato Grosso	4,000,000	100
1970	Pantanal (south)	2,200,000+	30
	Pantanal (all)	5,000,000	38
	cerrado	2,800,000+	38
	campo limpo	2,200,000	30
	Southern Mato Grosso	7,500,000	100
	Mato Grosso	13,000,000	100

Sources: Data from Mato Grosso, *Relatório do vice-presidente Dr. José Joaquim Ramos Ferreira devia apresentar à Assembléia Legislativa Províncial de Matto Grosso, 2ª legislatura de Setembro de 1887* (Cuiabá: n.p., 1887); Miguel Arrojado Ribeiro Lisboa, *Oeste de S. Paulo, Sul de Mato-Grosso: Geologia, Indústria, Mineral, Clima, Vegetação, Solo Agrícola, Indústria Pastoril* (Rio de Janeiro: Typografia do Jornal do Commercio, 1909), 141–145; Brazil, Ministério da Agricultura, Commercio e Obras Públicas, Directoria Geral de Estatística, *Recenseamento realizado em 1 de Setembro de 1920, vol. 3, parte 1 "Agricultura"* (Rio de Janeiro: Typ. da Estatística, 1926), 8, 26; Gervásio Leite, *O gado na economia matogrossense* (Cuiabá: Escolas Profissionais Salesianos, 1942), 13; Afonso Simões Corrêa, "Pecuária de corte em Mato Grosso do Sul," report delivered at Encontros Regionais de Pecuária de Corte, Brasília, 27 November 1984, 5, 8.

Notes: State population totals also include regions not dedicated to cattle ranching, which accounts for any apparent disparity between figures. "Pantanal (south)" and "Southern Mato Grosso" refer to populations within the present-day state of Mato Grosso do Sul.

"+" indicates numbers are uncertain but are likely higher.

TABLE 4.2. REGISTERED LAND OCCUPATION BY ENVIRONMENTAL
REGION

Region	Year	Millions of hectares	% of total regional land area
Pantanal	1920	6.4	46
	1940	6.8	48
	1972	12.8	91.5
Total Pantanal land area		14.0	100
Cerrado	1920	6.4	37
	1940	8.2	47
	1972	11.3	64.5
Total Cerrado land area		17.5	100
Campo Limpo	1920	3.1	88
	1940	2.9	85
	1972	3.5	100
Total Campo Limpo land area		3.5	100

Sources: Data from Brazil, Ministério da Agricultura, Indústria e Commercio, Directoria Geral de Estatística, *Valor das terras no Brazil, segundo o censo agrícola realizado em 1 de setembro de 1920* (Rio de Janeiro: Typ. da Estatística, 1924), 24–25; Brazil, IBGE, Recenseamento Geral do Brasil (1º de Setembro de 1940), Série Regional, Parte XXII, *Mato Grosso: Censo Demográfico, Censos Econômicos*, 193; Brazil, Ministério das Minas e Energia, Secretaria-Geral, *Projeto Radambrasil: Levantamento de Recursos Naturais, vol. 27, Corumbá*, 412–422, and *vol. 28, Campo Grande*, 376–388 (Rio de Janeiro: IBGE, 1982).

Note: The apparent decline in areas occupied in the campo limpo between 1920 and 1940 is probably the result of a disruption in the local economy during the 1930s, as smallholders suffered much more than large landowners from the world depression.

be exploited in mixed farming. This gradual subdivision and growth in cattle population guaranteed that ranching's relationship with its environment would become more intense over time.[7] (See table 4.2 for land occupation figures between 1920 and 1972.)

Compared with the Pantanal, the cerrado and campo limpo experienced the greatest expansion of cattle raising and human intervention. Visitors to Mato Grosso repeatedly were impressed by the lush native pastures of campo limpo (the Vacaria), seen as a natural area to raise cattle, while many were shocked by the apparent sterility of the cerrado. Yet, thousands of head of cattle were pastured in the cerrado. The occupation of the cerrado and campo limpo between 1910 and

1940 was feverish and eventually led to property subdivision as financial pressure and speculation entered the picture. By the 1940s, there was no easily exploitable land left in campo limpo, as the 1940 census reveals that 85 percent of all land (2,950,000 ha) was in private hands, with 21 percent (632,900 ha) of that covered by forest. Nearly 57 percent (1,984,500 ha) was pasture, only 5 percent (148,400 ha) under cultivation. Data from 1920 indicate that this had changed little over the previous 20 years. Such a static situation inevitably led to property subdivision.[8]

In the cerrado, ranching continued to expand during the period, albeit on poorer land, where ranches had to be larger, on average, to survive. Here, the same census shows some land availability, as 47 percent of cerrado was privately held (8.2 million ha), half of that covered in pasture (4.3 million ha). By comparison, the 1920 census already counted 6.4 million hectares officially under private ownership, 37 percent of the total cerrado area. (See table 4.2.) Occupation and purchase of public land continued on the same scale over the next few decades. By the 1970s, 35 percent of Mato Grosso do Sul territory was dedicated to cattle, and pastures made up 70 percent of all properties surveyed. Properties of less than 1,000 hectares made up 76 percent of the ranches, occupying only 14 percent of ranching land. The phenomenon of widespread in-migration between the 1940s and 1980s, and the division of many properties followed by eventual sale to large enterprises, produced a disequilibrium that significantly altered the regional environment, as will be seen.[9]

State Land Laws, Settlement, and Prices

The first step in bringing some order to the chaotic world of property ownership was the passage of the state land law of 1892, which became the basis of all subsequent legislation. It recognized all land entitled during the empire and accepted all claims to untitled land that could prove "habitual residence" and the use of at least one-third of the parcel. The law also sought to encourage the development of small property over large, following the expressed ideals of the nation. Land grants were discouraged, except for colonies (regulated by the colony law of 1895), while properties were restricted to 900 hectares for agricultural land, 450 hectares of extractive product land (erva mate and rubber, for example), and 3,600 hectares for cattle raising. All land occupiers, whether owners or squatters, were required to register their lands with the municipal mayor's office, which then passed on the information to the newly created State Land Office (Repartição de Terras). Lands were to be surveyed through the auspices of property judges and had to be legalized within four years by approval of the state president. With bureaucratic delays due to inadequate resources, the period was extended to 1909.[10]

As might be expected in a region as remote as Mato Grosso and subject to political passions, state presidents often interpreted the law in distinct ways. Regulations required that old sesmarias and squatted lands be charged excess fees if they measured over the legal stipulations, but presidents were less than consistent in applying the law. There was no legal pattern to each president's decision-making, since some properties were granted fee exemptions at the same time that others were forced to pay. Local political connections were key; smaller holdings of between 200 and 1,500 hectares usually were charged, but the big latifundios of more than 100,000 hectares seldom paid. Clearly, this favored long-time large ranchers and those who were either active or had contacts in regional and state politics.[11]

Large ranchers also were able to legitimate their holdings and get around the law by making separate *posse* (squatting) applications for each *retiro* (line camp) on the ranch. Some ranches had as many as 20 retiros. Also, lands exceeding the legal limit could be bought but were charged a fee per hectare of excess, on top of purchase and surveying costs. Ranchers were able to minimize their expenses by not buying the excess outright but by declaring the posse and paying the fee in installments. Abuses were recognized as early as 1896, when state president Antonio Corrêa da Costa, who applied the law more rigorously than his predecessors and most of his later colleagues, suggested measures to restrict the registration of latifundios. His government also passed a law in 1896 that gave free concessions of 50 hectares to small farmers. This complemented an 1895 law that offered the same to settlers, Brazilian or foreign, in the 66-kilometer border zone. The intention was to attract small farmers who could later expand their holdings by purchase. The 1907 homestead law also sought to limit property sizes. Such small land grants, however, were useless for ranchers, as even small ranching required more land than 50 hectares, which under the conditions of the day could support a maximum of 20 head. Besides the attraction of other regions of Brazil, agriculturalists were discouraged from immigrating to Mato Grosso by the chronic lack of facilities, financing, transportation, and personal security.[12]

Yet immigration did occur and landholdings were registered. Corrêa da Costa noted that by 1895 more than 3,000 holdings, close to 28 million hectares (14 percent of state territory), had already been registered, although only 1.5 million hectares had been legalized. All claims in Corumbá, Miranda, and Nioác were reportedly for cattle ranches, a total of almost 9 million hectares. He did stress that these figures should be read with caution, since property sizes were overestimated by the claimants, claims were often duplicated, and properties held by one person were frequently registered under different names. Nevertheless, these figures give a good indication of the importance of land possession in Mato Grosso at the time.[13]

With the entry of the many gaúcho immigrants and the beginning of railway construction from São Paulo to Corumbá, land occupation and registration increased and prices began to rise. Official figures show that between 1906 and 1910, state income from the sale of public lands and legalization charges rose 250 percent, from 188 contos to 470 contos. In the early years of occupation, land had little value in itself; witness the sale of a ranch in the Vacaria in the 1880s for 50 young steers, valued at 5 milreis a head. Sesmarias of 13,000 hectares were said to sell for 3–20 contos (US$1,500–$10,000) with cattle, only 240 reis to 1.5 milreis per hectare. By the late 1890s, legal posses of 3,600 hectares were being sold by the state for 700 reis per hectare along navigable rivers and 350 reis per hectare two or more kilometers from such waterways. Following a revised land law in 1902, prices for ranching land rose to 800 reis per hectare away from navigable rivers and 1.5 milreis a hectare along the rivers. These rates were the norm for Mato Grosso until the land boom brought about by the construction of the railroad and the purchase of large parcels of land in the south of the state by foreign syndicates. By 1909, land prices in the Vacaria had increased to between 800 reis and 2.2 milreis per hectare, and by 1914 some ranches were selling for as much as 6 or 7 milreis per hectare.[14]

Some of the reason for these dramatic price increases and accompanying speculation can be assigned to the Farquhar syndicate's Brazil Land, Cattle and Packing Company. In order to establish a ranch in the Três Lagoas region near the future railway route, the company bought more than 500,000 hectares from a total of seventeen settlers in 1911–1912 for more than 1,000 contos (US$320,000), 2 milreis a hectare. Until then, land in the region had been selling for as little as 280 reis per hectare. The number of large landholdings increased as well, with over 190,000 hectares destined for cattle raising sold by the state in 1909 alone. In the municipality of Campo Grande, one report claimed that in 1913 there was little good land left after latifundio expansion. Only more remote lands and those with no surface sources of water were still available. With the completion of the railroad in 1914 and the outbreak of war in Europe, the region's dramatic boom in the value of its cattle products initiated a corresponding inflation of land prices as suitable property became scarcer.[15]

THE SCRAMBLE INTENSIFIES: POST—WORLD WAR I

The prewar boom in land occupation and sales intensified through the war and into the 1920s, declining only slightly in the 1930s. In just five years (1914–1919) government income from land transfers more than tripled, while total annual area awarded titles rose from under 400,000 hectares to over one million hectares. The end of the war did not slow this fever. Detailed figures from government reports reveal that between 1920 and 1930, a total of 2,146 properties covering more than 7 million hectares were awarded with definitive or provisional titles, averaging 3,300 hectares per property. Many properties encompassed large expanses dedicated to cattle ranching, including purchases by foreign syndicates. Others were small plots engaged in family agriculture. Generally, the size of properties varied by region. In 1923, for example, only two properties totaling 42,000 hectares were issued provisional titles in Três Lagoas, but 129 properties totaling 200,000 hectares were registered in Ponta Porã. Both regions raised cattle. The difference was that Três Lagoas, by virtue of its location on the railroad and the local ecological conditions of the cerrado, required more land per animal than was required on the campo limpo in Ponta Porã, where a number of small farmers raised crops on properties as small as 50 hectares.[16]

Registration also depended on local and national economic and social conditions, and some years saw greater activity than others. After the temporary recession following World War I, the most dynamic years were 1924 and 1928, when 1,100,000 hectares (1924) and 919,000 hectares (1928) were awarded titles. Other years experienced extraneous constraints, including official land price increases in specific regions as valorization continued apace and, in 1925–1926, the armed rebel Prestes Column, which passed through the south of Mato Grosso and aggravated an economic crisis caused by reduced demand for cattle products in São Paulo. By 1929, the amount of land registered began to decline as the availability of public land in preferred regions was virtually exhausted.[17]

Political reorganization after the 1930 national coup led by Getúlio Vargas disrupted the collection of land data. Federal receivers (*interventores*) were sent to all state capitals under orders to reorganize bureaucracies and to govern in the national, as opposed to state, interest. Not only was collection of data disrupted but also the actual registration of land claims was put off for some years until the Vargas dictatorship could straighten out irregularities in land ownership. As might be expected, the general stagnation of the economy during the 1930s slowed land acquisition, public or private, even further. Between 1929 and 1934, for example, the number of property transfers dropped from a high of 1,560 (1929) to 922 in 1932, remaining at around 1,000 a year for much of the rest of the decade. Tax revenue on land transfers and sales in 1934 amounted to a little over 1.2 million milreis, comparable to 1920, when the postwar recession temporarily depressed all activity, but below the average for the previous decade.[18]

Still, in many areas state land was available into the 1940s, at least for the raising of cattle, and was often considered of excellent quality, especially when viewed from afar and compared with opportunities in other regions of the country. In São Paulo, for example, where public land had been scarce for at least ten years, prices of private land had skyrocketed beyond the means of most ranchers to expand. Land prices saw considerable change in Mato Grosso itself beginning during World War II.[19]

Land Prices

Speculation and the entry of foreign investors were the main reasons land prices rose significantly between 1900 and 1914, but they were modest compared with those of the postwar period. Prices of state-owned land were established in the new land law of 1918, and they differed depending on location. As might be expected, the south of the state, with its fertile soil, had the most expensive land: 3 milreis in 1918 per hectare for cattle and farming land within 6 kilometers of a navigable river or roadway. Land was cheaper by half if beyond the 6-kilometer zone. In other areas of the state, the government offered ranch land near watercourses for 1.5 milreis a hectare and agricultural land at 2 milreis per hectare. Private property sold at whatever price the expanding market could bear. Consequently, the price of some ranches in the Vacaria in 1914 was 6 to 7 milreis per hectare. In 1919, when allegedly there was no unclaimed public land left in the area, the price had gone up to as much as 14 milreis. By the mid-1920s, however, land registry records in Ponta Porã reveal that prices declined somewhat, probably in partial response to the availability of cheaper land elsewhere. Most sales noted during the decade were for improved ranches or farms of between 400 and 4,000 hectares, ranging between 5 and 12 milreis per hectare, depending on the quality and number of upgrades. Also of note was the subdivision of properties originally purchased from the state. This could be construed as a form of speculation, but it seems it was primarily a response to the high cost of maintaining land and the opportunity to generate some cash for property improvement offered by a seller's market. As time went on, the size of parcels began to shrink. In 1921 the average sale was 3,000 hectares, but by 1929–1930 this had dropped to 1,000 hectares.[20]

Overall, between 1919 and 1929, land prices in southern Mato Grosso rose from 5 milreis per hectare to as much as 18 milreis. Campo Grande and Nioác municipalities had the most expensive land, with land values calculated in 1920 between 18 and 22 milreis per hectare. Comparable values were recorded in Bela Vista and Aquidauana, but in the Pantanal, land values seemed to depend on proximity to markets. In Corumbá municipality, values were around 7 or 8 milreis per hectare, in Miranda 11 to 14 milreis, and in Poconé, near the capital city of Cuiabá, between 15 and 17 milreis per hectare. In the remote municipality of Santana do Paranaíba, where drought was common, values reached 7 milreis a hectare. Through the 1920s, prices across the state averaged between 3 and 8 milreis for vacant state land and between 10 and 20 milreis for private unimproved land, although land with planted pasture could fetch upwards of 50 milreis per hectare. This was at a time when prices in other regions of the country were 10 to 20 times higher. For example, in 1923 in western São Paulo, prices ranged from 50 milreis in the most remote areas of the state to over 1,000 milreis a hectare in the most heavily populated zones near transportation and with improvements. In the Triângulo Mineiro cattle-raising areas of Minas Gerais, prices averaged between 50 and 300 milreis per hectare[21] (see table 4.3).

As did most of the economy, land prices suffered from the contraction of the 1930s. Between 1930 and 1935, it was possible to pay as little as 6 milreis for a hectare, even along the railroad. Beyond the rail line and roadways, most prices had come down below 5 milreis. Better quality

TABLE 4.3. LAND PRICES IN SELECTED MUNICIPALITIES, 1923

State and Municipality	Average Price (in milreis per hectare)	
	With Improvements	Without Improvements
Mato Grosso		
Aquidauana	14	11
Bela Vista	18	15
Cáceres	4	3
Campo Grande	19	17
Corumbá	8	7
Coxim	3	3
Miranda	14	11
Nioác	22	18
Paranaíba	7	5
Poconé	17	15
Ponta Porã	14	9
Porto Murtinho	15	13
Três Lagoas	17	13
State average		
Vacant land	3–8	
Private for ranching/agriculture	10–20	
Minas Gerais		
Araguari	71	63
Araxá	27	24
Ituitaba	36	30
Monte Alegre	35	30
Monte Carmelo	48	42
Patos	58	51
Patrocinio	107	99
Uberaba	56	46
State average		
Central zone, ranching	60–200	
Triângulo Mineiro, natural pasture	50–200	
Triângulo Mineiro, artificial pasture	up to 300	
São Paulo		
Barretos	200	173
Bauru	180	148
Franca	237	201
Pennápolis	140	109
Ribeirão Preto	1,099	895

TABLE 4.3. (CONTINUED)

State and Municipality	Average Price (in milreis per hectare)	
	With Improvements	Without Improvements
State average		
Plains (far from roads and railways)	50–100	
Plains (near roads and railways)	100–300	
"Roxa" soils (far from roads and railways)	60–300	
"Roxa" soils (near roads and railways)	200–500	
Rio Grande do Sul		
Alegrete	121	107
Bagé	196	181
Bento Gonçalves	171	140
Cachoeira	111	93
Santana do Livramento	148	135
Santa Maria	97	85
São Borja	81	69
Uruguaiana	157	147
State average		
Coastal stock raising	80–200	
Prairie stock raising	130–350	

Sources: Data from Brazil, *Valor das terras no Brazil . . . 1920*, 24–30, 41–50; US Department of State, Report by A. Gaulin, US Consul in Rio de Janeiro, to Department of Agriculture, April 23, 1924 "Land Prices in Brazil," *Brazil*, entry 5, box 63, Record Group 166, USNA.
Note: In 1920 the milreis was worth US$0.21.

and improved lands naturally commanded higher prices, up to 36 milreis along the railway, but even those were low compared with prices in other states. By 1939, however, prices began to rise again, with the best unimproved land in the state selling for between 50 and 70 milreis, while poorer quality land commanded 10 milreis per hectare. Planted pasture, though rare, fetched up to 200 milreis per hectare.[22]

The prices rose partly in response to a gradually improving national economy, but some credit is due the nationalist Estado Novo's (1937–1945) promotion of settlement in Brazil's interior. Vargas's "Marcha para o Oeste" program to occupy Brazil's backlands as an integral part of the nation's development led to the establishment of a national agricultural colony at Dourados in southern Mato Grosso in the early 1940s, further valorizing land in that region and throughout the state. By 1952, essentially there was no public land left in southern Mato Grosso. With the exception of Dourados, until this time Mato Grosso offered opportunities only to those either rich or intrepid enough to risk settlement in the more remote frontier regions. Poor subsistence farmers were discouraged not only by prices but also by official action and the control of landholding by large and medium-sized ranchers. Land was cheap compared with other states, but it was largely destined for extensive ranching, seldom agriculture, except near major population centers like Campo Grande, Corumbá, and Ponta Porã. Cattle had played and would continue to play a principal role in determining land tenure throughout this period and well into the following decades.[23]

Ranching and Large Property

The expansion of ranching between 1914 and the 1940s involved a great deal of territory that before had hosted little human occupation. Ranches varied widely in size but tended to be large, particularly in the Pantanal and the cerrado. The presence of large ranches meant that the most successful of the original ranchers, plus new ranchers with access to a sizable amount of capital, dominated the raising of cattle. In the market of the time, small ranchers (with 500 to 1,000 head on 2,500 hectares or less) found it hard to supply even local demand. This was the character of cattle raising, naturally tending to large ranches, particularly in remote areas.

Many of these properties were foreign-owned. For example, Brazil Land was soon followed by a number of other foreign firms, from England, France, Argentina, and Uruguay. These became the largest landholders in Mato Grosso, with the most sizable herds. Throughout the war years and into the 1920s, they accumulated land at impressive rates. In the 1920s, Brazil Land already controlled more than 1.6 million hectares, and Sociedade Anónima Fomento Argentino-Sud Americana claimed one million hectares. Other major foreign ranching enterprises at the time included Fazendas Francesas (French), Miranda Estância (English), Brazilian Meat Company (English), Barranco Branco (Uruguayan), and Mate Larangeira (Argentine). These operations continued to be prominent over the next two decades, some accumulating even larger parcels. By the late 1930s, it was reported that nine foreign

TABLE 4.4. PROPERTIES OF FOREIGN COMPANIES IN MATO
GROSSO, 1941

Company	Municipality	Hectares
Brazil Land, Cattle and Packing Company	Paranaíba	190,000
	Três Lagoas	759,087
	Campo Grande	146,379
	Cáceres and Corumbá	763,508
Total		1,858,974
Brazilian Meat Company	Três Lagoas	311,010
	Campo Grande	250,000
	Aquidauana	5,000
Total		566,010
Fazendas Francesas	Miranda	246,456
	Corumbá	172,352
Total		418,808
Miranda Estância Co.	Miranda	219,000
Agua Limpa Syndicate	Três Lagoas	180,000
Sociedade Anon. Rio Branco	Corumbá	549,156
Sud-Americana Belga S.A.	Corumbá	117,060
Soc. Anon. Fomento Argentino	Corumbá	1,001,077
Cia. Mate Larangeira	Bela Vista	164,590
	Dourados and Ponta Porã	180,436
Total		345,026
All foreign holdings		4,878,191

Source: Data from US Department of State, Report "Foreign Holdings in Mato Grosso,"
by Roger L. Heacock, US vice-consul, São Paulo, 23 January 1941, no. 479, Brazil, entry 5,
box 20, Record Group 166, USNA.

companies owned more than 4.8 million hectares in eight municipalities in southern Mato Grosso, covering 13.6 percent of present-day
Mato Grosso do Sul. Brazil Land held the most, 1.8 million hectares in
18 ranches (though some of these ranked as retiros of large ranches).[24]
(See table 4.4.)

After the export boom of World War I and some uncertainties over

land control, Brazil Land was able to pursue its program of modernization of the ranching industry envisioned by its owner, Percival Farquhar. During the war years the company was placed under receiver-
ship by a US court, a result of overextension of Farquhar investments
just before the international economic recession of 1913. The company
did so well during the war that by 1919 control was restored to Farquhar
and his Brazil Railroad Company. Brazil Land then prospered through
the 1920s and 1930s. Comparable to the British experience in Patago-
nian sheep ranching and elsewhere, the company's intention was to
bring European and North American ranching techniques to Central
Brazil, where ranching was considered to be backward.²⁵ For this reason, Murdo Mackenzie (seen earlier, in chapter 2) was hired to oversee
all Farquhar ranching operations. Mackenzie brought a number of his
compatriots with him, hard-nosed businessmen who had learned the
cattle business in Texas and were not averse to some discomfort if the
salary was right. These men became managers of the many ranches in
the Farquhar empire. They immediately began stringing fences, planting exotic pasture grasses, and importing purebred English and Texas
bulls for breeding. These men defined ranching "improvement" as the
importation of technology and methods that had proved so successful
in the United States.²⁶

Only a few Brazilian entrepreneurs had the resources to purchase and maintain such large holdings. Of these, most owned ranches
in the cerrado and the Pantanal, including those whose families had
been in the region for generations. The Gomes da Silva family, for
example, controlled more than 180,000 hectares in the Pantanal, divided between four owners and pasturing more than 20,000 head.
The Alves Ribeiro family in the eastern Pantanal also held a considerable expanse of land, in excess of 200,000 hectares. In the cerrado,
particularly Campo Grande municipality, most ranch holdings were
not as extensive, though they still covered considerable territory. In
1919, Domingos Barbosa Martins controlled over 350,000 hectares, and
at the end of the 1920s three other members of the Barbosa Martins
family held over 85,000 hectares in five ranches, while a distant relative owned 57,000 hectares. Another landowner, whose primary occupation was organizing and leading cattle drives, held title to almost
50,000 hectares.²⁷

The average ranch did not approach such proportions. Instead, depending on the region, ranches ranged between 2,000 and 10,000 hect-

ares and pastured from 500 to 2,000 or 3,000 head. Despite the land rush previously described, by the 1930s and into the 1940s the number of ranches increased. This was the result of fragmentation of some large holdings rather than a continued influx onto unoccupied land, which is confirmed when census figures for 1920 and 1940 are compared. Out of a total of 2,914 rural properties in 1920, 641 were over 5,000 hectares (272 of those were over 10,000 hectares and 127 over 25,000 hectares). The 1940 census, from a total of 7,476 properties enumerated, noted 2,300 properties over 5,000 hectares. Despite considerably more registered properties, the proportion of large holdings had changed little between 1920 and 1940. In area, the 1920 census included just over 14 million hectares of property over 5,000 hectares, and the 1940 census recorded less than 14.2 million hectares in large-scale ranching (ranches under 5,000 hectares occupied just under 2.5 million hectares). The trends are obvious; although the number of large holdings in ranching had increased almost fourfold, the total amount of land grazed had not risen much at all. This suggests considerable fragmentation of large holdings, a pattern that continued into the 1950s. It is surprising that observers made only passing reference to the phenomenon. Apparently it was the result of several factors, including subdivision within families, previous speculation leading to sale, a need to rid oneself of some property during the lean years of the early 1920s and the mid-1930s, and the resolution of title disputes to the benefit of smallholders[28] (see tables 4.5, 4.6, and 4.7).

It is worth pointing out that the municipalities hosting the largest number of substantial holdings in 1920 were the same in 1940 and 1950. They included Três Lagoas, Corumbá, Campo Grande, and Aquidauana, all important cattle-raising areas. It is also important to note that small-scale agriculture was growing in Mato Grosso, thriving near urban centers like Campo Grande and Cuiabá, but most small holdings were dedicated to mixed farming, with an emphasis on ranching, either subsistence or providing limited produce and milk to local urban centers. This also was the case in and around Ponta Porã and Bela Vista, though by the late 1940s a statewide increase in urban population had encouraged more small-scale ventures. Still, these comparisons illustrate how land use had changed by degrees over the years. As the population grew, more land was brought into use for mixed farming, yet unquestionably cattle dominated in the Mato Grosso rural sector, in large and small holdings alike. This continued until the 1970s, when

TABLE 4.5. LARGE AND SMALL PROPERTY IN MATO GROSSO
MUNICIPALITIES, 1920

Municipality	Small Holdings (Under 5,000 ha)		Large Holdings (Over 5,000 ha)	
	Total Area	Number	Total Area	Number
Aquidauana	197,404	95	1,021,690	44
Bela Vista	274, 193	135	557,870	36
Cáceres	83,968	78	1,185,799	24
Campo Grande	593,026	473	1,479,474	110
Corumbá	108,470	87	1,880,086	45
Coxim	272,783	82	1,151,211	69
Miranda	174,142	87	643,677	37
Nioác	223,115	108	515,242	41
Paranaíba	296,451	237	1,218,909	60
Poconé	118,084	65	412,986	29
Ponta Porã	483,359	272	2,638,129	66
Porto Murtinho	66,370	31	495,386	15
Três Lagoas	252,715	256	878,559	65
State total	3,144,088	2,006	14,079,018	641

Source: Data from Brazil, Ministério da Agricultura, *Recenseamento realizado em 1 de Setembro de 1920, vol. 3, parte 1 "Agricultura"* (Rio de Janeiro: Typ. da Estatística, 1926), 150–153.

commercial agriculture came to play an important, though still secondary, role.[29]

Conflict over Land

Conflict over access to land was common in the wake of the pre-1914 speculative boom. The Pantanal experienced relatively little discord, perhaps because most of the land already had been occupied by several interconnected families since the mid-nineteenth century, and local environmental conditions limited opportunities for small holders. This was not the case in the cerrado and Vacaria, where long-time ranchers had not established a rigid regime of land control and where environmental conditions and access to markets were much more favorable than in the Pantanal. Significant conflict occurred along the Paraguayan border, especially between small ranchers and Mate Larangeira. Other claimants in the region also entered into disputes over jurisdiction. Much of the resulting litigation was by descendants of the original settlers, who had arrived in the region before the Paraguayan War, or involved squatters who had settled on lands and begun the process of legitimization on the supposition the land was public. Most disputes occurred between 1910 and 1915 and seem to have been prompted not only by real occupation but also by an attempt to take advantage of the speculative boom of the time.[30]

For example, a dispute arose in 1914 between the heirs of nineteenth-century claimants and some 3,000 squatter families near Bela Vista. The property, Fazenda do Apa (sometimes Fazenda São Rafael do Estrela), extended over 2.5 million hectares, along the Apa River. It was claimed

TABLE 4.6. LARGE AND SMALL PROPERTY IN MATO GROSSO
MUNICIPALITIES, 1940

Municipality	Small Holdings (Under 5,000 ha)		Large Holdings (Over 5,000 ha)	
	Total Area	Number	Total Area	Number
Aquidauana	138,813	80	1,520,069	187
Bela Vista	72,616	68	552,407	129
Cáceres	76,194	109	725,000	1
Campo Grande	70,633	1,127	1,885,016	321
Corumbá	114,204	52	2,172,899	159
Dourados	37,553	95	52,643	32
Entre Rios	74,237	650	451,371	86
Herculânea (Coxim)	195,891	29	759,500	127
Maracajú	58,238	95	449,245	137
Miranda	39,584	53	196,163	60
Nioác	59,561	43	189,754	68
Paranaíba	310,095	290	808,520	128
Poconé	281,068	1,372	540,678	140
Ponta Porã	424,041	863	662,415	144
Porto Murtinho	28,921	22	970,200	57
Três Lagoas	494,665	228	2,243,304	257
State total	2,476,314	5,176	14,179,184	2,033

Source: Data from Brazil, IBGE, Recenseamento Geral do Brasil (1° de Setembro de 1940), Série Regional, Parte XXII, Mato Grosso: Censo Demográfico, Censos Econômicos (Rio de Janeiro: Serviço Gráfico do IBGE, 1952), 196–197.

TABLE 4.7. LARGE AND SMALL PROPERTY IN MATO GROSSO
MUNICIPALITIES, 1950

	Small Holdings (Under 5,000 ha)		Large Holdings (Over 5,000 ha)	
Municipality	Total Area	Number	Total Area	Number
Aquidauana	385,789	221	1,243,360[a]	87
Bela Vista	335,490	266	307,289[a]	37
Cáceres	159,552	87	1,090,582	48
Campo Grande	644,928	717	2,284,273	82
Corumbá	366,317	198	3,587,404	165
Dourados	234,341	308	52,448[a]	15
Maracajú	238,179	147	246,365[a]	29
Miranda	92,234	60	1,112,218	19
Nioác	244,320	206	158,246[a]	17
Paranaíba	748,576	729	645,702[a]	60
Poconé	290,453	352	126,545[a]	28
Ponta Porã	304,756	416	437,814	23
Porto Murtinho	120,644	118	474,699[a]	24
Rio Pardo	211,551	100	1,160,188	70
Três Lagoas	380,260	250	2,133,194	143
State total	10,576,500	8,606	19,865,395	1,232

Source: Data from Brazil, IBGE, Conselho Nacional de Estatística, Serviço Nacional de Recenseamento, Série Nacional, Vol. 2: *Censo Agrícola do Brasil, 1950* (Rio de Janeiro: IBGE, 1956), 136–137.

[a]These numbers are undercounts since properties over 100,000 hectares were not recorded.

by the heirs of Dona Rafaela López, sister of the nineteenth-century Paraguayan dictator Francisco Solano López, and by the heirs of a Brazilian woman, also called Dona Rafaela, who had settled in the area with her husband before the Paraguayan War. The Brazilian Rafaela had been married to one of the original settlers in southern Mato Grosso, Gabriel Francisco Lopes, and had been in the region since the 1840s. The Paraguayan Rafaela, who claimed she bought the property from the Paraguayan government in 1863, had married a Brazilian after the war, which gave her claim legal standing in Brazilian courts. Although the area had always been claimed by Brazil, it was also accepted that Paraguay maintained an official presence in the region before hostilities. Allegedly, both families had settled and prospered in the area after the war, with no conflict between them. By 1914, both women were dead, and action was taken by their children. The claims were confusing, not only because the names were similar but also because documentation was in question. The civil upheavals between 1896 and 1910 destroyed many records, depriving the Brazilian Rafaela of proof of long-term occupancy. In the end, only documents filed by the husband of the Paraguayan Rafaela were found to substantiate the claims of her heirs. Questions involving other settlers, however, remained unresolved.[31]

Many of the settlers on the disputed property believed they had squatted on public land. Still others took advantage of the chaotic situation of land tenure in Mato Grosso to claim land as theirs by right of occupation. Apparently, the state was unclear what it owned, and it regularly awarded provisional and definitive titles to settlers on lands claimed by absentee landlords and old families, like the Fazenda São Rafael. Initially, there was plenty of land to go around, and many of these landowners had permitted squatting on their land without consideration of future consequences. Before they realized it, vast holdings that had been in the family for generations were occupied by land-hungry settlers. In certain ways the old families had themselves to blame, for they held more land than they reasonably could farm or ranch. Perhaps they were victims of their own traditions, since few made the effort to consolidate their holdings by subdivision among family and friends (as was done in the Pantanal) or by outright coercive measures (as was common in the Paranaíba and Três Lagoas areas). This was especially the case in the Vacaria and Bela Vista region, where many such fazendas were located. The haste of the government to legalize many dubious holdings, often facilitated by graft, only exacerbated the situation. In the case of Fazenda São Rafael, there is no evidence that the conflict was resolved to the satisfaction of the claimants, but considering the number of squatters and the decision of the state to continue issuing titles during litigation, it is unlikely that the claim was successful. In the ultimate analysis, Fazenda São Rafael was an example of land tenure transformations in southern Mato Grosso. By ignoring the trends, the owners of a large expanse of property were overwhelmed by a land hunger they didn't understand and were incapable of foreseeing. Fazenda São Rafael was simply one example of a large, minimally exploited holding established in the nineteenth century and forcibly broken up by the tide of settlement sweeping over the state. The fazenda's fate explains to a great extent why the average size of large ranches declined between 1920 and 1940.[32]

Brazil Land also found itself in dispute over land. Early in the acquisition process some settlers protested that the company was trying to buy land from the state that was not public but instead belonged to them. Apparently these objections were not heard, for from 1913 to 1918 the company found its efforts to survey and utilize one parcel blocked by groups of armed men. Employees were threatened and work was paralyzed until 1919, when federal police were sent in. Part of the problem was political, between the company lawyer and the leader of the protestors, and was not resolved until well into the 1920s. Still, most attempts by wealthy or powerful ranchers and syndicates to gain possession of land were successful. During the boom in the Três Lagoas area, for example, smallholders who were not willing to sell were soon forced to do so by well-placed politicians and large ranchers "backed by armed force."[33]

Uncertain property titles and dubious sales motivated demands for some sort of control. The state administration, however, was often forced by fiscal necessity to grant provisional titles to wealthy companies and individuals who could pay well for the privilege, regardless of who might have been living on the land at the time. This sometimes created problems with federal authorities. This was the case of Fomento Argentino-Sud Americano.

Two years after acquiring a private lease on more than four million hectares along the Paraguay River in 1908, Fomento Argentino proposed to the state government that the company buy more than a million hectares of the land outright, in return relinquishing the rest. The state government of the day was uncomfortable permitting such extensive land control by a foreign enterprise, as expressed by President Pedro Celestino in 1911: "It is not reasonable that the wealthy of today acquire [lands] for low prices, preserve them unutilized with prejudice to the State, [in order to] sell them tomorrow for exorbitant prices." Yet Celestino granted permission after the company promised to establish a ranch that could act as a stimulus for modernization by local ranchers. The question, however, did not involve solely the state government. Much of the land was within the 66-kilometer border zone under federal jurisdiction, and approval for foreign purchase first had to be granted by Rio de Janeiro. In the past the regulation had often been ignored or permission granted without question, but by this time, perceptions in the nation's capital were beginning to change. The speculation that accompanied the land boom in Mato Grosso drew federal attention to conditions in the state. As a result, Rio took over ten years to decide on the case, and in 1920 it refused authorization. Meanwhile, the state continued to process the company's application for title and, after payment in 1921, issued conditional title to 274,000 hectares outside the border zone.[34]

Such uncertainty guaranteed that the company would not fulfill the promises made to Celestino. The property was largely left to deteriorate, permitting the entry of speculators, rustlers, and *aves de rapina* (plunderers), who operated with impunity in the area. This, plus the property's proximity to the border, caused the Vargas government in 1931 to order its confiscation. Once the government acquired the land, little was done either to divide it into smaller ranches or to sell it off to large national operators. During the 1930s, the area was invaded by wealthy speculators from São Paulo and Paraná, who only intensified the climate of speculation, smuggling, and intimidation, to the detriment of small squatters. Finally, in the 1950s most of the land was sold to a São Paulo entrepreneur, who then directed the property's first real development, in ranching and agriculture.[35]

The other side of the coin was the fate of smallholders in Mato Grosso, most whom were squatters. They were often subject to abuse from large landowners who evicted them as "intruders," even though these new potentates frequently were themselves the intruders. As Virgílio Corrêa Filho emphasized, concerning the years immediately after World War I, in many cases wealthy syndicates took advantage of the land rush to legalize property with dubious titles. The land was then sold to foreign buyers, who were able and willing to pay prices far in excess of local market value. This speculative process encouraged the practice of *grilagem*, claim-jumping of property, common throughout Brazil. Some unscrupulous individuals made their living from the activity. In Mato Grosso, grilagem was practiced primarily in two forms: the falsification of documents from the years just after the 1850 national land law, which allegedly confirmed original and long-term occupation; and the theft from land offices of old blank receipts issued for payment of property transfer taxes. This latter method was done with the cooperation of land office clerks, and the stolen receipts were duly made out in the name of the highest bidder. It is easy to imagine the predicament in which many smallholders found themselves, trying to defend their rights either as squatters or even as legal owners. In some cases, if legal owners refused to sell out, despite high prices, intimidation was employed. Several wealthy speculators with political connections, especially in Três Lagoas and Paranaíba municipalities, acquired large properties through threat. Most smallholders were simply too weak to resist such pressure from powerful ranchers or companies, and as long as there was the possibility of healthy profits in land speculation, coercion prevailed. The government's uncertainty over the extent of its control of land facilitated intimidation. The downturn of the economy in the 1930s, coupled with some tough laws decreed by the centralizing Vargas dictatorship during that decade and into the 1940s, helped to diminish conflict, though it did not disappear and periodically plagued the rural sector of the state into the 1960s and 1970s.[36]

Credit

One significant element that helped to exacerbate the struggle over land was credit. Rural credit scarcity was not new in Mato Grosso or, for that matter, in Brazil. From the mid-nineteenth century, Brazilian agricultural lobbyists had made many attempts to encourage the national government to facilitate the formation of rural credit banks, most particularly to finance large property. While coffee was favored throughout the period, there were several calls for rural banks and cooperatives in other sectors by the Sociedade Nacional de Agricultura (SNA), an association of primarily agricultural elites who sought the agricultural diversification of the nation through promotion of regional rural development. Given the continued repetition of these concerns by the Sociedade in publications and congresses, it is clear Rio was not willing or able to satisfy most of their credit demands, though the government occasionally provided relaxed tariffs and subsidies in special cases. Much of the reason may have had to do with the accepted understanding of the role of banks, which until the twentieth century favored private institutions over publicly supported development bodies.[37]

Until World War I the only commercial bank in Mato Grosso was the Banco Rio e Mato Grosso, created in 1891 by the politically influential Murtinho family, in association with several important banking investors from Rio. The bank, however, did not function as a credit institution; instead it was more a holding company for the expansion of the erva activities of Mate Larangeira. As a result, its capital and expertise were not offered to smaller producers or even to large cattle operations. Its demise in 1902 was largely the product of state political rivalries.[38]

In the absence of other commercial banks, financial transactions were routed through local merchants, particularly those in Corumbá who had accounts in the major banks of Rio de Janeiro, Montevideo, or Buenos Aires. Since credit was private, access and interest rates were quite arbitrary, and often small producers or nascent industrialists were denied loans, except at usurious rates. For ranchers, collateral was understandably land or cattle, which meant that small ranchers ran a much greater risk than did those with vast holdings. Commercial banks would have imposed the same obstacles, but at least they offered an administration and fixed regulations, and the borrowers knew what to expect. Private lenders could call in loans at any time, and often did. Many ranchers had to rely on annual sales to pay off loans, frequently owed to the very drovers who bought their cattle, though some of the larger operators were able to contract credit directly in Rio or São Paulo, often from foreign bankers. For example, Percival Farquhar reported that just before World War I, the Swiss manager of the Brazilian office of the Banco Francês e Italiano, Louis Dapples, extended credit to Mato Grosso ranchers for their drives to Barretos in São Paulo. Dapples believed this was a much safer investment than either securities or real estate. Considering the subsequent boom, he proved to be a minor financial prophet.[39]

Most credit on goods was extended on expectation of successful seasonal animal sales, which put both ranchers and lenders in a difficult position, for local economic prosperity thereby depended on the buyers, severely limiting the independence of ranchers and their partners. If sales were less profitable than anticipated, either lenders were left without sufficient operating capital or, more likely, ranchers were forced to sell cows and heifers in order to pay their bills. These conditions led to constant calls for the establishment of commercial banks in the state, especially with the onset of World War I.[40]

After the Murtinho venture, the first commercial bank in Mato Grosso was a branch of the Bank of Brazil, established in 1916 in Corumbá, followed by some others responding to the boom of World War I. The Corumbá branch was not particularly useful to the region. It had little lending capital, even for urban entrepreneurs, reflecting the general outside perception of the region. Apparently, risks were too great when collateral value was limited. Despite the actions of the SNA, throughout the period rural lending in Brazil remained most intense in the state of São Paulo, primarily for the expansion of coffee cultivation. Remaining funds might then find their way to owners of winter pastures or slaughterhouses, but rarely to the source of the raw material, especially in remote areas like Mato Grosso.[41]

Some observers blamed the government for not acting to attract more credit to the state. In 1922, for example, a rancher decried that in a state with nearly three million head of cattle there were only five bank branches, three of the Bank of Brazil, in Cuiabá, Corumbá, and Três Lagoas, and two of the Banco Nacional de Comercio, in Campo Grande and Corumbá. Yet in a state that relied on agricultural production for its economic prosperity, none of these specialized in rural credit. Another observer, also a Mato Grosso rancher, noted that the federal government had issued regulations in 1922 for the provision of rural credit by the Bank of Brazil, but nothing had been done. He argued that considering the cattle crisis of the time, it was necessary that the government act on its laws in order to aid the industry in Mato Grosso.[42]

Little was done through the 1920s. President Corrêa da Costa's report for 1925 pointed out that the chronic lack of rural credit was a significant factor in the nearly stagnant agricultural and ranching sector. He suggested that the state offer concessions for the establishment of credit cooperatives in order to stimulate the rural economy. This idea apparently went unheeded, for a study undertaken in 1929 noted that Campo Grande hosted only one bank, a branch of the Bank of Brazil, since the Banco Nacional de Comercio had recently closed its doors. Ranchers lucky enough to be awarded credit by this institution were subject to transactions that took up to four months, with interest rates of 12 to 18 percent, and no guarantees. Farmers and small ranchers were shut out of borrowing, and because of the onerous terms, those ranchers who could preferred to borrow in Rio or São Paulo. This state of affairs continued through the 1930s and into World War II, when the federal government provided credit through the national Farm and Ranch Fund (Carteira Agropecuária), which helped to stimulate the cattle sector nationwide to meet wartime needs. In the postwar years, however, the fund was closed, forcing the sale of cattle below market value, as in the period after World War I, to the detriment of the regional economy. Credit availability in the rural sector quickly returned to prewar conditions.[43]

This constant drought in credit for ranchers reveals almost complete government inertia, both in Rio and Cuiabá. As has been common in Brazilian history, good ideas abounded, but limited concrete action was taken. Programs that did get off the ground were usually in response to crises, not geared to long-term development. It would be decades before serious agricultural planning was put into action.⁴⁴ Meanwhile, ranching retained much of its traditional character over the decades, although some changes did occur.

CONFLICT AND LOSS: RANCHING AND NATIVE PEOPLES

Throughout the Americas, the neo-European frontier encroached dramatically on the lives of aboriginal peoples, who frequently were forced to acculturate to the expanding new society, or face brutal consequences. A constant in Brazilian history, such treatment has continued in some form to the present day. Mato Grosso represented no variation in the customary Brazilian attitude toward native peoples, except that their treatment during the period stimulated the first serious national attempts to deal with conflict brought on by the expansion of Brazil into its remote interior. Ranching's relationship with indigenous communities in Mato Grosso was pivotal in prompting such efforts.

As explained in chapter 2, the natives of Mato Grosso were caught in regional rivalries, first between Spain and Portugal and later between Paraguay and Brazil. Some groups were able to exploit the situation to their advantage; others became victims of international and local expansionism. By the mid-nineteenth century, a modus vivendi had been established in the remote Brazilian province whereby some groups remained aloof of the Brazilian settlers, while others participated in a limited manner in the regional economy. The main activities of those groups who did interact with nonindigenous settlers were trading cattle, horses, crops, and handicrafts to the settlers and military garrisons, and working as guides, canoe handlers, and ranch hands on the larger ranches, particularly those near the Paraguay River.⁴⁵

Occasional conflict did occur between settlers and natives, mostly evident along the Cuiabá-Goiás road, and there were some successful attempts at conversion to Christianity, particularly of the Terena near Miranda. But for the most part, direct contacts with nonindigenous Brazilians until the 1860s were either businesslike or detached, with only occasional disruption of native society. This changed with the Paraguayan War, which was a traumatic experience for indigenous peoples and a watershed in their history and survival. Some of the smaller groups simply ceased to exist in the years following the war. Others found their traditional societal structures fundamentally challenged and eventually weakened. The expansion of ranching was a significant factor in this cultural onslaught, by design of some ranchers, unwittingly by others.[46]

For the invading Paraguayan forces in 1864, anyone living in Brazil was considered the enemy, including peoples who had only minimal contact with Brazilian settlers and ranchers beforehand. Some Indian groups allied immediately with Brazil, others played both sides. Still others tried to avoid the conflict. Eventually, indiscriminate treatment by the Paraguayans guaranteed active assistance in the weak Brazilian resistance by virtually all natives. War's end brought nominal recognition from Rio of the aboriginal contribution to the nation's defense, but few concrete rewards. The war had caused the dispersal of most tribes, and their return to traditional territory was sporadic and unorganized. By itself, perhaps, this would not have been as critical to aboriginal cultural survival, but with the postwar increase in attention to Mato Grosso from outside the region, native peoples soon felt the pressure of neo-European immigration. Experiences common to other regions of Brazil and the Americas were recorded in Mato Grosso, including population decimation, exploitation, violent conflict, and widespread acculturation.[47]

Incomplete censuses done soon after the war indicated a still relatively heterogeneous native population, including more than thirty tribes throughout Mato Grosso, and nine major groupings in the south. The estimates did not include all aboriginal peoples and thus provide no absolute idea of numbers, but an unofficial census made in the late 1870s counted between 8,000 and 9,000, while suggesting that there were probably some 25,000 natives in the entire province. The largest groups counted were the Terena, with 2,200, and the Kadiwéu, numbering 1,600.[48]

The 1872 report placed the natives in three categories: those living in villages (*aldeias*) with regular contact with Brazilians; seminomads who had irregular contact with the settlers and authorities; and "hostile" groups, who either engaged in sporadic raids against the settlers or tried to avoid any contact whatsoever. Following Brazilian imperial policy, the report suggested increased peaceful contact with the natives, including trade, the presenting of gifts (usually tools and agricultural implements), offers of education, and other attempts to secure both nonnative settlement in the province and the loosely controlled borders with Bolivia and Paraguay. The report also included a request for money and personnel to carry out the recommendations. This was a common policy of the imperial government, but it had mixed success in aiding settlement in the region. Imperial authorities seldom came up with sufficient financing, and there was a chronic lack of experienced personnel to carry out any peaceful "pacification" efforts on behalf of the government. Treatment of the native peoples in Mato Grosso was left largely to the settlers themselves, whose actions rarely conformed to the government's stated goals. The result was a predictable combination of accommodation, exploitation, conflict, and acculturation.[49]

Not all native groups had the same experiences, however. Some, like the Terena, were able to reach something of an accommodation, often involving labor in the ranching sector. Others experienced severe pressure on land, resulting in considerable conflict. This latter case was most notable for the Kadiwéu.

Guaikuru/Kadiwéu

Within only two or three years of the war, most districts had been assigned an Indian director, usually a local settler, rancher, or military commander, and frequently the interests of the director and those of the natives were polar opposites. One element of potential strain on relations between natives and neo-Europeans was the indigenous culture. Members of the Guaikuru culture in the southern Pantanal along tributaries of the Paraguay River, the Kadiwéu were seminomadic and traditionally expressed an unwillingness to respect Brazilian jurisdiction in the area, including concepts of property. This was evident before and even during the Paraguayan War, when the Guaikuru fought the Paraguayans not as allies of Brazil but as a separate invaded nation. Another factor was the reputation of the Guaikuru. Sometimes referred to as "Indian knights," for their equestrian skills, they offered some of the fiercest resistance to early Mato Grosso settlers in a highly effective alliance with the water-borne Paiagua. Probably most important was the location of the group, which occupied some of the finest pasture land along the Paraguay River, land that was coveted by several ranchers despite being part of a reserve of 600,000 hectares apparently awarded by the Rio government in gratitude for Guaikuru resistance to the Paraguayans.[50]

The most blatant example of native-rancher conflict over land involved the Portuguese immigrant Antonio Joaquim Malheiros between the mid-1880s and the early twentieth century. Malheiros began his career in Corumbá immediately following the war under the protection of the president of the province, the Baron of Maracajú. Like several immigrants of the time, he supplied ships and goods to the various expeditions passing through the region during the 1870s. As reward for his services in provisioning two government boundary expeditions, he was given the directorship of the Kadiwéu reserve. A more unfortunate choice could not have been made. Malheiros quickly took advantage of his position in 1878 by buying a vast territory, reportedly covering more than 200 square leagues (720,000 hectares). It included a large section of the reserve, for which he paid the natives a paltry 300 milreis (US$625). Malheiros established a ranch with several thousand head and used his position to force the Kadiwéu to work for him with minimal compensation, employing corporal punishment and imprisonment as his coercive tools.[51]

In 1885, Kadiwéu chief Nawila traveled to Cuiabá to complain about the treatment. Supported by testimony from the Brazilian captain of the steamboat on which he traveled, Nawila was well received by the provincial Director General of Indians. The director then recommended that Malheiros be replaced, but this advice was not followed, probably thanks to Malheiros's contacts in Cuiabá. Malheiros remained in control of the Kadiwéu for another fourteen years, provoking greater conflict with the tribe. Until he was removed from his post the Kadiwéu lived in a state of periodic war with the man who was officially responsible for protecting their interests. At the behest of Malheiros, federal troops were deployed against the Kadiwéu, particularly during the frequent state rebellions of the 1890s and in response to occasional rustling practiced by the natives. The problem for Malheiros was not just that the Kadiwéu resisted his attempts to enslave them but also that they refused to become sedentary farmers or passive ranch workers. They continued to live their hunting and foraging lifestyle, sometimes at the expense of his and others' herds. Naturally, Malheiros was able to enlist the support of the growing number of ranchers in the region, who considered the natives to be an affliction on their livelihood. Rustling brought retaliation from the ranchers, and the natives were accused of harboring Brazilian deserters and other "adventurers." The ongoing conflict and resulting chaos eventually led to the demission of Malheiros as director in late 1899 and his replacement by Marianno Rostey.[52]

Rostey's subsequent investigation revealed that the encroachment of ranches onto Kadiwéu land was more organized than had been assumed. In early 1900, the chief of the tribe, Captain Joaquim Timoteo, presented a detailed report to Rostey on the situation in the Kadiwéu reserve. He painted a dismal picture. Lands awarded by the imperial government had been occupied by several ranchers; native cattle and horses had been stolen, homes burned, crops destroyed, and Kadiwéu firearms confiscated by the federal military force called in by Malheiros. This constant onslaught had caused the Indians to move into more remote regions, which was seen as a threat by the ranchers, who misinterpreted migration as organization for attack. Apparently no ranchers had been assaulted, with the exception of Malheiros, who was the driving force behind the encroachment on Kadiwéu lands. Settlement expanded so quickly that it led to claims on land as public, but which was actually part of Indian territory. As a gesture of goodwill, Captain Timoteo had even offered to supply native labor free of charge for the relocation of one ranch under construction on the Kadiwéu side of the Nabileque River reserve boundary. There is no indication of whether the offer was accepted, but it seems unlikely, considering subsequent events.[53]

The removal of Malheiros from his post did not eliminate conflict between the Kadiwéu and the rancher and his allies. In fact, the conflict only intensified in the following years, as civil war swept across southern Mato Grosso, directly involving the Kadiwéu. In 1902 and 1903, Malheiros took advantage of the revolutionary upheaval. Allied with local bandits or revolutionaries, he financed a number of expeditions against state police forces and the state government. He used a force of more than 200 mercenaries lodged on his ranch to intimidate the Kadiwéu. The violent times and imminent threat led the natives to initiate an attack against the ranch, resulting in the death of a number of mercenaries, the sacking of the ranch, and theft of cattle and horses. Malheiros himself was forced to seek safety on the Paraguayan side of the Paraguay River, although this temporary exile had as much to do with his involvement in antigovernment activities.[54]

The ongoing discord led Rostey to urge Cuiabá to settle the land issue by granting definitive title to much of the land claimed by the Kadiwéu. He wished to guarantee them a secure territory in order to establish a colony and create conditions for them to settle into a sedentary way of life, free from territorial dispute. He based his petition on a survey made of the territory in 1899–1900, which recommended the recognition of a reserve encompassing much of the area the Kadiwéu claimed had been granted by the imperial government. In 1903, a reserve of more than 370,000 hectares, including pasture and agricultural land, was approved by the Cuiabá government, though legal and other arguments for encroachment on native territory and the violence that characterized the previous two decades did not completely disappear.[55]

The conflict had taken a severe toll on the Kadiwéu. The population had been forced to disperse even more than is normal among semi-nomadic peoples. Of the alleged twenty-eight to thirty villages that had existed before the Paraguayan War, by 1900 the natives had been concentrated into only three, two of which were inhabited primarily by the elderly. The total population had declined from more than 1,500 in the 1870s to 850 by 1914, and by the 1930s there were only 150 Kadiwéu resident on the reserve. This was the result of warfare, disease, malnutrition, migration, and infanticide, as well as neglect on the part of Mato Grosso authorities. The Kadiwéu had practiced infanticide before the arrival of the settlers, but its incidence from the 1880s surpassed Kadiwéu norms. The constant fear of attack from ranchers and the subsequent need to keep on the move in a rapidly shrinking territory led to deteriorating health, greater exposure to disease, particularly yellow fever, malaria, and tuberculosis, and the perception of few options for the future. The familiar scourges of alcoholism, prostitution, and chronic malnutrition became a significant part of Kadiwéu life. In addition, the long-running conflicts siphoned off a number of workers to the surrounding ranches and led to divisions within the community. This aspect will be discussed in more detail in chapter 5.[56]

Although the Kadiwéu now lived on territory guaranteed by law, the thirst for land in the 1920s throughout Mato Grosso led to the influx of claim jumpers (*grileiros*), who attempted to survey land within the reserve at least three times during the decade. They also launched a propaganda campaign against the newly established Serviço de Proteção aos Indios (SPI), which had established a post in the reserve in 1912. The campaign questioned the size of the reserve, which was considered far too large and thus restricted opportunities for immigration and "development" in the region. The SPI was created in 1910, through the efforts of Matogrossense Cândido Mariano da Silva Rondon, who became its first director. It furthered Indian land claims throughout the state and helped to consolidate Kadiwéu rights to the region. Only through concerted effort by the SPI and the Brazilian government did the Kadiwéu survive complete obliteration, a situation that only gradually improved from the 1930s.[57]

The land issue simmered periodically over the next several decades and was not fully resolved until as late as 1984. Kadiwéu protests against the constant encroachment by ranchers on their lands, which included holding a group of ranchers prisoner, led to the definitive awarding of 540,000 hectares to the tribe in that year and marked a new beginning for the Kadiwéu.[58]

Terena

Violence and exploitation were the norm suffered by the Kadiwéu, but other native groups in southern Mato Grosso had varied experiences, each demonstrating distinct responses to their predicaments. The best organized native group at the time and to this day has been the Terena, members of the Guaná-Aruák culture. After the Paraguayan War, several Guaná tribes survived in Mato Grosso. They included the Terena, Laiana, and Kinikinao, numbering over 2,000 persons. They were the most integrated of local Indians at the time, already engaged in trade with ranchers and other settlers and employed in navigation and later as cowboys on some of the newly established ranches. Generally a sedentary people, they had experienced significant evangelization before the Paraguayan War by Italian Capuchins and, before that, had served the Guaikuru in what has been called a form of serfdom. Some have suggested these experiences helped the Terena to enter a similar work regime brought to Mato Grosso by neo-Europeans and thus survive intact as a group. Perhaps forming religious and economic alliances with settlers also was a strategy to seek opportunities away from Guaikuru control, though the Terena relationship with ranchers and other settlers was hardly one of mutual benefit and respect.[59]

The spread of cattle ranches into traditional Terena territory and the employment of community members brought a process of physical disruption and subsequent cultural breakdown. In many cases, the Terena had won title to the lands they occupied, but since they did not understand the Brazilian land tenure system, it was easily usurped by ranchers. For example, in gratitude for their service in the Paraguayan War, the imperial government had issued a number of land grants to individual groups through their chiefs, who also were granted honorary titles of "captain" in the National Guard. Terena custom allegedly required that when a chief died his possessions were burned on his grave. Since land titles were considered the chief's responsibility, observers concluded that these went up in smoke along with the rest of his belongings, though this may be more of an outside justification for lack of titles than an actual reason.[60] But land was considered a common resource, and as such, titles had little intrinsic value. Ranchers, then, were able to take advantage of these conventions to take over land where the natives could show no title or to buy up large areas of Terena grants at cheap prices.[61]

The ultimate result for the Terena was dispersal. A shrinking land base, limiting opportunities for hunting and subsistence agriculture, forced a growing number of young people into wage labor, which accelerated the cultural disintegration of the tribe and made it even more susceptible to exploitation by ranchers. This was a trying period in Terena history, known as "the time of captivity." The captivity was a reference to debt service but also to the native reliance on the goodwill of ranchers in permitting them hunting and cultivation rights on the ranches. By the early 1900s, the Terena population had grown, perhaps to as many as 5,000, but when visited by Rondon in 1904, their condition was abysmal. The leader of an expedition to string telegraph lines across southern Mato Grosso, he lamented that he had never seen such misery:

> This whole population lived in the midst of the greatest misery. Dislodged from their lands, reduced to a bitter servitude that was not even disguised; without the least support in the laws that seemed to be made solely to protect the rights, real or imaginary, of their truculent oppressors; abandoned by the authorities who did not even condescend to hear their grievances, nor to curb outrages against their persons and their wives and daughters—life for these Indians was a heavy burden, as sad and unfortunate as that [regime] under which the cruel Spartans punished their miserable slaves.[62]

Rondon was wrong on one point, for grievances were heard by some authorities, particularly the state General Directorate of Indians. Between 1894 and 1902 this office received Terena complaints, and at least one delegation, describing rancher invasion of their traditional lands. The official response was positive, and the tribe was awarded title to land in the Miranda area comprising 7,200 hectares, and survey teams were recommended to mark off the territory and thus guarantee title from settler encroachment. What Rondon encountered, however, was the result of years of delay on the part of the government in Cuiabá, a product of bureaucratic indolence and the series of civil conflicts in the region at the turn of the century, which involved the very ranchers to whom the Terena had fallen victim.[63]

Rondon offered the Terena deliverance from these conditions, as he publicized their treatment nationwide, scandalizing public opinion and the government in Rio. He then used his contacts in the state government to win a special legislative act in 1904–1905 recognizing Terena lands. Two villages, encompassing 5,400 hectares, were established. He also used his influence to convince some local ranchers, who had moved their property markers into Terena territory on the grounds that the Indians would not be prejudiced, to withdraw their claims. By 1940, thirteen villages supervised by SPI officials had been established, involving more than 25,000 hectares and 4,000 people. With time the ethnicity of the settlements broadened as Terena and other tribes moved from the towns or other reserves into those run by the SPI. In some, Brazilians and Paraguayans settled and intermarried with the natives.[64]

Rondon's campaign on behalf of the Terena was aided by the character of the group's way of life, which was not markedly different from that of nonindigenous settlers, and its peaceable nature. He reported that the Terena raised cattle and horses; cultivated manioc, sugar, bananas, and cotton; and had planted citrus and guava trees. They were the ideal of acculturation so desired by Brazilian authorities.[65]

Rondon also intervened in the case of conflict between the Ofaié (Chavante) Indians and rancher José Alves Ribeiro in the eastern Pantanal. In that case the Indians, who were semiresident on Ribeiro's ranch, had slaughtered cattle for consumption and cut fencing wire for arrows, allowing a number of animals to escape into the surrounding Pantanal. They also used semiferal donkeys and horses, animals that Ribeiro considered part of the ranch. Ribeiro had been appointed director of these people by the Cuiabá government in 1896, but his relationship with his charges was clearly not close. Increased tensions between the two parties led to armed conflict and the death of some Ofaié, which brought Rondon into the picture. Through his influence, Ribeiro and the natives came to an understanding without further bloodshed, and peace returned to the area. The Ofaié then left the region. Subsequently, they were persecuted throughout Central Brazil and even as far south as Rio Grande do Sul, as they made futile attempts to locate a region where they would not be in conflict with expanding settlement.[66]

Up to 1900, land in Mato Grosso had very little intrinsic value. It was only considered useful for what could be extracted from or produced on it. This meant that property wealth was judged not by territorial extension, which often was quite impressive, but by the estimated number of cattle on it. With the impressive increase in the state population over the almost four decades between World War I and 1950, and the intense interest in ranching that accompanied this growth, land quickly took on a value of its own, often to the detriment of long-term ranchers. An influx of investors, foreign and national, and especially speculators caused land prices to escalate to levels never before imagined. Original ranchers often experienced an erosion of their properties, and native peoples suffered loss of traditional territories. The state government was caught in the middle as it tried to regulate the sale of public land while also earning a healthy income from the sale and legalization and surveying fees. In several cases, exceptionally large parcels destined for cattle ranching were given title, and up to the 1930s and 1940s conflict over land rights was common.

In addition, access to credit throughout the region was extremely limited. Even relatively wealthy ranchers and processors had trouble borrowing significant amounts of money, and when they could there were no regulations that protected debtors. This created an unhealthy dependency on creditors, most of whom were local merchants or cattle drovers. The result was apparent in the slow economic development of the state, a situation that improved only in the 1950s and 1960s as private banks began to appreciate the potential of Mato Grosso, especially its cattle sector.

The experience of competition for land was a particularly onerous one for native peoples. As in the cases of native peoples throughout the Americas, ranching in Mato Grosso often represented the first sustained contact between Indians and neo-European society. In most instances it cleared the path for accelerated settlement that did away with aboriginal society as it had existed. Native groups regularly challenged encroachment but were often overwhelmed by the numbers of settlers, who tended to view original peoples as obstacles to development, and the indifference or ineffectiveness of governments. Loss of cultural specificity and at least partial incorporation into nonindigenous work regimes, including ranching, were the results, as will be seen in the next chapter.

FIVE

Cowboys, Hands, and Native Peoples
LABOR RELATIONS

R anching has been less labor-intensive than other agricultural pursuits, though the degree depended on the ranch and time period. As in other ranching regimes, work functions on Mato Grosso ranches tended to be divided into cattle care and ranch maintenance. Some ranches also hosted tenants or squatters, who had an interdependent relationship with ranch owners and were important to a ranch's economic success. Cattle drives, with their need for skilled drovers, were vital to ranching, providing the means to integrate ranches into the external market and offering employment opportunities to itinerant cowboys. There were ethnic dimensions at play as well, particularly in relation to native peoples, though not always to their detriment. These were the human forces that made up the running of a ranch and the cattle economy and reveal the faces of regional ranching.

HIERARCHIES ON THE RANGE: DIVISIONS OF LABOR

Unlike better-known ranching regions, such as North America, in Mato Grosso a hierarchical division of labor developed between those who directly worked with cattle and those who did not. Cowboys (*vaqueiros*) tended to concentrate on the cattle, which involved roundup, branding, castration, slaughter, and other tasks. Ranch hands (*camaradas*) worked on fence stringing and repair, building maintenance, tanning of hides, and, where applicable, gardening. Management tended to be the responsibility of resident ranch owners or a trusted foreman (*capataz*) who had been in the ranch's employ for some years. Distinctions were sometimes blurred, depending on the organization of a particular ranch, but on the whole a tiered structure developed between cowboys and ranch hands, to the point that cowboys often refused to string fences or work on construction because they considered such chores beneath them. This was especially the case among permanent cowboy employees, necessitating the hiring of temporary workers to do jobs not directly involving cattle.[1]

Ethnicity also played a role. In the Pantanal a notable number of foreigners, usually Paraguayans or Argentines from the province of Corrientes (Correntinos), worked as hired hands. Cowboys were often Brazilians of diverse origins. Ranch labor in Mato Grosso, as in virtually all areas of the interior of Brazil, was predominantly made up of men (and occasionally women) of mixed race, usually mestizo but sometimes mulatto. Some were indigenous, especially from the Kadiwéu and Terena peoples, although native presence depended on the location of the ranch.[2]

Distinctions developed between cowboys and ranch hands based largely on job, horsemanship, and ethnicity, though there was little to choose from in terms of quality of labor. For example, Paraguayans often were considered to be superior horsemen, but they also were seen by some as unreliable because of an alleged penchant for drunkenness and gambling; this did not mean that ranchers would refuse to hire the immigrants on principle. Periodic labor shortages forced some ranchers to accept them, while others went against the common belief and considered Paraguayans and Correntinos to be the best vaqueiros in the region. The experience they brought from similar ranches in Paraguay and northeastern Argentina stood them in good stead among many Matogrossense ranchers. There was a sizable Paraguayan community in the state throughout the period, mostly landless peasants from the interior of that neighboring nation. The majority worked in the Mate Larangeira forests. Many others arrived in Mato Grosso fleeing political upheavals or as economic refugees from the expanding latifundios in their homeland. They undertook several kinds of employment, as ranch hands, cowboys, charqueada workers, tenant farmers, hired guns, and bandits. Initially there were few women among them, except in the port town of Corumbá. As the population grew, increasing numbers of immigrants came as families, while the cattle towns, particularly Campo Grande, attracted young women for work in domestic service, as street vendors, or prostitutes. Local customs in southern Mato Grosso were permanently altered by the immigrants, and today certain foods, musical preferences, and speech reflect that long-standing presence. In fact, although Portuguese was the lingua franca in ranching, often the language was enriched by Guaraní expressions and words, and the Paraguayans invariably spoke their native language among themselves.[3]

Another distinctive group in the region were Gaúchos, immigrants from Rio Grande do Sul, who became a significant sector of the southern Mato Grosso population after the 1893–1895 Federalist civil war in their home state. Mato Grosso received thousands of Gaúcho immigrants who brought experience in ranching with them. They became the backbone of small ranching in the state, defending their rights vigorously against large ranchers, the state, and particularly Mate Larangeira. Gaúchos settled the southern portion of the state, from Campo Grande to the Paraguayan border, establishing a presence through continued immigration that has endured to the present day. Bringing an independent spirit and some capital, they developed small ranches and farms but seldom provided labor for the large ranches. Their style of ranching was more intensive because their holdings were smaller, and they readily adopted innovations like wire fencing, planted pasture, and stabling when capital was available. They also employed fewer workers, relying on mutual cooperation during roundups and in emergencies. The Gaúcho contribution to Mato Grosso ranching was less spectacular than that of the large ranches, but no less important, at least in the long term. These small ranchers contributed relatively few cattle to the industry until into the 1930s, but they helped to expand the border region, especially around Bela Vista, as an important source of cattle for the Paraguayan market. With the completion of the railroad from Campo Grande to Ponta Porã in 1953, they exported their cattle to São Paulo frigoríficos and, eventually, to the slaughterhouse set up in Campo Grande in the 1960s.[4]

With the exception of native peoples (discussed below), other groups tended to be considered part of an amorphous whole, such as Brazilian mestizos. In the years after the Paraguayan War, when slavery was still an important institution in Brazil, there were some slaves on cattle ranches, but they tended to be few in number, and most African or mulatto employees were freedmen. With abolition, those who could sought new opportunities. Other immigrants were from Minas, Goiás, and São Paulo, but besides dietary preferences and accent, they exhibited no predominantly distinctive characteristics to differentiate themselves from one another.[5]

Class Structure and Ranch Labor

Most labor obligations followed breeding cycles and involved a number of different activities. The number of vaqueiros employed at one time depended on the size of the herd and property; the average around the turn of the century was one man to 200 to 600 head. More were needed on ranches where animals were geographically most scattered. This did not change much over succeeding decades, until greater enclosure of pastures in the 1930s and 1940s reduced the need for as many cowboys per head.[6]

Cowboy responsibilities are specialized work and were considered as such. A typical working day on the range during roundup started at daybreak, with no breakfast except hot mate tea (*chimarrão*), served in a wooden, metal, or gourd cup, sipped through a metal straw, and passed from hand to hand. This custom was a gift from both Paraguayans and Gaúchos and is part of Mato Grosso ranch ritual to this day. Horses were saddled and ridden out to the herd, with a provision of erva mate for cold mate (*tereré*), dried beef, and farofa (manioc flour) carried for the day's meal, to be heated or consumed cold if necessary. Roundups occurred at specified times of the year. Once the herd was encountered, and if the animals were not skittish, young animals were separated into groups of yearling calves to be branded and two-year-olds to be castrated. Other animals were inspected for fading or damaged brands, injuries, and signs of illness or debilities. Vaqueiros ate their lunch in the saddle or together, if the work was not heavy. Labor ended near sundown, and they returned to camp for a hot meal of salted beef, beans, rice, and farofa prepared by a camp cook, often swapping stories of the day's endeavor before turning in. Vaqueiros commonly slept in hammocks strung between trees, in makeshift shelters, or on the ground when on the open range.[7]

Branding and castration occurred at roughly the same time, between April and June, although in the Pantanal this was from September to January, before floodwaters rose beyond workable levels. Branding usually involved an ear mark, more rarely the hot branding iron. Castration was always by the knife, the scrotum slit and the testes tied off or removed, the wound then cauterized with a hot iron. In Argentina, it was common practice to castrate yearlings, but in Mato Grosso, where cattle grew more slowly, this was done at two years of age. Before the construction of corrals and containment pens at the retiros (line camps), this work was done on the open range, one animal at a time, and in keeping with traditional agricultural beliefs, when the moon was on the wane. Once these jobs were completed, and if the ranching regime was not too extensive, young calves were weaned from their mothers by separation for a period of time in a makeshift corral, and adult animals were inspected for injuries and illness. Throughout most of the period, treatment of injury, disease, and parasites was rudimentary, often involving local remedies like pitch, papaya oil, and horse dung. Vaccines and other preventive medicines, as explained in chapter 6, were only employed on a regular basis beginning in the 1940s and 1950s, partly because of availability, cost, cowboy resistance, and inadequate understanding of application techniques.[8]

Vaqueiros were often colorful characters. Their dress was similar, borrowing a great deal from traditional gaúcho styles, but with regional variation. Paulo Coelho Machado listed the following items of clothing and accessories, although these appear to be more formal attire than daily wear: pants were the baggy bombachas, shirts were checkered, and accordion-type boots reached halfway up the calf. The cowboy wore a wide-brim hat made of felt or plant fiber (depending on the season), a bandolier belt studded with bullets, long and noisy spurs, thick locally tanned hide chaps to protect the legs when riding through brush, and a deer-skin belt decorated with coins. He carried a long knife tucked in his back belt, a whetstone, a pistol at his side, and a rain poncho for inclement weather. Other observers also noted that the vaqueiro used a long lasso, between 20 and 40 meters in length, made of twisted deer or steer rawhide, and in the Pantanal some often rode barefoot, using a toe stirrup. Unlike in North America, cattle were usually lassoed by the horns rather than the neck, probably a consequence of the enormous appendages sported by the original Mato Grosso animals. Certainly not all cowboys were as well-equipped as described by Machado, especially those who relied on seasonal employment. Seasonal workers were more likely to have no boots, only a straw hat, one pair of pants and shirt, a knife, and no firearm. This did not make them any less effective, however, as most observers were duly impressed by their skills.[9]

The cowboy's work was somewhat more specialized in the Pantanal than in other regions. Regular flooding commonly forced cowboys to brave rising floodwaters to rescue stranded cattle. This meant leaving their horses behind and taking to the water, and Pantaneiro cowboys were as adept with canoes as they were with their steeds. Typically, canoes were hewn from one log and manipulated with a long paddle, 2 to 3 meters in length, or a stout pole if floodwaters were sufficiently

Cattle crossing a river, 1928, near Rancho Jeroquya. Courtesy of Elza Dória Passos.

shallow. Cattle were guided across wide streams or through flooded regions using a combination of canoes and horses. Cowboys also were not
averse to swimming with the cattle in order to round them up. Sometimes a daring vaqueiro took charge of a small herd by clinging to the
horns of the lead animal, guiding it to dry ground where other cowboys waited. Here, the vaqueiro was also a fisherman, often catching his
own food when on roundups. Of all the cowboys in Mato Grosso, the
Pantaneiro was certainly the most versatile, and as such was wooed by
ranches to remain in the region, if not always on the same ranch.[10]

Ranch hands normally were seasonal workers contracted for up
to seven months at a time, depending on the region and ranch needs.
Among these workers, Paraguayans were the most numerous, appreciated by ranchers for their hard work and willingness to take orders.
After the ranching season, they worked as lumberjacks or at odd jobs
in the towns. Some returned temporarily to their homeland. However,
they were also known for gambling and drinking, which often guaranteed their return to the ranches the following season, after exhaust-
ing their savings. These workers sometimes were called *peões*, the term
likely brought into the region from Paraguay and Argentina. Some
were not temporary, living on the ranch year-round with their families,
but the bulk were seasonal. It was rare for a camarada to work at the
same ranch every year, thus workers were contracted in nearby towns
by the ranch owner or trusted permanent employees.[11]

Hands generally did not handle cattle but instead strung and re-

paired fences, constructed buildings and corrals, tanned hides, and occasionally broke broncos, although this latter labor was normally reserved for cowboys. Other tasks included digging wells, clearing land,
and making sugar from cane grown on the ranch. Camaradas normally
lived in rustic open-sided shacks in remote areas or in wattle-and-daub
huts near ranch headquarters. Food included the normal fare of most of
Brazil, though some ranches had small garden plots, often cared for either by the rancher's family or by selected hands. Ranchers and their
employees usually shared the same diet. Some ranches allowed seasonal
workers to plant small gardens and to raise animals, usually cows, especially if they had families, but this was not common, largely because
the work was so transitory. These same ranches often hosted tenants
and squatters in return for similar labor obligations. Even into the mid-
1950s, these basic living conditions had changed little.[12]

Housing was provided free of charge to cowboys and ranch hands
alike. Wages varied over the years. Until World War I, they tended to
be relatively high, reflecting a chronic lack of labor. In the late 1870s, a
camarada earned 60 milreis (US$28) a month, comparable to cowboys in
the United States at the time, while paying 15 milreis ($7.05) for a bushel
of rice and 6 milreis ($2.80) for an arroba (15 kilos) of sugar. Sugarcane
alcohol (*cachaça*) cost 700 reis ($0.33) a liter, and a *vara* (1.1 meter) of
rope tobacco 7.5 milreis ($3.50). Aboriginal workers earned significantly
less, 8 to 10 milreis ($3.75–$4.70) a month, meals included. Part of the
reason Lúcia Aleixo gives for these minimal wages for indigenous cowboys was competition from Paraguayan immigrants, who flooded the
market with experienced ranch labor. Whatever the reason, it is apparent that a single camarada could survive relatively well, although
the ability to support a family on these wages was restricted. Over the
following decades, prices and wages fluctuated but remained relatively
comparable to the early years.[13]

At the turn of the century, ranch wages had declined to about
15 milreis (US$2.25) a month, plus room and board. Such low income
probably contributed to a labor crisis in the state's rural sector, as many
workers were attracted north, where a rubber boom had developed.
On ranches that Miguel Lisboa visited in 1907, camaradas were paid for
piecework, for example, 500 reis ($0.15) to string a *braça* (2.2 meters) of
fencing, with materials and tools provided by the rancher. By 1919, the
expansion of the industry and the completion of the railway provided
many more opportunities for cowboys. Wages for indigenous camaradas in the Pantanal had risen, to between 2 and 3 milreis ($0.50–$0.80)

a day, with or without food. Nonindigenous camaradas earned 3 milreis per day plus food, while cowboys earned 50–90 milreis ($13.00–$23.50) a month, including food. Specialized laborers like carpenters commanded 12 milreis ($3.10) for eight hours of work, but apparently they were hard to find because the railroad and the towns had absorbed so many. Many ranches suffered a consequent deficit of skilled labor, since they simply could not match the wages offered by these expanding sectors.[14]

That had stabilized by 1922, when the regional economy slowed down and the milreis was devalued by half. Ranch hands earned 70–120 milreis (US$9.10–$15.50) a month and cowboys 9–12 milreis (US$1.20–$1.60) a month, with food and other privileges not extended to ranch hands. By comparison, cowboys in neighboring Paraguay earned between $5 and $8 a month. Sometimes payment would be in kind, particularly sugar and cachaça, which were not part of the daily food allotment, and some clothing or fabric and occasionally bullets. Those working at clearing land usually were paid for the area cleared, 150 milreis ($18) per alqueire (2.4 hectares). A ranch foreman earned 100–200 milreis (US$13–$26) per month, and day laborers took in 3–5 milreis (US$0.40–$0.65) daily, depending on whether meals were included.[15]

Despite these low wages, the corresponding cost of living was fairly high, particularly in the towns. In Campo Grande municipality in 1922, an arroba of sugar cost 15–23 milreis (US$1.95–$3.00), the same amount of beef jerky was 17–21 milreis (US$2.20–$2.75), a kilo of fresh beef 500–700 reis (US$0.06–$0.10), and a chicken 3 milreis (US$0.40). A 60-kilo sack of corn fetched 11–18 milreis (US$1.45–$2.35), beans 25–40 milreis (US$3.25–$5.20), and rice 26–40 milreis (US$3.40–$5.20). Building materials also were expensive; the cost of building a house was far beyond the salary of an agricultural worker, which explains in part why so many lived in the most rustic housing, either on the ranches or in the towns. Under such living conditions, it takes little imagination to visualize how difficult it was to raise a family with just one salary, especially for temporary workers.[16]

Ranch Relations

Relations between ranchers and their employees varied depending on the individuals involved and also on the time period. In the years just after the Paraguayan War, there were several incidents of illegal slavery, physical abuse, and extreme exploitation. In 1873, a newspaper in Corumbá expressed concern over a Brazilian deserter who was purportedly conducting a group of Paraguayan prisoners of war to the Corumbá area for sale as slaves, probably on the remote ranches then being reoccupied. There is no information on the fate of these unfortunates, but the attitude that contributed to such acts was clearly prevalent, particularly in the most remote regions. For example, the interim Chief of Police for Mato Grosso made a trip through the Pantanal in 1877 to investigate conditions. He found that debt peonage, indiscriminate corporal punishment, and the organized hunting down of workers who had fled brutal conditions were common on the ranches of the region. Such conditions were echoed by other observers in Mato Grosso in later years.[17]

Notwithstanding these appalling circumstances, toward the end of the nineteenth century the greatest labor abuses in Mato Grosso occurred in the Mate Larangeira erva mate stands and in the sugar plantations near Cuiabá. Owing to the semi-independent nature of cattle raising, where the means for flight were readily available and there were other employment options, cowboys and ranch hands seldom suffered physical abuse. The practice of debt service, however, was fairly common, particularly in the Pantanal, and in many cases became a permanent part of a ranch worker's life. By restricting workers' purchases to ranch stores, ranchers could tie laborers to the region. In some cases, workers could take advantage of the relative flexibility in debt contracts, and they frequently moved from one ranch to another, where the new boss would pick up their debts. It was not uncommon for ranchers to rescue a cowboy or camarada by assuming his debt accumulated with another rancher. This was not the case with new workers, however, as it applied only to those who had proven themselves to their employers. At the same time, it was not unusual for a camarada simply to disappear from the ranch without settling his debt. Allegedly, this was one of the risks of employing Paraguayans, who did not have far to travel to reach home. This phenomenon reflected a high degree of mobility common in the Mato Grosso ranching scene.[18]

Labor mobility was constantly lamented by ranchers, who suffered chronic labor shortages throughout the period. Several observers called the local cowboys, especially in the Pantanal, nomadic in character, and ascribed this to their ethnicity, Paraguayan or aboriginal, ignoring the unstable character of the work itself. Some believed this was a major factor in the slow development of Mato Grosso, although others did not consider mobility to be so onerous. Paradoxically, for some a solution was to contract more Paraguayans, who were seen as less nomadic, but often this was a last resort, again because of their alleged propensity for drinking, gambling, and fighting, despite a contradictory reputation as hard workers. Stereotyping was clearly prevalent in the thinking of Mato Grosso ranchers and outside observers.[19]

Apparently, no solution to the supposed problem of labor mobility was ever encountered, since in the Pantanal even as late as the 1980s up to 25 percent of cowboys were permanently mobile. Today ranchers and other observers in that region, which has best preserved the traditions of the Mato Grosso vaqueiro, report that the typical Pantaneiro changes ranches several times in his working life, almost in a cyclical fashion, periodically returning to ranches he has worked on before. Frequently, a cowboy will tell the rancher to "save an opening for me," expressing his intention to return at some future date. In this way, Abílio Leite de Barros suggests that the Pantanal cowboy wasn't an employee of a specific ranch but rather of the zone. With few possessions and personal responsibilities, the cowboy was free to cultivate an independent spirit. Ranchers have observed that first among the cowboy's priorities was his horse, allegedly followed in order by his saddle, his wife, and his children. Outsiders typically exaggerated these priorities, again ignoring the region's skewed labor relations, limited educational opportunities, and the willingness of ranchers to accept such behavior since it kept wages manageable.[20]

Women, with the exception of the wives and children of tenants and ranchers, were a small fraction of ranch populations during the period. Even as recently as the 1980s, ranches in the Pantanal were over 80 percent male, indicating that the majority of vaqueiros were bachelors or lived in loose matrimonial unions. Those unmarried women living among the cowboys were traditionally considered common property, at least as long as there were no children. They were sometimes prostitutes from cattle towns like Campo Grande. Indeed, life on the ranches was not easy for families, even for those of ranchers. Medical care was necessarily minimal, usually dependent on local remedies, which were often insufficient. A ranch's remoteness meant that injuries could be more serious than they would have been in the towns, and the risks of childbearing to both child and mother were proportionately higher. As a doctor in Corumbá during the period observed, mortality among men in many remote regions was frequently the result of gunshot wounds, but women died in childbirth. Where possible, ranch owners and some trusted hands and tenants sent their children to board in the towns; ranch schools were seen only on the largest and wealthiest establishments.[21]

Violence in the Brazilian Far Oeste

As in virtually all cattle economies in remote regions, arbitrary violence existed in Mato Grosso. In a study of banditry, Valmir Batista Corrêa characterizes two types of bandits operating in the region: those in the employ of local bosses (*coronéis*), and the independents. For the most part, Corrêa observes that there were few independents during the years 1889–1930, when the power of the coronéis was at its height. Political rebellion, especially between 1891 and 1906, required personal armies and acted as both a source of employment and a stimulus to banditry, whether sanctioned by a *coronel* or not. A large number of mercenaries came from Paraguay, and that nation was used as a natural hideout and launching pad for cross-border attacks and cattle rustling.[22]

Since ranchers often were involved in these rebellions, ranch employees were expected to ride with their bosses' private armies. Many included the Paraguayans who had come to the region to escape similar upheavals in their home country. For some young men, this may have provided an opportunity for adventure, and it did attract a number of adventurers to the state, but for the most part it did nothing but disrupt regional life and undermine any semblance of the peaceful existence most settlers were seeking. During the military rebellions of the early 1920s, which briefly spread from São Paulo into Mato Grosso, many Paraguayan workers fled, creating a temporary labor shortage on many ranches and further restricting an already weak local economy.[23]

This was what might be called organized violence. There were also several independent bandit gangs in the region throughout the period, beginning soon after the Paraguayan War. Most of the gangs were short-lived, using Paraguay or Bolivia as bases of supply, refuge, and market for their stolen goods. They were most often involved in cattle rustling, since the relatively large number of animals and the general lack of fencing and close attention made this by far the easiest and most lucrative illegal activity around. The smallest ranchers, lacking political influence and the resources to defend themselves, were the most frequent victims. During the most severe years of bandit predation, several observers reported that the only law that prevailed was the "law of article 44," that is, caliber .44. Paraguayans were invariably involved in the business, especially in the areas of Ponta Porã, Bela Vista, and Porto

Murtinho, and these cross-border bandits and others were immortalized in local folk songs. One song recounted: "The rustler comes from afar/from the confines of Paraguay/he sprang from a bloody Guarani revolution." State authorities found them difficult to control, nevermind eliminate, as there were always too few resources available for police forces, and distances were great. Besides, Cuiabá only began to appreciate the economic potential of southern Mato Grosso once the railway was under construction. Bandit depredations diminished with the authoritarian and centralized rule of Getúlio Vargas in the 1930s.[24]

Despite such a climate, violence was limited on the ranches themselves. The growing town of Campo Grande was known as the Brazilian Far West, for its periodic shootouts and assaults, but cowboys and ranch hands on the ranches, even when carrying weapons, seldom resorted to violence. The prohibition of alcohol on the ranches may have been largely responsible for the relative peace on the ranches. Many of the ranches were "dry" out of this very concern. A rigid moral code among cowboys also restricted the playing out of rivalries and vengeance to the towns, reinforced by the threat of dismissal. This was not unlike ranches in North America during the nineteenth century, where a ranch could not function if its employees were bickering, and any sign of discord usually meant the discharge of the most divisive employees. The presence of debt service probably had some effect in controlling employee excesses too, particularly the honoring of debt carried from one rancher to another, whereby a cowboy could leave a confrontational situation for another ranch without losing his livelihood.[25]

The Insecurity of Security: Tenants and Squatters

Tenants (*agregados*) and squatters were often an integral part of ranch life, particularly in the Pantanal, where many ranches were so huge that the owner had no way of controlling all of his animals without permanent help. These people tended to be quite stable, apparently staying six to eight years, on average, some as long as twenty years. They were normally Brazilian mestizos, from Mato Grosso or neighboring states, but Paraguayans also made up a significant proportion. They often had families with them, and for the most part agregados were given the right to live on the ranch, usually at line camps, and raise animals and crops. In return, they cared for cattle in the vicinity at their own expense and fulfilled whatever other obligations the ranch owner required. Tenant women and children also worked, cooking, laundering, or taking care of gardens and barnyard animals like chickens and pigs. If they had no animals of their own, they were frequently permitted to slaughter one steer per month for personal consumption.[26]

One notable benefit, at least for the more permanent ranch employees, had its roots in the colonial era. As long as cattle were well cared for, on top of their salary cowboys and sometimes tenants were granted one calf in four from the annual herd increase. This meant that in a herd of 10,000 animals with an annual increase of 20 percent, as many as 500 calves might be acquired by employees. Under ideal circumstances, a cowboy or a tenant could take possession of up to 25 animals per season, a number that could be built into a small herd in a relatively short time. This was a strong incentive to take good care of all cattle. To prove their claims, cowboys commonly dried the umbilical cords of newborn calves, a practice honored by ranchers. In fact, this system was an important part of the relationship between ranchers and these employees.[27]

Sometimes ranchers signed over title to the retiro to a long-term agregado. This, plus the cattle earned through the custom of one calf in four, permitted some tenants to become small ranchers. Squatters occasionally benefited from this process as well, although much less frequently. Some of the tasks required cooperation with other tenants and camaradas, a form of teamwork called *mutirão* or *muxirão*, whereby workers from different sectors of the ranch would be brought together to raise a building, construct a fence, or work on other projects. The rancher usually provided food, drink, and music as incentive, and the job could develop into a festive occasion, something akin to barn raising in the North American west. Over time, however, mutirão gradually declined, allegedly because of the cost of providing food and entertainment for the workers. The decline signaled the general breakdown of the traditional paternalistic structure of relationships among ranchers, employees, and agregados, as loyalty and dependence began to dissipate and the more competitive values of the outside world filtered into the region.[28]

Both agregados and squatters always were quite vulnerable. If a rancher decided to expand his operations, he could force tenants to vacate land they had looked after for years, and evict squatters entirely. Such evictions were not too frequent, as there were other forms of dealing with employees that did not necessarily involve dismissal. In later years, some agregados were permitted to occupy land under the obligation to plant pasture, but without payment in cattle or agricultural goods. Though not as common, this system was similar to one used in Argentina some decades before. Overall, tenants usually were not forced to leave the land after a fixed period of time, though this provided them no real stability, since they always had to rely on the goodwill of the ranch owner. Labor relations became more complicated with time, however, as the state and the cattle industry expanded economically and socially.[29]

As land was occupied and became more expensive, particularly in the Vacaria and along the Paraguayan border, some ranchers lacking sufficient pasture or other resources signed contracts, called *parcerias*, to rent out cattle to another rancher in return for a percentage of production. This system benefited both parties, as the rancher renting out the cattle eliminated the expense of raising his own animals, while the renter started a herd with little initial capital investment. There were several types of parceria contracts in Central Brazil. The most common stipulated delivery of a given number of animals for a period of three to seven years, to be raised at the renter's expense, with payment of half of the calves born by the end of the contract and return of the same number of cows originally received. A variant contract delivered a given number of cows to be returned at the end of the contract along with an equal number of additional cows; all other production and bulls were then the property of the renter. Other forms of parceria, such as agreements seen in western São Paulo, involved shared raising for fattening and for milk cows.[30]

In Mato Grosso, the extensive nature of the cattle regime made it difficult to know the exact number of animals born each year. Consequently, the 50-50 system often was modified: an agreed-upon number of animals, bulls and cows, were rented out at the expense of the renter, with the obligation to return 15 to 20 percent of the annual increase of bulls every year. At the end of the contract, the same number and proportion of animals originally rented out were to be returned. Sometimes this system was slightly modified by stipulating a fixed number of bulls to be returned annually. Such contracts assumed an annual increase of bulls between 15 and 20 percent, a reasonable assumption in a regime in which natality reached 50 or 60 percent. This general information is confirmed when one looks at rental contracts in specific areas. Records from the local land office in Ponta Porã, for example, reveal that between 1917 and 1928, contracts required either a fixed number of young bulls in payment annually or 20–25 percent of bulls born each year. Contracts covered periods of between three and five years. It seems reasonable to assume that similar contracts pertained in other regions of the state, particularly in the south, where properties were generally smaller and pressure on land and production consequently greater.[31]

These contracts reveal a certain mutual dependence among ranchers in Mato Grosso, a situation that often was extended to relationships between ranch owners and their employees. Relations on the ranches, within the ranching family, and between ranch owners and their workers were similar, although external influences gradually modified behavior. In the first decades traditional, almost medieval, practices were observed, including within the ranch owner's family. For example, on campaign during the Paraguayan War, Alfredo d'Escragnolle Taunay described his visit to a ranch near Coxim, where the owner was a well-read man but refused to present the women in his family to the visitors, including his wife, who was hidden behind a screen. This experience gave rise to Taunay's classic nineteenth-century novel of the Brazilian backlands, *Inocência*. The rancher in this case may only have been protecting his family from the dangers posed by several hundred soldiers passing through the region, but a glance at other regions and reports of similar behavior in later years suggests a more firmly rooted custom. Corrêa Filho noted that in the Pantanal, ranchers, though polite, were commonly wary of strangers unless they carried a letter of introduction from a trusted relative or friend. This was likely a product of the isolation of most ranches, since self-protection was essential for survival, for rancher and cowboy alike. Such an attitude was reinforced by the authorities' apparent lack of interest in the region; they seemed to surface only to extract taxes. The draw of the frontier to adventurers and fugitives from justice only reinforced a profound distrust of outsiders, who at best represented a potential imposition on the insular lifestyle of ranch inhabitants, at worst a threat to women or children.[32]

ROMANCE DISPELLED: THE CATTLE DRIVE

One of the most romantic scenes in the literature of ranching in the Americas is the cattle drive. Images of thousands of steers crossing the open prairie, kicking up immense dust clouds visible for tens of miles, have emblazoned themselves on the popular consciousness. Most of our perceptions come from descriptions and film depictions of the famous Texas cattle drives of the late nineteenth century, but no less

Cattle drive in eastern Pantanal, 2001. Photograph by the author.

dramatic were the drives of Central Brazil. Indeed, they were the life-blood of the business in Mato Grosso until very recently. (See map 3, p. xiv, for routes of cattle drives.)

Drives were similar in North America and Mato Grosso, and broad consensus affirms that they were conclusively unromantic. On the contrary, they usually were very tedious. Hundreds, even thousands of steers were herded under the cracking whips of tough trail riders who faced arduous conditions, including the chaotic fording of streams and rivers, rustling, and regular exposure to scorching sun and torrential rains. Despite these physical difficulties, probably a key feature of the drive was the relatively mundane activity of contracting the sale of cattle from the ranch to drovers (*boiadeiros*). To the ranching industry, this was a key activity that could make or break a year's worth of toil, and it was commonly dependent on the state of the market far beyond the control of either the rancher or even drover.[33]

In Mato Grosso, the character of the drives dictated the quality of cattle bought for the markets of São Paulo and Rio de Janeiro. Throughout the period, cattle were driven east to Minas Gerais and São Paulo, journeys that usually took two to three months, though occasionally six, traveling up to 3 leagues (12 miles) per day. The drives ran between 1,000 and 2,000 head, occasionally more, at one time. These enormous undertakings required animals hardy enough to withstand the rigors of such a trek. Only tough and wiry cattle, usually weighing 200–250 kilograms, were selected by the drovers, as these had the best chance of surviving the drive. Once at the destination, the cattle were fattened in pastures provided by the slaughterhouses for a period of six to twelve months, until they reached their original weight and were ready for slaughter.[34]

Besides the usual Mato Grosso cattle drives to Minas Gerais and São Paulo, there was also irregular traffic to Paraguay, especially from the ranches near the border. In the years immediately following the Paraguayan War, this was the favorite destination for Mato Grosso cattle. Drives were attracted partly by the market, but they also avoided the distance and the time consumed in trailing animals east. As a result, cattle from eastern Mato Grosso, particularly Paranaíba, Coxim, and parts of Campo Grande, went to Minas, and those from the Vacaria and border regions went to Paraguay, encouraged in the 1920s by the construction of salt beef and beef extract plants in the Chaco. Pantanal cattle were largely traded within the state, though improved communications permitted ranchers to send their animals to Paraguay and the east, and when the railway arrived during World War I, an influx of cattle buyers stimulated the expansion of sales and drives to the voracious Paulista slaughterhouses.[35]

The loss of cattle along the drives depended on specific conditions, but ranged between 5 and 10 percent, sometimes the result of disease but more often of exhaustion, injury, or escape. That percentage changed little from before the Paraguayan War, improving only when drives became short marches to local slaughterhouses or jerky plants. Weight loss ranged between 40 and 60 kilos per animal, depending on the length of the drive, and figured significantly in the prices paid by buyers. For much of the period, a drover's profit margin remained constant. During the boom between 1916 and the mid-1920s, it was approximately 30 milreis per head, hardly a substantial amount in light of the risks run by boiadeiros during the drives. Various observers complained that proper facilities for the transport of cattle, including adequate trails and access to pasture and the completion of a bridge over the Paraná River, were essential for the reduction of costs to drovers and increased profits to the producers. The completion of the railroad bridge over the Paraná River at Três Lagoas was an attempt to resolve these issues, since a passageway for cattle was incorporated into the design; however, it proved insufficient to replace the crossings at Taboada and XV de Novembro.[36]

One result was that drover expenses rose in the 1930s. By 1935, it could cost up to 60 milreis a head to drive animals from Campo Grande to Barretos in São Paulo. No one enjoyed a booming profit. Ranchers received between 110 and 140 milreis per head, and drovers got roughly 180 milreis at drive's end. The slaughterhouses sold the meat for 225 milreis per rendered steer, not including hides, fat, and viscera. Animals arriving in Barretos averaged between 180 and 200 kilograms, and eight to twelve months of fattening brought the weight up to between 250 and 275 kilograms at slaughter, rendering roughly 100 to 150 kilograms of fresh meat. Meanwhile, consumers in São Paulo city were charged 1.3–1.7 milreis (US$0.11–$0.14) a kilo, depending on the grade of the meat. These figures suggest that drovers normally earned around 10 to 20 milreis a head, less than 15 percent, and the rancher's take was no better. Clearing comparatively more, slaughterhouses still received only roughly 40 milreis per head on the meat alone, a return that remained relatively constant throughout the period. Limited infrastructure and processing in Mato Grosso narrowed the economic margins for ranchers and drovers. Only when the frigorífico in Campo Grande came on stream in the 1960s did Mato Grosso ranchers begin to gain better than subsistence income from their operations.[37]

In the decades immediately following the Paraguayan War, most boiadeiros and their hands came from Minas, often driving breeder bulls or horses into Mato Grosso to help pay for the purchase of thin steers and cows. Drives occurred in the spring, between September and December, when light rains created conditions for forage along the routes, particularly the normally dry trail from Campo Grande to Paranaíba. The drover usually arrived in the region during the winter and traveled from ranch to ranch collecting suitable animals to drive to Minas. Because the activity was concentrated into a relatively short time span, ranches sometimes accommodated two or three drives at a time, putting great strain on ranch resources, particularly corrals and pasture. There was always the danger of adventurers, as well. Some drovers independent of the traditional organized drives offered their services at better prices, but they also took advantage of the general lack of fencing, increasing the size of their drives by plucking the odd unbranded animal from ranches as they passed through. Once ranchers became aware of such rustling, they became directly involved in the initial stages of the drive. Ranch owners or trusted cowboys then accompanied the drive to the end of their property, passing on vigilance to their neighbors, and so on along the drive route.[38]

Most boiadeiros were honest, however, and ranchers had little reason to complain about their services. For the most part, they did not pay outright for the cattle. After receiving payment in Minas or São Paulo, they then returned to the ranches several months later to reimburse the ranchers. In this way reputations were made, and some drovers established small ranches in the region. This meant, however, that only the richest ranchers could afford to send their cattle out without receiving immediate payment. Even for the better off, it was necessary to join together in order to finance a drive, as the demand in Minas and even São Paulo for Mato Grosso cattle was not urgent until World War I. Then, an intensified interest in Mato Grosso cattle brought cattle buyers representing the São Paulo slaughterhouses directly into the region, particularly as rail transport facilitated their access. Cattle were seldom exported by rail, but the EFNOB greatly improved the efficiency of the drive purchasing process.[39]

Initially, cattle were conducted over open range, but the creation of corridors and later roads facilitated the herding of the animals and reduced time to market. The majority of drovers financed themselves, with capital usually saved after several drives as a cowboy. One important exception, mentioned earlier, was the manager of the São Paulo branch of the Banco Francés e Italiano, who during the boom of World War I became heavily involved in financing drives organized by Mato Grosso ranchers, as he perceived that cattle were a better risk than securities or even real estate. That type of investment was unusual, however. Credit remained tight in Mato Grosso, and most ranchers and drovers survived from drive to drive.[40]

It was no easy task to conduct up to 2,000 head over 1,000 kilometers for sixty or more days. The number of cowboys depended on the size of the drive, requiring four to six per 1,000 head and the same number of mounts per rider to keep steeds fresh. This average remained constant down to the end of the drives in the 1990s. Salt had to be carried or found en route, and water was always a concern. Boiadeiros normally owned everything used on the drive, from horses and mules and oxen to cooking utensils. In the early days they carried all their food with them, relying, as in the North American west, on a cook wagon to travel ahead of the herd and set up camp at an agreed-upon site. The normal fare included the traditional foods already described and, according to Machado, "a lot of erva-mate." In São Paulo, there were rest stops where food for the cowboys and forage were provided for a price. The drover paid all costs.[41]

In Mato Grosso, mules often were more common on the drive than horses; allegedly they were more resilient on the long journeys. Oxen also were common, particularly for pulling the cook wagon and carrying some articles, and apparently herd dogs accompanied the boiadeiros. Cattle were led by a cowboy who blew periodically on an enormous steer horn to keep the rhythm of the drive, while the rest flanked the herd, cracking whips in the air as necessary to keep animals together. Drive trails were no more than 20 meters wide (much narrower than the 70 or 80 meters in Rio Grande do Sul). Periodically, streams or rivers had to be swum, until the road system expanded in Mato Grosso, beginning in the 1930s. These tropical waterways teemed with fish, including the infamous piranha, whose feeding frenzy is legend, if exaggerated. Cows could be seen with parts of their udders missing after having encountered these voracious aquatic carnivores, and where there was an extreme concentration of the fish, especially when streams were low, a weak or old animal was sacrificed in order to draw them away and permit a safer crossing for the herd.[42]

Though always a precarious business, drives during the boom period were better served and gradually became more efficient as infrastructure, such as the bridge over the Paraná, was improved. The character of the drives changed as well, developing into four phases that carried into recent years. These included fattened cattle for local consumption, rainy-season drives to town slaughterhouses and charqueadas, drives of hardier cattle to the traditional markets of São Paulo, and truck drives, which began in the 1950s, to transport cattle to slaughterhouses in the towns and into São Paulo. The regular raising of zebu cattle in the Pantanal from the 1940s allowed that region to participate directly in the commercialization of cattle to São Paulo, instead of relying on sales to the charqueadas. The new breed was important because the hooves of Pantaneiro cattle, well-adapted to the marshy Pantanal terrain, were too soft to withstand long drives. Zebu did not have this problem. Meanwhile, the introduction of truck drives helped to expand the state's export of live cattle. From this time, drives took on a new character, and the long and arduous journeys east gradually came to an end. Most recently, cattle drives of any length are seen only in the most remote zones of Mato Grosso do Sul, such as parts of the Pantanal.[43]

A KIND OF ACCOMMODATION: NATIVE LABOR

As outlined in chapter 4, indigenous participation in the ranching world was largely centered on land access, which often degenerated into conflict. At times, there was a form of accommodation, usually through involvement in labor opportunities, but sometimes through migration. As might be expected, experiences differed between groups and among individuals.

For example, indigenous Guarani, who lived along the Mato Grosso–Paraguay border, found their way of life challenged by the encroachment of the erva mate industry. These people, frequently called Kaiowá or Ñandeva, were said by Darcy Ribeiro to be descendants of the Guarani from the Jesuit missions, who, after the Spanish Crown's expulsion of the Jesuits in 1767 and the gradual disintegration of the missions, chose to return to their original way of life as hunter-gatherers and subsistence farmers rather than integrate into mestizo Paraguayan society. In 1880, the Guarani were reported to number as many as 6,000. They worked periodically for Brazilian ranchers and settlers, but only in order to earn enough to purchase goods needed for their people, at which point they returned to their villages. They were initially recruited to work in the organized extraction of erva since, along with the majority Paraguayan workers, they spoke Guarani. Eventually, the work regime and lack of respect for their culture, which had little in common with that of the Paraguayans, led to a messianic movement in the early 1900s. A small group of Guarani, disturbed by the theft of their lands and the exploitation of their people, followed several of their shamans to the east in search of a "Land without Evil." When they arrived on the São Paulo coast south of Santos, they established a thriving community based on an economy centered on the sea and eventually tourism.[44]

Other Guarani/Kaiowá remained in Mato Grosso, numbering 4,000 to 6,000 in 1924. Through the efforts of the Serviço de Proteção aos Índios (SPI), the state government awarded them three parcels totaling 3,600 hectares near Dourados in 1926, on which they collected erva mate and engaged in farming. Their lands were also placed off-limits to nontribal members by the SPI, to restrict exploitation of erva in their territory by freelance operators and nearby settlers. This also helped limit the spread of diseases like tuberculosis and alcoholism, a problem particularly acute among those members who maintained regular contact with the settlers.[45]

For the Kadiwéu, increased settlement and the group's relative poverty led many young people to seek work on the expanding ranches. By the beginning of the twentieth century, Kadiwéu cowboys were a common sight on local ranches, while others were contracted as professional hunters and day laborers. The irregular migration of those members who normally would have supported the tribe through hunting and fishing guaranteed that the Kadiwéu experienced a process of slow acculturation, if resisted by many members of the tribe. The land issue had not been fully resolved either. The development of ranches awarded to the large foreign conglomerates Fomento Argentino and Fazendas Francesas in Kadiwéu territory during the empire meant there were jurisdictional disputes, although these did not degenerate into violence as they did with Malheiros. Most of the conflict came from the occasional act of rustling and the forcible removal of any encampment set up within the ranches' boundaries. Not surprisingly, ranch administrators and their employees expressed a lack of respect for native culture. According to Claude Lévi-Strauss, who visited the Kadiwéu in the 1930s, his peaceful interactions with the tribe opened the eyes of the administrator of the French-owned ranch and led to closer communication between the natives and several ranchers.[46]

Nonetheless, because definitive jurisdiction over land was not fully resolved until the 1980s, the ability of the Kadiwéu to expand and flourish was highly limited. Ranching continued to expand into their territory, and hunting and fishing were subsequently restricted. The creation of FUNAI in the 1960s brought greater scrutiny of relations with native peoples, after which the population increased to roughly 1,600 by 1998 and economic opportunities expanded. A handicraft industry organized by the Kadiwéu themselves is an important source of income, especially for women. Beginning in the 1980s, ranching became a significant activity through mutual partnerships (parceria) with outside investors.[47]

The Terena were the most visible on ranches throughout the time period. Allegedly, members of the tribe were prized as employees by the Brazilian settlers, largely because they were hard workers and exhibited a passive nature. Yet, until later in the twentieth century, permanent employment was not that common, in part because relations were strained between the workers and ranch owners. As ranch workers, the natives were not spared the burden of debt service. Unlike servitude under the Guaikuru, the European concept of debt service did not include any freedom of movement whatsoever, forcing ranch hands, agricultural workers, and erva laborers to remain in one region until they had paid off their debts. Since income was so low, especially for indigenous laborers, the possibilities of escaping debt were minimal at best. For the cowboy, the nature of whose work offered some freedom of movement, this was not so serious a problem, but for other ranch workers conditions were highly onerous. The ranch store with its high prices was common. In one case, a rancher exploited the debt process by lowering individual debts at the beginning of the year, then encouraging celebration by selling alcohol to the native workers, increasing their debts once again. Some ranchers followed the lead of the erva industry by utilizing corporal punishment; because the ranchers were close to local authorities (if not holding the positions themselves), it was guaranteed that flight to avoid debt was restricted. The expansion of ranches not only affected individuals but also had a deleterious effect on indigenous communities.[48]

As noted in chapter 4, Cândido Mariano da Silva Rondon's experience with several tribes in Mato Grosso and violence against natives in a number of states led to the founding of the SPI. That momentous decision by the federal government fundamentally altered Brazilian-Indian relations throughout Brazil. It was an innovation in the official treatment of native peoples, for it removed jurisdiction over Indians from the states, where local politics and influence had frequently undermined attempts to provide equal rights to the aboriginals, and placed the service's direction under a man of part native extraction himself, deeply committed to the betterment of all tribal peoples in Brazil.[49]

The creation of SPI villages was not the end of exploitation, however, since the natives were seldom permitted full control over their daily lives; legally they were considered adolescents and thus incapable of making their own decisions.[50] The concept within the service was that aboriginals needed care and control in order to survive the onslaught of settlement and attempts to take over Indian lands and to enter civilized Brazilian society. This meant that community activities, labor, and religious beliefs all came under its supervision. In some cases, this was to the benefit of the natives and received their support, especially if it involved restriction of alcohol; in others, it created tensions with the service and divisions within communities.[51]

The SPI took over many of the traditional rights and obligations of the chiefs, placing this authority under the local Brazilian directors, some of whom were not particularly scrupulous. Native labor often had been contracted out by the chiefs, particularly to the ranches and the railroad. With the establishment of SPI posts, this responsibility passed to the director. In some cases exploitation occurred, as ranchers made deals with individual directors for cheap labor and customarily paid wages directly to the directors, who would then control native cowboy access to their hard-earned wages. Nonetheless, the presence of the SPI helped to uphold the interests of the natives against those of local settlers and authorities. Reserve boundaries were defended, if not always as effectively as necessary, and labor relations were irregularly monitored for abuse.[52]

Still, for most groups, relations with the SPI were fragile. Until the reorganization of the service into FUNAI, all economic activities on reserves were dictated by the Indian agents. Though generally honest men, these bureaucrats held views of Indian abilities and culture that frequently conflicted with the self-perceptions of the natives. Indeed, by its nature the SPI only accelerated a process of integration that had begun with the dispersal of native peoples in the first place.

For the Terena, wage labor and Brazilian customs, habits, and diet contributed to the acculturation of many tribe members. By the 1920s, few still engaged in hunting and gathering. The extended family structure and village communality, weakened by the need to seek work outside the community, had been replaced by individual family units, further eroding traditional culture and permitting greater control from outside. This breakdown was significant in opening the reserves to missionary groups, supported by the SPI. The experience of the Terena with both Protestant and Catholic missionaries was not a happy one, as communities split over religious differences and conflicts ensued between missionaries and the SPI. In some cases, growing antagonism forced several families to seek other villages or to leave the reserve system altogether.[53]

The Terena experience with the expansion of ranching and settlement in Mato Grosso was different from that of the Kadiwéu, but it was still harsh. That the Terena managed to survive and to multiply to a population in 1999 of nearly 16,000 in several villages probably has as much to do with their adaptability to Brazilian society and their integration, if culturally altering, into the rural proletariat of the nation. In this way, and with the initial support of the SPI, they were able to survive an abusive process of frontier expansion while preserving, at least in part, some customs and lifestyles. Like the Kadiwéu, Terena culture and handicrafts have gained a stature today that would have been unthinkable during those years when, at best, the group was looked on with "tolerant disdain," at worst as an expendable obstruction to the spread of modern civilization.[54]

Labor in Mato Grosso's ranching sector was not dramatically different from that encountered in other ranching regimes throughout the

Americas. It involved working cattle, constructing and repairing fencing and buildings, as well as tanning hides, tending to gardens, and the like. Also like many cattle regimes, the labor force was made up of an ethnically heterogeneous population, including foreigners and native peoples. This generated inevitable prejudices but also led to outside influences that helped form characteristics of the Mato Grosso rural sector that endure to the present day. Many of the chores done in ranching today are identical or similar to those of the past, indicating the durability of its work regime.

Perhaps distinctive to Mato Grosso was the division of labor between vaqueiros and ranch hands, whereby the former dealt only with cattle and refused to work at other chores. A certain cowboy aristocracy developed as their treatment and compensation surpassed that of ranch hands. Also, they were a permanent fixture, whereas ranch hands were normally hired seasonally. This did not mean, however, that cowboys were permanent employees on one ranch, for there was continual migration among ranches. Nor were ranch hands necessarily dismissed after every season, as some, particularly those who were married, were able to stay on the ranches as camaradas, with their own responsibilities and living quarters. The relationship between ranch owners and their employees, then, could vary from mutual trust and friendship to severe exploitation. The latter circumstance was generally more common when ranches were remote, especially in the early years, or when there was a surfeit of labor, as during the 1910s and 1920s.

Other labor essential to the ranching economy involved the cattle drive. Here, the regional and national economy determined the character and costs of the drive, and ranchers in Mato Grosso were at the mercy of cattle buyers in Rio de Janeiro or São Paulo or local charqueadas. This dependence was particularly acute during the boom of World War I and the 1920s, when the São Paulo slaughterhouses controlled most of the market and consequently the national industry. Because appeals for improved rail services or local slaughterhouses tended to fall on deaf fiscal ears, Mato Grosso was forced to rely on the drive well into the 1950s. The drive may seem romantic to folklorists and the general public, but was an important factor in slowing expansion of the regional cattle industry beyond its traditional role as a purveyor of cheap animals to the national market.

Ranching also had a profound effect on the original inhabitants of Mato Grosso. Competition for land, deep cultural misunderstandings, and abusive exploitation of Indian labor all contributed to the decimation of tribal populations. Those groups most resistant to rancher encroachment onto their traditional lands suffered the most. Groups who were integrated, at least partially, into the ranches survived in greater numbers, although suffering the loss of culture and their previous way of life. The presence of the SPI did permit, at least in part, the preservation of some vestiges of native culture. It also was an instrument in the expansion of Brazilian society that led to the assimilation of many native groups, a significant number of whom became part of a rural working class that found employment in ranching.

SIX

The Dynamics of the Mundane

EVERYDAY RANCHING

The everyday activities of ranching usually are similar across regimes, but in every region they take on a particular character rooted in the local human and environmental conditions. For Mato Grosso, an added feature was the question of ranching efficiency. Economic demands imposed from outside the state drew attention to what was frequently characterized as sluggish traditionalism versus innovative modernization, the latter usually expressed as "rational" ranching. Throughout the period under study, observers repeatedly worried about the backward state of the local industry, yet ranchers' responses were not as simple as they were often portrayed.

THE TRADITIONAL AND THE RATIONAL

Despite the growth of animal and human populations and a gradual expansion of production and exports, Mato Grosso's so-called backwardness remained a major concern for the region's presidents. In an 1872 report, the provincial president blamed "lazy Matogrossenses" for the virtual absence of agricultural and ranching development, but President Cardozo also cited the lack of manpower caused by the recently terminated war, the irregular transportation facilities, a lack of capital, and disease. Cardozo thus insisted that Rio must accord greater attention to the strategic border area if it was to emerge from its "lethargy." Rio responded by exempting most Mato Grosso cattle products from export taxes, and in 1874 it offered a prize of 5 milreis per head for the import of purebred breeding bulls.[1] Little appears to have come of these efforts, as João Severiano da Fonseca, traveling through the area just a few years later as part of a Brazilian commission demarcating the Brazilian-Bolivian border, remarked on a profound lack of initiative among Mato Grosso ranchers, who were involved only in the buying and selling of their animals, otherwise allowing the semiferal cattle to fend for themselves. He argued that most Brazilian ranchers were aware of the value of managing their animals, providing water and salt, and planting pasture, yet in Mato Grosso such practices were ignored, to the detriment of the industry's future. As we have seen, a decade later Vice President Ramos Ferreira lamented the wretched state of ranching, emphasizing its disorganization and hinting at a lack of interest in improvement on the part of ranchers.[2]

The same sentiment was expressed over succeeding decades and continued well into the twentieth century. To be fair, the criticism was not confined solely to Mato Grosso. Ranching throughout the country, particularly in the Center-West, occasioned continual reproach for its "rudimentary" character, largely compared with more lucrative ranching in Argentina and North America. In 1903, a study of ranching in Minas Gerais concluded that the problem was a combination of insufficient capital investment, inadequate technical skill and training, and lack of government interest. The author's recommendations—more attention to better and cheaper transportation, construction of local frigoríficos and charqueadas, and organization of marketing through cattle fairs—were all equally applicable to Mato Grosso. Yet, rudimentary ranching was still a concern in Brazil ten years later, when Eduardo Cotrim, a distinguished Brazilian veterinarian and agricultural expert in the São Paulo state government, reported that Brazilian ranching had hardly developed in recent years:

> We do not have any method whatsoever. Up to today, our cattle have lived in complete abandonment. The industry does not exist because the system adopted as the most convenient is that of perfect savagery [*selvageria*].

Speaking primarily of the Center-West, Cotrim believed that only with strict care of cattle and concerted support of the industry could ranching emerge from its perceived stupor.[3]

In Mato Grosso this critique was routine. In 1907, Virgílio Alves Corrêa listed the reasons for ranching's failure to progress in the state. He mentioned the ease with which ranching could be undertaken, as well as government inertia in recognizing the importance of the industry to the region and to its own income, despite its reliance on taxes from cattle production. He also cited high import taxes and freight rates, which discouraged imports of fencing wire and breeder animals, plus inadequate transportation. He did not spare the form of ranching, however, and especially condemned the practice of communal cattle raising and field burning, which exhausted pastures and added to care costs and losses of animals. Echoing earlier recommendations, Corrêa urged the federal government to support local ranching with subsidies, tax exemptions, and free training. These solutions came slowly as time-tested methods continued to dominate.[4]

THE ESSENTIAL APPLICATION OF FIRE

Despite Corrêa's concerns about the use of fire to clear fields, tropical cattle raising relied on this historic, if controversial, practice. Burning to improve pasture for grazing was common in Europe before the conquest of the Americas, and burning was widely applied by native peoples living in the plains of the Americas as an aid to hunting. The arrival of ranching only expanded the practice. By the 1820s, neo-Europeans and their cattle were a permanent presence in Mato Grosso, and the regular burning of fields soon had a noticeable impact.[5]

The application of fire for clearing in savanna today is a complex issue. Environmentalists and some scientists argue that the practice should be abandoned, while most ranchers supported by other scientists have urged some form of controlled burning. Annual burning is the quickest and most cost-effective method to renew grasses that have lost their palatability or nutritive value because of age or overgrazing. This is especially the case where extensive ranching predominates. The intention is multifold: to clear old and coarse grasses, brush, and shrubs; to leave a mineral-rich ash layer (calcium, magnesium, and potassium) for the growth of new grasses; to kill parasites and insects like ticks; to form protective clearings against accidental fires; and to attract animals to an area otherwise avoided because of old and unpalatable grasses. Employed over the long term, however, and especially in conjunction with overpasturing, fire causes permanent damage, a serious matter when the area affected is restricted, as it is on small ranches. Soil nitrogen is consumed through the burning of surface litter; soils lose humus and eventually become hard and mineral poor, leading to surface erosion; invader grasses, which are often tougher and less pal-

Burning the cerrado, in the 1920s. Courtesy of Elza Dória Passos.

atable, push out species common to the area; and the entry of termites and other damaging soil insects is encouraged, further impoverishing the soil. In addition, conditions become propitious for the entry of cattle parasites such as ticks and berne fly. One writer described the result in Mato Grosso as "the law of the least effort employed in squandering, in the short term, landed capital."[6]

In Mato Grosso, and in Brazil as a whole, the economic rationale won out almost exclusively, including to the present day. Burning had been practiced in the country for centuries. In eastern Mato Grosso, Alfred Taunay of the 1865 expeditionary force noted the widespread use of fire by ranchers from western Goiás to Coxim near the headwaters of the Taquari River.[7] For the most part, there was no attempt to control burning, as fires were simply set and left to burn for weeks. More than the pasture was destroyed, as flames were dependent on the serendipity of the winds. In 1903, agricultural expert Joaquim C. Travassos described a scene witnessed throughout Brazil during the winter months:

For centuries the entire surface of Brazil, in the months of July, August and September, [has] blaze[d] like an ocean of fire, the atmosphere full of parching smoke and cinders, darkened, stifling us under a blanket of dense smog that even makes navigation difficult.[8]

At roughly the same time, Franz Van Dionant, the Belgian administrator of Descalvados, described burning in Mato Grosso that went on for weeks:

The curtain of flames, from one to several kilometers wide, and occasionally of several leagues, often rises to great heights. The columns of opaque smoke surge even higher and with the clouds sometimes completely obscure the sun. The heat is intense, the air thins; animals flee in terror, man feels an uneasiness, an anxiety analogous to what one feels before a severe storm, but even more intense.[9]

The impact on the environment triggered doubts quite early. Toward the end of the nineteenth century, some Brazilian observers began to question the usefulness of indiscriminate burning. Botanist J. Barbosa Rodrigues, on a trip through southern Mato Grosso in 1896, described dry fields ruined by cattle and indiscriminate burning, which he felt was rapidly destroying the region. In 1918, Moises Bertoni wrote that in the Paraguayan highlands, part of which borders on Mato Grosso, the practice of regular burning had diminished productivity, and native grasses had already disappeared or were disappearing in response to the combination of fire and indiscriminate grazing. His assessment of conditions in Paraguay was a portent of what was to come in Mato Grosso. By the 1920s, observers were noting a significant amount of deforestation caused by felling and burning, even along rivers and streams, where such clearing was ostensibly prohibited by state law. Also noticed in the same period were numerous termite mounds between Três Lagoas and Campo Grande, an indication of soil degradation caused by the combination of fire and overpasturing. Studies done in 1941 revealed that the nutritious native mimoso, *Heteropogon villosus*, was fast disappearing where it was burned and cattle were allowed to graze soon after. This combination severely compromised the survival of native species, which generally have shallower root systems than the hardier invaders. During his journey, Taunay observed the spread of the fire-hardy and coarse barba de bode in western São Paulo, the Triângulo, western Goiás, and eastern Mato Grosso. He attributed this to uncontrolled burning. In addition, indiscriminate use of fire contributed to a rise in cattle pests.[10]

The proliferation of ticks and berne and warble flies appears to be the result of regular burning and overpasturing. As soils are degraded and tougher, woodier vegetation invades, conditions improve for the insects. Fires retard their development for a short time, but since fields are usually burned during the insects' dormant cycle, fires are not especially effective in eliminating the pests. The increase in insects, especially berne flies, was noticed in Mato Grosso in the 1930s and apparently became a problem for cattle. With time, prophylactic measures kept the flies in check throughout most of the region, although not all ranchers could afford to treat cattle regularly. Despite these negative results, not all observers were decidedly against the use of fire.[11]

Rodolpho Endlich saw fire in a different light. He did not consider the practice destructive per se, and he emphasized that forest was not destroyed, since trees were generally resistant and acted as firebreaks. He acknowledged a negative return, however, if burning occurred outside the season of August to October. The benefit of new green shoots would soon be offset by an invasion of tough grasses and weeds. This opinion was echoed some years later by Cotrim, who recommended the use of fire throughout Brazil, since he saw the need to renew grasses in the most economical manner possible and believed that ticks were permanently eliminated as well. Such burning did have to be controlled, preferably through division and rotation of pastures, and control of this sort required the expense of fencing and extra labor, which simply were not within the budget of many ranchers. This was particularly the case among smaller ranchers, who had limited space in which to pasture their animals, a situation common on the campo limpo.[12]

More recently, Ana Primavesi has accepted that in extensive ranching, annual burning has been the cheapest and most effective method of rejuvenating pastures in remote regions. As such, she does not condemn this practice outright but perceives burning as somewhat rational. Where a form of "minimigration" occurred, as ranchers took advantage of seemingly unlimited territory to rotate their animals, environmental damage was limited, at least as long as the cattle were not restricted to one area for too long. As the sector expanded and competition for space increased, such practices lost their validity. Strictly speaking, Primavesi is correct, although the environmental consequences of ranching over time reveal the complexity of cattle's interaction with their surroundings, especially given the distinct ecological zones of Mato Grosso.[13]

In the Pantanal, fire was and is less important than in the cerrado. Although seasonal flooding takes the place of burning to rejuvenate pasture, fire was used periodically, largely in the drier eastern region bordering the cerrado, with local impact. The use of fire appears to have started as soon as cattle were introduced to the region, though it became common only in the years following the Paraguayan War. Burning was most frequent in the northern sectors of the Pantanal, where population pressure was greatest. During the postwar reorganization, many new ranchers set fires in an attempt to round up the unbranded feral cattle of abandoned government ranches. In other sectors of the Pantanal, some burning was common during the initial occupation of the territory in order to clear land of brush and trees for the construction of ranch buildings and for easier pasturing. Renato Ribeiro relates how his ancestors in the 1840s and 1850s set fires that lasted weeks and covered hundreds of hectares. For the most part, however, burning in most of the Pantanal was done only to remove scrub where cattle had not grazed and when pasture was needed in times of emergency, such as after an excessive flood year or if a rare frost killed existing grasses, and in the periodic hunting of jaguars.[14]

Nonetheless, there were some detrimental effects of fire specific to the Pantanal. It was recognized that burning produced a great deal of ash that could be harmful to fish when it entered the river system, disrupting the aquatic life cycle. For a time, it was believed that excessive potassium from the ash killed fish directly, since residents noted many thousands of dead fish after burning. It appears, however, that it is the excessive growth of aquatic vegetation, forming floating islands and triggering decomposition, which increases levels of carbon dioxide in the water and deprives the fish of oxygen. Locally called *diquada* or *dequada*, and in other regions *toque*, this has been a recurrent problem in various regions of the Pantanal. It is not clear how much burning contributes to this situation. At the same time, ranchers noticed that animals grazing on the new grasses developed a nutritional deficiency if they ingested potassium through the ash while maintaining a low-salt diet. Potassium chlorate combines with sodium chlorate in the blood to eliminate sodium through urination.[15]

As late as the 1930s and 1940s, observers continued to critique similar rudimentary practices, as well as limited fiscal investment, isolation, deficient transportation, inadequate capital for improvements, and labor shortages. The problem was primarily one of extensive ranching. Until land took on a value other than as a medium for animals to go forth and multiply, ranching relied on nature for its survival. There was a certain logic in this, for if the climate is sufficiently benign, as it was in most of Mato Grosso, cattle could survive on their own with limited inputs from ranchers. There was little competition from wild herbivores, predators were a nuisance but not a major threat to herds, and there was plenty of forage and even natural salt to sustain significant numbers of animals. Any losses incurred either went unnoticed or were seen as the cost of doing business.[16]

IMPROVEMENTS AND THEIR CONSTRAINTS

Thanks to such rudimentary conditions, ranching in Mato Grosso compared unfavorably with Rio Grande do Sul and Argentina, but the state's cattle sector shared this weakness with most of the rest of the nation. Cotrim argued that although cattle in the Center-West lived in "complete abandonment," the nation's ranching sector had a potentially bright future if "rational" methods such as veterinary care, wire fencing, planted pasture, and cross-breeding were introduced. Those were serious concerns for Brazilian veterinary scientists well into the 1950s, and their apprehensions were reflected in proposals for Mato Grosso.[17]

The main concern was the quality of cattle for meat production. Mato Grosso herds were considered inferior to animals from more controlled ranching in Argentina or Rio Grande do Sul. A semiferal animal is usually lean, tough, and wary. Roundup of such animals could be a difficult process, resembling a hunt more than a harvest, and the quantity and quality of meat and hides produced generated only minimal income. The result was close to subsistence ranching, despite the vast expanses of territory claimed by some ranchers. Unless the inputs necessary for improvements were compensated by higher cattle prices, little attempt could be expected to improve the situation. Here, the isolation of Mato Grosso was decisive, for the cost of raising an animal any other way was seldom recovered in its sale. Poor transportation, scarce rural credit, high taxes, and competition from other ranching areas of Brazil combined to prevent Mato Grosso ranchers from modernizing their operations, even when there was the will and knowledge necessary for a modest beginning. Under the circumstances, extensive ranching was not the product of a simple, backward view of the world. Rather, it was the only available response to environmental and economic conditions that limited the scope of action.[18]

Veterinary Care, Experimental Farms, and Cattle Fairs

These conditions were understood by several observers, including some presidents of the state. Official attention was not absent, but it was limited because government resources were scarce and politicians were further restrained by competing political demands. As noted in chapter 3, transformation came with the boom of World War I, when the value of Mato Grosso cattle inextricably drew the state into the broader national and international economies, though official recognition of the possibilities began a few short years before. In 1912, President Costa Marques suggested the purchase of breeder-age cows and heifers, the subdivision of large ranches, introduction of fencing on a wider scale, improvement of transportation facilities, and the development of an effective medicine to combat mal das cadeiras in horses. He suggested that the state offer assistance in finding a cure and that it establish an experimental farm or ranch (*campo de demonstração*) to improve cattle breeds. The suggestions, though not particularly new, fell on more responsive ears with the wartime demand, and eventually some concrete measures were taken to support the sector.[19]

In emulation of initiatives in São Paulo beginning in 1895, a state agricultural school and a campo de demonstração were established near Cuiabá in 1912–1913. World War I stimulated the creation of a *feira de gado* (live cattle market or fair) in 1919 at Três Lagoas, modeled on feiras established during the colonial era and in Minas Gerais at the end of the nineteenth century. The measures were intended to encourage the application of modern ranching methods, as practiced in other nations, and to aid in the marketing of Mato Grosso cattle to the São Paulo slaughterhouses.[20]

Yet these facilities were insufficient and poorly maintained. Just one year after it opened, President Manoel Faria e Albuquerque declared the experimental farm expensive and overly bureaucratic. His report suggested that itinerant extension agents might be cheaper and more effective. Nevertheless, the farm continued to function, and by 1920 some grasses had been planted for study and three purebred bulls purchased. In 1918, the federal government also proposed setting up a *fazenda modelo* (model ranch) in Campo Grande on land provided by the state, but Cuiabá delayed acquiring a site. By the mid-1920s, the experimental farm had been shut down, ostensibly for financial reasons, although the state president favored the concept. In his 1925 report, President Mario Corrêa da Costa suggested that its grounds be used as a "Patronato Agricola" to teach traveling instructors. The following year he reported that the model ranch at Campo Grande was not yet functioning, even though the state had bought land and donated it to the Ministry of Agriculture, and a director had been appointed in 1924. This delay could have been attributed to the civil upheaval of 1924–1926, for the director reported in mid-1925 that construction of the "first works" would begin shortly. Nevertheless, the ranch was still idle in 1929.[21]

The following year, the Campo Grande ranch had only planted a few hectares of pasture and no buildings had yet been constructed. The new federal government of Getúlio Vargas halted even that minimal activity, which was not surprising, since Vargas soon replaced most public officials. Eventually the Vargas regime paid more attention to establishing Brazil's hold on its remote regions than had previous governments, and by 1936 the ranch was in operation, engaging in breeding experiments under the direction of the Ministry of Agriculture.[22] World War II and the postwar transition to democracy distracted most legislators, causing the ranch's work to be uneven, and ranchers paid it little attention. Years later, attempts were made to formalize animal health (for example, a state government laboratory in Aquidauana distributed vaccines in the 1960s), but little was developed until 1975, when the *fazenda modelo* became part of the national center for the study of beef cattle under the auspices of the federal Empresa Brasileira de Pesquisa Agropecuária (EMBRAPA), officially created in 1973.[23]

The feira de gado in Três Lagoas experienced similar obstacles. Under private concession, the feira had operated since 1920, but by 1925 it had contributed little to stimulate either local business or the Mato Grosso cattle industry in general. Letters written to a Três Lagoas weekly in mid-1925 complained that the best efforts of the directors had achieved nothing because the state government offered the feira little support. Bridges between the cattle center of Campo Grande and Três Lagoas were still not completed, relegating the cattle drives to continued use of the Paranaíba and Porto XV de Novembro trails. The feira offered no pesticide baths, developed pastures, or zootechnical station; there was virtually no infrastructure, and water and electricity were insufficient. The letters also noted that the demand for cattle was greater than the supply; hence, animals were purchased the traditional way, on the ranches themselves, bypassing the feira altogether and undermining the promise of competition. President Corrêa da Costa's address the following year bowed to the critics, emphasizing that as long as bridges between Três Lagoas and Campo Grande and Paranaíba were not finished, and given the traditional nature of cattle raising in the state, the feira simply was irrelevant. He might also have added that the taxes charged by the feira and the still incomplete rail bridge across the Paraná River made Três Lagoas an illogical export center for Mato Grosso ranchers.[24]

Little improved in subsequent years, as a letter from the mayor of Três Lagoas in 1930 indicated. The mayor pointed out that the feira was closed, its pastures inactive, and a bridge over the Sucuriú River on the trail to Paranaíba had been constructed of wood instead of the iron stipulated in the feira's contract with the government. As a result, the municipality was forced to repair it constantly, described as a "real calamity." Considering that the bridge over the Paraná was now complete, opening up opportunities for rail transport to the São Paulo winter pastures, he saw no excuse for the failure of the feira. State government intervention was requested. Yet the feira appears to have been abandoned, for no more reference was made in subsequent reports and newspaper articles through the 1930s and 1940s.[25]

There was some hope, however, as reported in another Três Lagoas newspaper. At the start of 1927, a veterinary post was opened to ranchers there. Under the direction of the federal Ministry of Agriculture, it offered information, vaccines, and other medicines. Apparently it was part of a rudimentary program undertaken by the federal government, one that had been requested by rancher organizations and the state government for years. Similar to the *fazenda modelo* in Campo Grande, the post limped along with limited financing and attention until the development of the EMBRAPA network, which superseded its utility. Such posts were only part of the recommendations made during the 1920s for continued improvement of the cattle industry. Other important measures that required implementation included organized imports of breeder bulls, breeding stations, disease control, pasture selection and instructions in rational utilization, and cheaper imports of fencing wire. Observers argued that most of these measures could be best addressed in the model ranches; however, the arrival of modern methods of ranching came from other sources.[26]

Planted Pastures

When Antonio Simoens da Silva wrote that "[c]attle are the mirror of the land" in the early 1920s, he was referring primarily to what he perceived as nutritious grasses supporting remarkably healthy cattle.[27] Certainly, quality grasses make for stout, robust animals. However, because herd populations were often difficult to ascertain with any accuracy in Mato Grosso, others observed vast territories set aside for relatively few animals. In the 1890s, Rodolpho Endlich estimated that local conditions demanded 3 hectares per head, much more than the 2.3 to 3 hectares in Paraguay and 1.8 hectares in Argentina. Yet he also reported that a ranch of more than 850,000 hectares hosted some 80,000 animals, a sizable number by any standard but at a density over 10 hectares per head. Clearly, not all land was suitable for grazing, but later observers noted similar land-to-animal ratios. Miguel Lisboa, who participated in the initial railroad survey across southern Mato Grosso, pointed out that ideal land-animal ratios in Mato Grosso depended on local environmental and geographic conditions. In the Pantanal, he believed, the land could support one head per 2.2 hectares; in the higher elevations, where the soil was less fertile and rainfall less frequent, the ratio decreased to one head per 4.5 to 6 hectares. In the Vacaria, conditions were slightly more favorable, 3.6 to 4.5 hectares for each animal.[28]

With time, and as more studies were conducted throughout Brazil, recommendations ranged all over the map. Depending on the region, estimates in the 1920s suggested that Mato Grosso natural pastures could support one head on 2 to 5.5 hectares. Comparisons were made with Argentina, which in 1916 was operating on a much more lucrative ratio of one steer per hectare on planted pasture. Little had changed by the mid-1930s. As an example of general perceptions, the Vargas government calculated territorial taxes based on 6 hectares per head. Observations in the early 1950s estimated that the cerrado could support one head per 2.4 to 3 hectares on natural grasses, but up to 2 head per hectare on planted pasture. In the Pantanal, the ratio was between 2.9 and 3.3 hectares per animal on natural pasture, while the Vacaria was said to support 1.3 head per hectare on its natural pasture.[29] Given these variations, the conclusion of most outside observers was that the most productive and efficient method of raising cattle in the region was through the introduction of exotic grasses.

The idea was not new, for as early as the 1850s, the rotation of pastures had been recommended to maintain the quality of forage. In later decades, observers urged the use of imported grasses, although in the 1890s some ranching circles emphasized the value of native species. One aspect of animal nutrition that tended to be ignored, however, was the potential for plant toxicity in free-range ranching. It was an important

Jaraguá grass, in the 1910s. Roberto Cochrane Simonsen, The Meat and Cattle Industry of Brazil: Its Importance to Anglo-Brazilian Commerce *(London: Industrial Publicity Service, 1919), 12.*

consideration, but for most ranchers such risks had never been a major concern; that was simply the cost of raising cattle. Hence, little was done on the ground, reinforcing arguments by observers who saw an urgency for regular access to more nutritious forage. Ultimately, the perceived pressure to emulate the success of Argentina led to an almost absolute faith in the value of introduced grasses.[30]

The most serious attempts to introduce planted pasture to Mato Grosso were made by the foreign consortia. Brazil Land first planted exotics after serious frost in 1917–1918 contributed to the loss of all of its imported breeding stock. The example was followed by all the other foreign operations and even some Brazilian ranchers who had sufficient capital for the investment. African-origin grasses like jaraguá (*Hyparrenhia rufa*) and gordura (*Melinis minutiflora* or *M. multiflora*), plus rhodes grass, sorghum grass, and even alfalfa, were planted. The most successful were jaraguá and gordura, and following long experience with these species among ranchers across the interior of Brazil, they were the principal species recommended, not only by Brazil Land but also by agronomists in São Paulo and Rio. Planting followed the pattern set in São Paulo, where winter pastures near the slaughterhouses of Barretos, Osasco, and elsewhere were sown with jaraguá and colonião (*Panicum maximum*, guinea grass).[31]

The exotic grasses served only as fattening pasture near frigoríficos, and under the circumstances, the demand for planted pasture in Mato Grosso was limited. Little need was seen in the Pantanal, and some observers promoted the extensive tracts of native pasture in campo limpo. For ranchers other than the well-capitalized foreign operations, the costs of stringing fences and planting and maintaining pasture were simply beyond their means. Only with gradually increased wealth and diminishing access to native grasses did ranchers turn to planting. For the most part, planting did not occur on a significant scale until after World War II, and it only became common (though not in the Pantanal) in the 1970s.[32]

Fencing

An important aspect of livestock and pasture care involves fencing. The need to confine one's cattle to a specific geographic area for reasons of control and supervision is a common consideration for the rancher. In Mato Grosso, however, competition for land, either for agricultural farming or between ranches, was limited, and so ranchers there infrequently concentrated on enclosure, despite recommendations from livestock specialists. Diverse climatic conditions created other constraints. In the Pantanal, for example, only a few ranchers used any type of fencing, generally in the higher, drier zones, where natural barriers like streams were rare. The main use was only to separate properties, and ranchers did not rotate pastures; the practice is still infrequent today. At the beginning of the twentieth century, outside observers believed that fencing would lower the cost of managing cattle by limiting their range, but it soon became evident that under extensive ranching conditions the cost of a horse and rider was cheaper than fencing wire. Although some trees were felled in the region, few were easily accessible, and the cost of labor to cut and carry wood for posts or rails was not justified unless land rights were in question. Once wire fencing entered the region, however, ranchers usually used the carandá palm. The palm's ubiquity and rapid growth reduced deforestation, and the environmental stress caused by the concentration of animals through confinement was limited. The one significant value of fencing in the region was to keep neighboring horses off one's property, above all during epidemics of mal das cadeiras.[33]

Only as the changes of the 1920s began to affect all of the state did ranchers in the Pantanal consider enclosure. In the rest of southern

Mato Grosso, the arrival of the railroad led to soaring land values and widespread speculation. The Pantanal, largely isolated from the railroad, experienced very little of its effects. Some ranchers began to build fences in the hope that enclosures would increase their land value and diminish their work load. Nhecolândia was the first area where fencing was established in any significant manner, as it experienced the first wave of subdivisions. Indeed, only when large properties began to subdivide in the 1930s and 1940s did fencing expand.[34]

In the rest of Mato Grosso, particularly the Vacaria, where smaller ranchers often lacked natural water or forest boundaries and thus ran the risk of disputes over calf ownership with neighbors, enclosure was more common. Still, the widespread use of fences took some time to develop, for several reasons, including a lack of wood, above all in the semiarid cerrado, and the high cost of imported wire.

The invention of barbed wire in the United States in the 1870s revolutionized organized ranching worldwide. Barbed wire was one of the major elements in the settlement of the American West, and it had a similar impact in Brazil.[35] The wire was introduced into Mato Grosso comparatively late; it seems it was not noticeable until after 1900. Unfortunately, because imports from São Paulo were never recorded, it is difficult to quantify the amount of wire entering the cerrado by way of São Paulo and Minas. Corumbá and Porto Murtinho customs records indicate that until the twentieth century, wire imports were irregular and in small quantities. Records from 1905 on document that barbed wire was one of the more common imports, along with salt. Clearly the imports were destined for the ranching industry, probably in parts of the Pantanal and the Vacaria, but quite possibly as far as the cerrado. Regular imports noted through Porto Esperança after 1918 indicate that this Paraguay River terminus for the railroad was more than likely used to move wire into the Campo Grande region, as it expanded with the cattle boom.[36]

Wire was never cheap, and so it was restricted to those few ranchers who could afford it. The federal government lowered its import taxes on barbed wire to stimulate its use, but even as late as the 1950s, wire in Central Brazil cost four times what it cost in neighboring countries. In Mato Grosso, wire remained expensive throughout the period. It appears that after 1914 virtually the only ranching operations that could afford the heavy investment required to fence off extensive territories were large foreign concerns. Where ranchers did adopt wire, though, the combination of fencing and planting pasture damaged the savanna environment and ultimately the rancher. That outcome was more often the case where there was little continuing investment, which was common in Mato Grosso with its tradition of allowing cattle to fend for themselves. The result could be disastrous for future productivity. Dolor Andrade noticed as early as the 1930s that in some areas of the cerrado, pastures had declined significantly where there was fencing. He attributed this to a lack of attention on the part of ranchers. His observation indicates that fences were gradually becoming more common, though their spread was slow. Even into the late 1940s, observers noted the absence of enclosures on many ranches.[37]

Stocking Rates and Salt

Even under extensive ranching conditions, attentive ranchers had to tend to their animals. Breeding, birthing, castration, roundup, and disease control were fundamental parts of the ranching routine. While officials and visitors repeatedly lamented the lack of attention paid to cattle, the concept of closer care for one's herd was not absent from the consciousness of Mato Grosso ranchers. Severiano da Fonseca observed in the late 1870s that one rancher, Joaquim José da Silva Gomes, had managed to introduce planted pasture, permanent water holes, corrals, orchards, and gardens on his Pantanal ranch. The son of the original southern Pantanal rancher (the Baron of Vila Maria), Gomes was a leader in the region, but he was murdered before his efforts could bear fruit. The ranch, Palmeira, apparently returned to its original unimproved state after his death in 1876, although his younger brother, Joaquim Eugênio Gomes da Silva (Nheco), settled an adjacent area in 1879.[38]

As the opportunities for sales to outside markets improved, ranchers increased the supervision of their animals, including the factors mentioned earlier. Above all, they paid attention to birth and mortality rates. Endlich noted in the early 1900s that ranchers throughout South America guided themselves by a herd growth rate of 26 percent per year. He judged that figure too high, especially in Mato Grosso, where he saw more arduous natural conditions than elsewhere. Depending on the ranch, he estimated a doubling of herds within six years, as opposed to three years in Argentina. Endlich based his assessment on the difficulty of commercial sale, ranching consumption, and other losses not experienced in Argentina, such as political revolution involving ranchers and their cowboys, rustling, snakebite, and jaguar attacks. A few years later, Lisboa assessed annual herd increase at a maximum of 18 percent in the Pantanal and 21 percent in the Vacaria. He based his figures on herds of 30 percent cows, with births ranging between 60 and 85 calves per 100 cows. At those rates, a herd could double within five years, though some ranchers reported that the time span was frequently much shorter. Under extensive ranching conditions, those figures are impressive. Despite perceived neglect, commentators consistently marveled over the favorable environment for ranching in Mato Grosso.[39]

Later years saw little change in this scenario. By the mid-1930s the birthing rate ranged from 50 to 70 percent in the Pantanal and 45 to 65 percent in the cerrado. Andrade attributed the slightly lower figures to increased fencing in the cerrado without corresponding improvements in care and pasture control. Often, fenced pastures became exhausted because ranchers either lacked or refused to apply the necessary resources to improve pasture productivity. That underlined another paradox of ranching in Mato Grosso: the quantity of animals increased, but quality either remained the same or even declined.[40]

The quality of a herd was believed to depend on the monitoring of breeding and of calf mortality. Endlich reported that natural breeding months varied in Mato Grosso; in the Pantanal, breeding was normally between August and October, before the rainy season. He also noted that ranchers did not engage in any controlled breeding whatsoever. Lisboa agreed, explaining that in his experience there seemed to be no specific breeding season at all. Procreation depended on the region, although in the Vacaria he witnessed birthing between the months of May and September, suggesting that breeding occurred between August and February, roughly comparable to Endlich's observations. The breeding times changed very little over succeeding years, but a key element was calf survival.[41]

Neither Endlich nor Lisboa cited mortality rates; such figures were probably difficult to obtain. Lisboa mentioned in passing that calf mortality was generally higher in the Pantanal than in the Vacaria, which was hardly surprising, considering the more extensive nature of ranching in the Pantanal. Endlich estimated that out of roughly 250,000 head in the Pantanal, some 5,000 head were lost annually to the elements. One can only guess the percentage of young animals in that number, but it must have been high. In 1918, Jacomo Vicenzi figured that calf mortality in the Campo Grande area did not exceed 2–3 percent, and he attributed the low mortality to the general healthiness of the region.

Studies in the 1930s and 1940s revealed a calf mortality rate for the entire state of 3–5 percent annually. The rate increased considerably after weaning, however, to 12–15 percent. That rate is consistent with information collected in the 1970s, when calf mortality was 5 percent until weaning, jumping to 15 percent thereafter. The higher rate clearly reflects increased vulnerability of the calf after maternal care is withdrawn. Most mortality was attributed to wildcats, climatic conditions like flooding or drought, injury, disease, and insects.[42]

Routine precautions taken to improve the future marketability of the animals greatly reduced such risks. They included controlled weaning, branding, separation, and castration, among other activities. Such measures were applied during regular roundups. Endlich noted that most ranches monitored and cared for cattle at retiros, distant from the main ranch house. Cattle were rounded up and held for periodic branding and castration. When they were sold to visiting drovers, the animals were either collected at retiros, selected, and then transported to the main ranch corrals, or they were picked up by the drovers directly as they visited the ranches. In the cerrado and Vacaria, there could be six or more roundups annually, even as many as twelve. Annual flooding could reduce the number to two to four a year in the Pantanal. To bring in the most reluctant animals, cowboys mustered at night when the herd was sleeping, since the presence of humans invariably sent an entire troop of semiferal cattle into nearby brush. The herds were usually attracted to an area with salt, which would keep them there for as long as it took to carry out the work.[43]

Retiros and roundups also served to provide salt. Salt is a dietary essential for all mammals, but cattle need large quantities. Some areas in Mato Grosso are endowed with natural sources of salt, such as barreiros and salinas in the Pantanal, which provide natural mineral salts mixed with earth or in the water, limiting the need for human intervention. Salinas are still used by cattle, and until recently, it was possible to come across cattle in the Pantanal with their snouts buried in dirt or mud banks, ingesting copious amounts of earth in what seemed to be a ravenous hunger. According to the famous Brazilian backwoodsman and Matogrossense Cândido Rondon, the barreiros and salinas were the true wealth of the Pantanal. He believed they made the region a virtual paradise for ranching and even ensured a better quality of meat than that from areas lacking natural sources of salt. The consequences of eating the earth, however, included the condition called toque. Cattle that rely exclusively on earth-held salts, without a regular balanced diet of good grasses, lose weight. The cows suffer miscarriages and can easily die if not moved to better feeding grounds. Toque was especially a problem in the arid sertão of northeastern Brazil, but it also occurred in the semiarid cerrado regions of Mato Grosso.[44]

Reliance on salinas and barreiros carried another risk in the Pantanal. During abnormal flooding, cattle could be deprived of their normal salt sources, and if ranchers were unprepared to provide supplements, many animals succumbed to malnutrition. Such deaths could be almost as numerous as losses from drowning or starvation. In addition, a recent study of the Pantanal has revealed that most barreiros and salinas do not provide sufficient salt for animals. Outside observers were awed that steers often licked one another's backs in order to ingest salt exuded through sweat. Desperate animals had even been known to attack sweating men. Theodore Roosevelt, traveling through the region in 1914, lost his recently removed sweaty underwear to a salt-starved steer. The general salt deficiency eventually was recognized by some ranchers, who by the 1930s began to provide commercial salt and minerals through licks, part of a greater effort to raise animals more scientifically.[45]

In most of Mato Grosso, natural salt licks are absent, and so ranchers must distribute the mineral to their animals. As mentioned previously, salt was imported from Spain, both for processing jerky and for cattle consumption. Therefore, ranchers in Mato Grosso faced not only the usual expense shared by ranchers in other regions of Brazil but also the additional cost of transporting the product into the interior, usually by river. The cost varied over the years, but ranchers usually paid what would be considered exorbitant prices elsewhere. Even within Mato Grosso, prices varied. In 1909, Lisboa noted that a 75-kilogram sack of salt sold in Corumbá for 3 or 4 milreis, while the same amount commanded 12 milreis in Campo Grande. Predictably, ranchers tended to import their salt clandestinely from Paraguay, where it was transshipped through the river port of Concepción. A state presidential tour of southern Mato Grosso in 1912 noted the extensive smuggling of salt from Paraguay, a traffic that responded to the needs of ranchers, whose cattle required salt supplements at least four times a year.[46]

By 1922, the cost had risen considerably, to 20–25 milreis for a 36-kilogram sack. By comparison, in 1917 ranches in Rio de Janeiro state were paying less than 4 milreis for 50 kilograms. In the 1930s, a ton of salt in Mato Grosso sold for between 500 and 800 milreis, comparable to 1922. This price stabilization can be attributed in part to restrictions on imported salt imposed by the Vargas dictatorship and the Spanish Civil War, which subsequently made Brazilian salt marketable, ensuring its acceptance by national consumers.[47]

Over the years, opinions varied as to how much salt an animal needed. Endlich recommended around 30 grams per head per month in the Vacaria. In 1916, it was estimated that animals there required considerably more, just over 80 grams per head per month. In the cerrado, the same amount had to be provided every two months. Another study, done in 1918 and covering all Brazil, suggested 25 to 50 grams per head per day. By 1929, a publication promoting Mato Grosso as a region for settlement recommended that calves be provided with 10 grams daily and steers 15 grams. Clearly, there was no consensus on how much salt animals needed to stay healthy, although more detailed studies today conclude that local conditions, particularly the quantity of minerals in the soil and grasses, determine salt and other mineral supplement requirements. In general, experts argue that beef cattle in Brazil need roughly 50 grams of all minerals daily over the two years of growth to adulthood. Unadulterated natural salts have always provided more than just sodium to the cattle diet, but sodium tended to account for the highest percentage until recently; now all commercial salts contain other essential minerals. During the period under study, then, cattle received salt rations that were inadequate for animal needs.[48]

A number of ranchers saw salt not only as a dietary supplement but also as a means to tame their semiferal animals. Salt placed in the corral of a retiro attracted the animals to human habitation. Gradually, steers grew accustomed to the presence of cowboys and their families and grazed closer to the retiros. That made care of the animals much easier, especially the regular roundups. A common way to attract the beasts was to call them with a horn and throw salt into the wind blowing toward their grazing area. It soon attracted the tamest cattle, which were usually followed by the others. The technique was so effective that often it was unnecessary for cowboys to ride out very far to collect the herd, as reflected in the saying, "Salt is the best cowboy." In addition, some ranchers believed that salt softened the hide and provided resistance to ticks.[49]

Nevertheless, there was some reluctance among ranchers and their cowboys to provide salt. Some of them, for example, believed that feral cattle had a tougher hide that was resistant to insects because they consumed less salt. The expense was always an important element in ranchers' decisions to supply salt irregularly, although as costs declined that rationale became less valid. The greatest resistance was seen in the Pantanal. Renato Ribeiro observed that on his family's ranch in the drier eastern Pantanal, little salt was provided before the 1940s. Others noted the same, and until recently barely half of Pantanal ranches regularly furnished salt, although today the practice of providing minerals is widespread. Still, into the 1980s only 20 percent of the region's ranchers regularly supplied mineral supplements in addition to basic salt rations.[50]

The story of salt is indicative of the contradictions in Mato Grosso ranching between suggested improved methods and realities on the ground. Ranchers were perceived to be resistant to change, but as we have seen, their behavior was based more on resources and need than on stubbornness. Similar circumstances determined the regional access to horses.

HORSES ON THE RANGE: AN ENDURING DILEMMA

With the devastation of the Paraguayan War, horses in Mato Grosso were invariably imported, and prices were not cheap. As early as 1869 imported horses were selling for 100 to 120 milreis each, at a time when the monthly salary of a cowboy was 30 milreis. This was a difficult time, since the war had not yet ended and horses were understandably hard to find, but prices varied little over the following decades. One Pantanal rancher recalled that during the early 1890s, stallions cost 100 milreis a head, while mares went for 40 milreis. At the time, average cattle prices ranged between 15 and 30 milreis per head. The most common way to acquire horses was through barter; bartered horses were usually poorer quality than those bought with cash. In such circumstances, during the first decade of the twentieth century, a horse was worth ten young steers, themselves valued at roughly 10 milreis apiece. Around 1905, Descalvados paid between 150 and 250 milreis a head for horses from outside Brazil, and 40 to 80 milreis of that was import tax.[51]

Given such high costs, it would seem logical to raise horses in the region, and in the past that was common. The presence of mal das cadeiras, however, made horse ranching virtually impossible. Endemic to the region, it regularly decimated the horse and mule population. The disease (surra, or equine anemia; *Trypanosoma evansi* or *T. equinum*) originated in Africa, arriving in Brazil in the 1820s and in the Pantanal in the 1850s, and caused havoc in the region into the 1930s. It is a severe form of anemia caused by a protozoan bacterium carried to the victim by blood-sucking flies or possibly bats, affecting not only domestic animals but also indigenous wildlife like the capybara. In the later stages of the disease, the animal loses control over its hindquarters. It is always fatal if not treated. In the Pantanal, mal das cadeiras usually struck at the end of flood season as waters began to recede, particularly if flooding had been more extensive than normal. Mortality could reach as high as 80 percent, and during one outbreak near the end of World War I, some ranches lost up to 600 horses. Ranchers were helpless to combat the scourge, principally because they had no idea of its origin, and so they were forced to import horses annually from outside the region at considerable expense. For many ranchers, particularly the smallest, the drain on income made impossible any development beyond subsistence ranching. Most, in fact, could not afford to import horses regularly and so resorted to using saddle steers as mounts, a common sight in the Pantanal into the 1940s. Obviously, such plodding beasts were incapable of rounding up semiferal cattle, underscoring the role of mal das cadeiras in inhibiting expansion of ranches and the industry as a whole.[52]

The disease ravaged the Pantanal almost annually. The worst attacks were noted following severe floods, as in 1901, 1905, 1919, and 1927. A study done in Miranda in 1940 by the Instituto Oswaldo Cruz, Brazil's most prestigious medical research establishment, concluded that extensive flooding contributed to the weakening of the horses because they had to search more assiduously for scarce forage. As floodwaters receded, the bot and warble flies proliferated, further debilitating the animals while increasing the chance of transmission of the disease through their bites or larvae.[53]

The state government showed early concern for the situation. In 1872, it offered a prize of 10 contos, a small fortune at the time, to whoever could rid Mato Grosso of the disease. In 1879, it bestowed on a scientist from Rio Grande do Sul, Carlos Berg, a monthly subsidy and a ranch where he could study the problem. Although he correctly hypothesized that the disease was passed on by an insect vector, he achieved no further success and in 1881 his subsidy was terminated. Endlich reported that in 1903 the cause of the disease had been isolated by an unnamed researcher in Paraguay. Yet, in 1919 the French administrator of a ranch in the lower Pantanal noted that despite the development of a treatment, the study of mal das cadeiras had ceased at the

Instituto Oswaldo Cruz in Rio de Janeiro. He emphasized the need for continued research if a cure was ever to be discovered.[54]

The cessation of research probably was only temporary, since by that time a prophylactic was in use, although it was very expensive. In 1916, the Mato Grosso government contracted for the purchase of some 8,000 doses of Protosan, produced by the Instituto, at the cost of 14 milreis apiece. The medicine was to be distributed to ranchers in the Pantanal at cost, but the state president soon admitted the program was a failure due to lack of training and follow-up. Many ranchers were not able to apply the remedy properly, breaking the vials or needles, and often gave up in disgust. The government did not set aside funds to send trained veterinarians to the ranches to instruct ranchers and their cowboys, thus providing no real motivation for ranchers to change their traditional methods of care or to risk the extra expense themselves.[55]

By the 1930s, however, an antiparasitic called Naganol was in use. It was similar to a medicine used in Africa to combat nagana, a parallel disease of horses and cattle transmitted by tsetse flies. This prophylactic was reported to have had a discernible effect in reducing the incidence of the disease, but it was very expensive. Five ampules cost as much as a third of the value of an animal. Nevertheless, the continued expense of purchasing horses from outside the region helped to make the campaign for its use a success, and by the 1950s no cases had been noted in the Pantanal for several years. The disease had been a drain on the Pantanal and no doubt had helped to stunt the growth of ranching in the region. Eventually, access to effective prevention, despite its cost, motivated local ranchers to make an investment in their future denied them previously.[56]

For decades, horses had to be imported into the Pantanal and into Mato Grosso. Incomplete data for Corumbá at the turn of the century indicate that 300 to 500 animals a year were imported. Probably as many or more entered from Paraguay through Bela Vista and Ponta Porã, while Minas and São Paulo likely contributed some animals. Various observers estimated that the average ranch in the Pantanal needed at least one horse per 100 steers. At that rate, some ranches might have required as many as 600 horses, although most could get by with fewer. Considering the mortality from mal das cadeiras, it is clear how draining the constant purchase of horses could be to the ranch economy. Renato Ribeiro wrote that his father reserved between 200 and 300 head of cattle every year for the purchase of horses. Mules also were uti-

Troop of horses at Miranda Estancia, 1910s. Reprinted from Cardoso Ayala and Simon, Album Graphico do Estado de Matto-Grosso, *286.*

lized, particularly in caring for newly born calves, which required long and sometimes difficult sojourns into breeding areas of the region, a time when horses were rested. Despite that alternative, horses generally were preferred, because they were better suited to specific work demands and to cowboy pride. Thus, one solution frequently proposed was to establish a horse-raising industry in Mato Grosso, where local production could guarantee a secure and cheaper supply. Yet, in 1907 Lisboa reported a chronic deficit of horses in the Pantanal but no horse ranching in the rest of the state. He seems to have overlooked some minimal breeding, but he did underline a serious deficiency in local ranching. That situation changed only with the expansion of the industry just prior to World War I.[57]

According to Paulo Coelho Machado, regular breeding of horses in the Planalto of southern Mato Grosso was started by his grandfather, who brought a herd from Argentina to his ranch in the municipality of Nioác around 1900. Whether the Rodrigues Coelho family was the first is not confirmed elsewhere, but all evidence indicates that horse ranching was thriving in the region by World War I. A national animal census in 1916 reported 140,000 horses in the state, 40,000 in Nioác and 25,000 in Campo Grande. The slightly more reliable 1920 census counted 152,000 horses in Mato Grosso, primarily in Ponta Porã, Bela

Vista, and Campo Grande municipalities. In 1929, a report on Campo Grande counted some 80,000 horses in the municipality, a mixture of breeds from Minas Gerais, Goiás, Paraguay, and Corrientes. The animals were being bred for the Pantanal market. They were offered at 150 to 200 milreis for a ranch horse and 80 milreis for a brood mare, prices not too much different from those demanded for beef cattle. By the mid-1930s, some 200,000 horses were counted in Mato Grosso, more than 140,000 of them in the southern municipalities. With the increase in numbers, prices had gradually dropped, and two adult steers could buy a good horse by that time. The figures indicate that the horse population in Mato Grosso during the period was growing at the same rate as the number of cattle. Consequently, it appears that sufficient mounts were for sale to the Pantanal, and by the 1930s, the increased use of Naganol against mal das cadeiras had stimulated tentative breeding of horses within the region itself.[58]

There were parts of the Pantanal where horses had been raised for decades without fear of the malady. It is uncertain why it did not strike in the north as in the south, but its absence in that area permitted the growth of a horse-ranching industry based on a local animal

Pantaneiro horses in the northern Pantanal, 2015. Photograph by the author.

called the pantaneiro. The horse was small, muscular, robust, and agile, ideally suited to the Pantanal, with a good temperament and great
endurance. In one case, a pantaneiro allegedly traveled over 1,000 kilometers in eighteen days. In addition to its ability to thrive on the natural Pantanal pastures, it also was well adapted to annual flooding, including a capacity to swim across swollen rivers. The pantaneiro was
probably descended from the Iberian horses that accompanied explorers and settlers into Mato Grosso from eastern Brazil and Paraguay. After the turn of the century, the modern breed was developed further
by crossbreeding horses from Poconé and Cáceres. Today, it is the most
common horse used in the Pantanal, but before recent breeding programs carried out by EMBRAPA began to supply a sufficient number
of animals to satisfy demand, the pantaneiro remained in the north,
providing little relief to the expanding southern ranching industry.
Even as late as 1949, Octavio Domingues and Jorge de Abreu observed
in their study of Nhecolândia that they encountered no pantaneiros in
the region. The horses they saw were mixed-breed English or Arabian,
probably imported from outside Mato Grosso, though there was a local horse called the pelo duro, likely purchased from suppliers in the
planalto. Over time, the availability of horses became a minor problem,
and by the 1950s Mato Grosso was producing sufficient mounts to supply ranching.[59]

MICROBES AND PARASITES

The story of the horse in the Pantanal raises the issue of cattle disease and parasite control. Although Mato Grosso was relatively free of
disease, compared with other regions, ranchers there still needed to pay
attention to fly infestations and the occasional pathogen brought into
the state. The risk became more acute with time, as greater communication between Mato Grosso and the rest of Brazil and the world led to
the introduction of pests and diseases unknown previously.

Concerns included foot-and-mouth disease (*febre aftosa*), Texas fever (transmitted by ticks), bot flies and warble flies (berne), carbuncle
or blackleg, rabies (*raiva*), and of course mal das cadeiras. Most were
minor annoyances or completely absent from the region until the 1920s
and 1930s, although local names for diseases or parasites were distinct
from the terms used today, making exact identification difficult. Ob-

servers of the cattle industry in Mato Grosso were almost unanimous in their declarations that the region was surprisingly free of most diseases.

Endlich noted an ailment called mancha negra, or manchilla, in the Mato Grosso–Paraguay region, but he considered it quite benign, as the symptoms were usually mild and it seldom caused death. The ailments were probably mild incidents of symptomatic carbuncle (blackleg or black quarter), also known in Brazil as manqueira, which is caused by a bacterium (*Clostridium chauvoei*) transmitted through contaminated water, forage, or skin abrasions, or as perceived at the time, possibly by blood-sucking flies. It can be devastating to young cattle, which have been known to die within 24 hours of contraction. The disease, only identified by name in 1919, apparently was present in the region from at least the late nineteenth century. Cases of blackleg were registered in São Paulo as early as the 1880s, and the disease was endemic to Minas Gerais by 1903. It probably spread to Mato Grosso through cattle imports and became a more serious problem with time. The region that has been the most susceptible is the Pantanal, where the bacterium found a suitable habitat in marshy areas. Vaccines were introduced in the early twentieth century, though they were only moderately successful, probably because of faulty application or resistance to use among ranchers. The disease continues to be a concern in Mato Grosso, as I encountered a case while conducting fieldwork, although effective modern vaccines have kept it in check.[60]

Another disease was aftosa, or foot-and-mouth disease, which first appeared in Brazil in 1895 near Uberaba, Minas Gerais, possibly arriving in a shipment of zebu cattle imported from India. A viral disease, it is considered endemic in southern South America, causing lesions around the mouth and on the hooves that restrict an animal's mobility and ability to eat. In extreme cases the afflicted die of starvation, though that is quite rare in South America, where the disease is relatively benign in character and cattle are resistant. It spreads very easily with contact and can be carried by droplets in the air, and so incidences were common in Central Brazil, especially when precautions were absent and live cattle were crowded together on drives. Apparently it struck every two or three years; though it manifested itself in a benign strain, the frequency gave the animals no opportunity to build up immunity.[61]

Foot-and-mouth disease first appeared in the cerrado area of Mato Grosso not long after appearing in Minas. From there it gradually spread throughout the state. The early outbreaks, though alarming, generally did not cause many deaths. It was considered endemic in the cerrado by the 1920s and in the Pantanal as late as 1975, where the infection rate was estimated to be over 40 percent in the late 1980s. The common method of dealing with aftosa was to let it run its course. If an infected animal was detected on a drive, drovers either stopped until no new instances occurred or, more commonly, left the sick beast behind. Because there are several strains of the disease, no one vaccine was successful until recently, and vaccine efficacy was usually short-term. The vaccines did not even enter Brazil until the 1950s, although today they are an integral part of the Brazilian beef industry's sanitary policy. Nevertheless, Brazil has had sufficient stocks of vaccine for its national herd only since the early 1980s. Until that time, Mato Grosso ranchers were hindered in their efforts to apply such preventive measures systematically.[62]

For the most part, the impact of foot-and-mouth disease in South America is economic. The presence of the disease led to sanctions against the continent's beef exports by European and United States authorities. Since eradication of the disease from the United States in the 1930s, government and animal health experts have been zealous in restricting the entry of beef products from the region; their assumption is that a reintroduction of the virus would devastate the marketability of American beef. The virus can live up to four months in frozen liver or kidney. Hence, today South American beef exported to the United States must be either canned or dried, and only beef frozen longer than four months reaches Europe.[63]

The ban also extended to breeder animals. By the 1940s, Brazil had become a major breeder of zebu cattle and was beginning to attract attention from other tropical areas of the Americas. In 1946, Brazil exported some 300 zebu breeding bulls to Mexico as part of a program begun in the 1920s. Coincident with their arrival at Veracruz, there was an outbreak of foot-and-mouth among Mexican cattle. Because one of the zebu imports had been diagnosed as well, the bulls were cited as the probable source; however, the animals had been inspected in Brazil before embarkation and again in Mexico, and authorities uncovered no other trace of the disease. A later assessment concluded that the one infected animal had contracted the disease when it was pastured briefly with Mexican cattle later known to be carriers. Fearful that the disease would cross the border, the United States government pressured Mexico to destroy the imported animals. In order to make its point more forcefully, Washington refused to permit the importation of any Mexican cattle into the United States until the animals were destroyed. In the end, the animals were not slaughtered but were eventually cleared by veterinary inspectors, and exports of Mexican cattle to the United States were resumed. Even some of the Brazilian bulls made their way into Texas, but further exports of Brazilian zebu to Mexico were suspended permanently. Not surprisingly, Brazilian ranchers and legislators interpreted the US action as a response to competition in zebu breeder markets, not contamination. If that was indeed the ulterior motive behind the US action, it was successful. Brazil was shut out of the business for some decades, as the Texan Brahman and Santa Gertrudis zebu-Shorthorn cross proceeded to dominate the market, including exports to Mexico.[64]

Another common problem in Mato Grosso involved warble flies (*Hypoderma bovis* or *H. lineatum*) and bot flies (*Gasterophilus* spp.), both called berne in Brazil. These flies appear to have had a long history in Brazil and neighboring countries. During the colonial era, observers considered "the fly," called ura, the curse of Paraguay, a condition that had not changed well into the twentieth century. In Brazil, a newspaper report in 1858 suggested that improved pasture could not only provide better nutrition to cattle but would also act to keep noxious weeds down and thus eliminate the "verme" fly. Mato Grosso suffered along with the rest of the country, for at the end of the nineteenth century, Endlich noted that ura was an affliction on both sides of the Mato Grosso-Paraguay border.[65]

Cattle are hosts for warble flies, which lay eggs on animal hairs, beginning a cycle in which emerging larvae pierce the skin, burrow along the surface, and then create an opening through which to breathe. Though the larvae seldom endanger a steer's health, unless they are extremely numerous, they do cause considerable discomfort and milk and weight loss as the animal attempts to relieve its pain and itching. Bot fly eggs enter cattle and horses through the mouth or nasal passages, obstructing the digestive system and causing weight loss. Eventually, the larvae are excreted with the feces and develop into adults, which then swarm around and worry the cattle. In economic terms, meat is considered unpalatable where larvae have burrowed, and their breathing holes ruin the hide for leather.[66]

Roy Nash reported in the 1920s that berne was absent in the Pantanal and south of Campo Grande. The latter observation may have been an error, since others had noted its presence as far south as the Paraguayan border, but it appears the Pantanal was indeed free of the insect. Most likely, periodic flooding made the difference, severely limiting the fly's life cycle, which requires dry pasture for the transition from pupal to adult stage.[67]

Traditional methods of treatment included the application of ashes, pepper, papaya oil, bacon fat, or tar to the sores; cauterization; the burning of pasture to destroy unhatched larvae; use of creolina (a creosote-based disinfectant) in corral and stable areas; and manual removal of nits from the animals. More recently, commercial parasiticides have been mixed into salt or mineral rations to destroy the internal bot eggs, and arsenic dipping has been used to combat warble larvae. In addition, several observers noted a lower incidence of berne in regions where the short-haired zebu predominated, although Octavio Domingues speculated that the lesser propensity of zebu to seek shade may have reduced infestations of flies, which remain in shaded areas. Either way, an increase in the zebu population in Mato Grosso from the 1920s may well have helped to reduce the incidence. Nevertheless, the flies presented constant problems for most Mato Grosso ranchers, and like many of the other diseases and parasites in the region, they were not systematically dealt with until the 1970s and 1980s, when sanitary controls expanded.[68]

Beginning in the 1920s, there was scattered mention of bat-carried rabies. A present dilemma for the region's cattle industry, it was not considered particularly serious during the period under study since a relatively small percentage of cattle pastured in the forest zones where bat colonies were found. Observers believed the bats became carriers of the disease through an unknown source in the 1920s, though concerns were not expressed until much later. The disease only began to be addressed in the 1950s, largely through poisoning of bats in their habitats and vaccination of cattle. Rabies was an issue, but it seems that because only individual animals were affected, ranchers accepted the disease as a price of raising cattle. State and federal governments paid attention once the cattle range expanded into forested areas and the disease affected larger numbers.[69]

A final malady that had only minimal impact in the Mato Grosso ranching world was Texas tick fever. Unlike other regions, Mato Grosso was not affected by the fever, although its presence in Brazil had an effect on the state's cattle industry. Known locally as tristeza, cases of Texas tick fever were not as common in Brazil as they were in North America or Europe, nor did the few cases registered develop into serious problems. Part of the reason was the type of animal affected. The tick (carrapato, *Rhipicephalus microplus*) usually comes into contact with animals by attaching itself to hairs as cattle pass through infected pasture; it feasts on the blood of its host for three to nine weeks, after which it drops off to lay eggs in the grass. The tick carries a parasite (*Babesia bigemina*) that lives in the cattle's bloodstream, causing severe anemia as red blood cells are broken down. Once an animal is infected, death results within three to ten days.[70]

It is possible that tristeza was present in Brazil as early as the 1820s, but the first confirmed incidence was at the beginning of the twentieth century, among cattle imported from Europe and Argentina. Of 315 animals affected, 298 died, prompting the São Paulo state government to begin a study of the disease at its Central Zootechnical Post in Campinas. Over time, it became clear to Brazilian veterinarians and ranchers that cattle breeds differed in their resistance to the disease. Almost invariably, the European breeds, especially those imported directly from Europe or the United States, became infected and succumbed quickly, while the Indian-origin zebu suffered little. Recent study has confirmed these observations. Experiments undertaken in Brazil, Australia, and the United States show purebred zebu to be highly resistant, crosses between zebu and European animals less so, and purebred Europeans most susceptible. Although the reasons are still uncertain, the zebu's success may have to do with its hereditary resistance, greater sweat secretion, and tougher hide.[71]

In Mato Grosso, ticks were common but Texas fever almost unheard of. The tick was unknown in the Pantanal, where flooding denies the insect a secure habitat. Furthermore, breeds long resident in Mato Grosso were noted for their resistance. Again, breed and environment may have determined such resistance, for in neighboring Argentina and southern Paraguay, where the climate was generally drier, grasses shorter, and the breeds primarily European imports, ticks were a serious problem. Over time, the introduction of regular dipping has reduced tick infestations throughout South America and virtually eliminated the possibility of an outbreak of tristeza in Mato Grosso.[72]

CATTLE AND NATIVE FAUNA

Although not directly an influence on ranching practices, the relationship between ranches and native fauna is an important part of the story of raising cattle in Mato Grosso. Ranching's impact on local fauna has depended on the degree of competition between cattle and wildlife for natural resources. Most reports indicate that until recently, ranching put little pressure on wildlife. In the Pantanal, impressionistic accounts from the years of initial occupation noted an impressive amount and variety of wildlife, particularly birds and fish. Fish were important to the ranches, since they provided food and oil, the latter for lamps and in the crude curing of hides. Professional fishermen were an integral part of the workforce on Pantanal ranches during the formative years. Wildcats, particularly jaguars, were cited as a constant hazard to the ranch animals. The predators, sensing an easy meal, would approach corrals and ranch buildings, demanding constant vigilance from ranch hands at certain times of the year. The jaguar hunt became an integral part of ranch life and provided an arena for cowboy manhood rituals, as well as a welcome diversion from daily routine.[73]

For a variety of reasons, however, cowboys and their bosses did not normally hunt. They held a vague conservationist attitude along with a cultural distaste for game meat; faced with restricted opportunity and the expense of ammunition, they had little incentive to hunt. The need to control wildcats and feral pigs provided an exception. The pigs, introduced in the nineteenth century with the first ranchers, caused problems for ranchers and disrupted the local environment by digging up the ground around baías with their snouts in search of food, restricting pasturing and endangering horses. They also contributed to the modification of vegetation. Cowboys dealt with them partially by creating a sport in which immature males are caught, castrated, and marked, to be hunted when adults, or captured for fattening on the ranch. Their population is not threatened today. Poisonous snakes, dangerous to both humans and animals, and armadillos, which burrow in the grasslands, putting horses at risk, occasionally had bounties placed on them by ranchers in an attempt to diminish ranching risks. Recent studies have noted population declines among both, though it appears that the greater concentration of cattle, rather than the bounties, is the main contributor.[74]

Thanks to generations of coexistence with local fauna, many ranchers today support conservation and try to protect wild animals. Certain species of wildlife, however, are fast disappearing in the Pantanal, most visibly wildcats, South American caimans (jacarés), several species of monkey, otters, and the giant anteater. All have suffered from hunting by licensed professionals from outside the region and poachers. Some birds, including several parrot species, also have shown population declines in recent years. Lately, such persecution has increased, particularly as the world market for skins and live animals for collectors has attracted smugglers into the region, often linked to the illicit drug trade. Such activity is not new. During the 1910s, one traveler reported "piles" of jaguar and deer skins at Descalvados. He surmised that as many as 1,000 animals were being killed annually. Another observer noted in 1916 that herons were being hunted for their feathers. The market paid a substantial US$1000 per kilogram, and he estimated that it was necessary to kill 200 birds to collect that weight in feathers. In the 1930s, deer, capybara, and giant toads were hunted for their skins, though apparently cowboys were seldom involved. All these species suffered population declines in specific localities, especially where they were easily accessible, such as along the margins of rivers. Thanks to a weak market herons, capybaras, and toads are no longer hunted, though deer skins are still sought illegally.[75]

Travelers' reports throughout the period continually referred to the impressive number of jacarés found basking in the sun along riverbanks, oblivious to the presence of man and his river steamers. Unfortunately, such apparent indifference often was fatal, as travelers and ships' crews used the defenseless reptiles for target practice, a sport eagerly engaged in to alleviate the tedium of long river journeys. Birds were victims as well. Still, those losses were not a direct consequence of ranching, but of the opening of Mato Grosso to the world and its civilizing influences. On the ranches, jacarés and birds were seldom disturbed, primarily because they did not harm the cattle or hinder ranch activities or expansion.[76]

Other animals, however, have been directly affected by ranching. Capybaras, important elements in the life cycle of the Pantanal, are herbivores. The rodents do not compete with cattle for forage as they are very adaptable, usually consuming aquatic vegetation but easily changing grazing habits with the seasons. They do suffer, however, from disease introduced with cattle. Capybaras are especially susceptible to mal das cadeiras, and in years of high equine mortality, the same was noticed in the rodents. Study of the phenomenon is still in its infancy, but the introduction of effective treatments for horses does not seem to have had any effect on outbreaks among the capybara, suggesting that the disease has become endemic among them. Fortunately, the disease's cyclical nature and the rodent's high birth rate have allowed capybaras to remain ubiquitous in the Pantanal.[77]

Hoof-and-mouth also claimed victims among the deer of the eastern Pantanal, as well as in other regions of the state. As early as the 1930s, ranchers noticed that when their cattle were affected, there was a corresponding decline in deer populations. Since the disease is endemic in the region, deer and possibly other animals are considered to be hosts. That is one of the reasons all four species of deer native to the Pantanal had been placed on the endangered species list by the late 1980s.[78]

Pantanal ranchers today assert that attention paid to wildlife in recent years has led to increases in the populations of several species, including capybara, armadillos, and even deer. Renato Ribeiro wrote in the early 1980s that he had not seen so much wildlife on his ranch in sixty-five years. One reason posited was that from 1974, climatic conditions had been ideal in the Pantanal, with adequate rainfall, no excessive flooding, and no drought. That could well be a factor, for as seen, survival in the Pantanal is determined more by climate than by human intervention. Nevertheless, it appears that in several areas, ranching has displaced local fauna, cattle have competed for resources, and wildlife stocks have been disrupted, especially where disease and predatory hunting have been introduced. Yet ranching's share of responsibility is still difficult to determine, as little past study was made of the region's wildlife. More research is required before definitive conclusions can be drawn.[79]

The plight of wildlife in the cerrado and campo limpo exhibited a slightly different character. Before 1870, travelers in Mato Grosso consistently marveled at the quantity and variety of wildlife. Tapir, deer, agouti, peccary, capybara, and many species of birds and fish were observed by an expedition in 1865, and professional hunters were recommended for future endeavors. The expedition was in an area of the cerrado similar to the Pantanal, where local fauna coexisted with numerous ranches that raised cattle on the open range. As long as ranching was of this type, there was little impact on local wildlife, since competition for resources was minimal. With an increase in domesticated animal populations, however, and the introduction of regular burning, exotic grasses, and animal diseases, the burden on fauna became quite pronounced.[80]

The most dramatic influence was fire. Though fire kills few large animals directly, it can be disastrous to the largely unseen small mammals, snakes, and soil insects. The small animals face a double danger: even if they escape burning, they become easy prey for birds attracted to the area. In the early 1900s, Van Dionant observed eagles patrolling a fire area, while vultures feasted on the semicarbonized victims of the flames. Recent studies have reported reduced populations of snakes, rodents, and insects after burning, though it is still uncertain whether the losses are due to death or flight. Certainly, breeding and migration cycles are disrupted, and birds with ground nests suffer even greater risks than reptiles and small mammals. On the other hand, the resulting tender forage growth attracts grass-eating mammals, including deer, contributing to population increases under the right conditions. The elimination of tall native grasses can also expose deer to predators and hunting by humans. Studies of deer in the cerrado indicate that ultimately fire has been a contributing factor in their decline.[81]

Other factors in the reduction of deer populations are habitat alteration, through overgrazing by cattle, and overhunting. Throughout most of the region, cowboys and ranchers preferred beef over deer meat, but hunting accompanied the increase in human and cattle populations, attracting professional or sport hunters and often leading to the unregulated killing of deer and other wild animals. Clearing and the introduction of exotic grasses also led to unexpected results. Although clearing automatically eliminates large species, it can provide conditions for the proliferation of smaller animals. Observers in other regions of Brazil noted that plantations of gordura and jaraguá attracted rodents, armadillos, poisonous snakes, leaf-cutting ants, and disease-carrying insects and mosquitoes. In the 1970s, this was observed in the campo limpo region of Mato Grosso. At the time it was a major concern, not only ecologically but also economically, since the increase in pests endangered the maintenance of pasture and cropland. These issues came to be addressed only in the latter decades of the twentieth century.[82]

All economically successful ranching operations rely on a combination of toil, ingenuity, vision, capital, and luck. Those elements existed in varying amounts in Mato Grosso during the period under study. Cattle raising on the Mato Grosso frontier after the Paraguayan War was rudimentary and plagued by obstacles, as many observers lamented, but the demands of national and, eventually, international markets pushed and dragged Mato Grosso into the world economic arena. As the scope of ranching in Mato Grosso was transformed, most ranchers welcomed the opportunities and responded as best they could. Despite some attempts at improvement, public institutions and authorities were poorly equipped to lend effective support to scientific innovation and economic expansion. Only the demands of the market helped to introduce "rational" thinking and methods into Mato Grosso ranching, and though most innovation began with private initiative, government support eventually became crucial. The process was slow and uneven, as most ranchers adopted innovation only when it was economically feasible. Once that threshold was reached, the structures and ultimately environment that had previously determined the character of ranching were permanently altered. Knowledge and opportunity combined to prod local ranchers into reexamining their businesses, leading to the introduction of ranching methods unknown or unprofitable before. The new methods included grass selection and salt provision, but as we will see in the next chapter, among the most influential was scientific cattle breeding, which despite overwhelming success led to considerable controversy throughout the Brazilian cattle sector.

SEVEN

National Breeds and Hindu Idols

From the early twentieth century, cattle breeds provoked passionate debate among Brazilian cattlemen. In the drive to modernize ranching, researchers and veterinarians engaged in a dispute over the suitability of specific breeds to Brazilian tropical and semitropical environments. The discussions were particularly heated regarding the introduction of zebu cattle from the Indian subcontinent. Mato Grosso was involved early in the arguments because the region served as a laboratory for testing the efficacy of the zebu against that of other breeds, especially European imports and nationally developed strains. The outcome was fundamental in establishing Mato Grosso's credentials as a major cattle-raising region, not only within Brazil but also on the South American continent.

THE LOGIC OF CATTLE BREEDING

Whether bred for purposes of draft, dairy, leather, or meat, the types of cattle developed around the globe over the millennia were genetically similar, if externally distinct. Much of the difference derived from local environmental conditions, such as climate, pasture availability, and overall range conditions, plus human intervention. The latter accelerated dramatically beginning in Europe in the late eighteenth century, as the new application of science to agriculture was extended to animal breeding. Over the course of the nineteenth century, various well-known breeds were developed in Europe for the production of milk and beef. Animals became larger, carried more flesh, and produced more and healthier calves.[1]

Originally exclusively European, the endeavor was soon extended to the Americas. Little deliberate selective breeding was done in regions where there were other varieties of cattle, most specifically India. There, the principal cattle breed was the humped zebu, initially considered a separate species from the European-origin animal. Throughout the period, confusion presided over scientific categorization of these visibly distinct animals. In part, the uncertainty involved the perception that one animal represented the future while the other, allegedly little changed for centuries, symbolized an unchanging and unproductive past. This perspective provoked a debate that spread across southern Brazil at the beginning of the twentieth century.

Modern taxonomy is not consistent, but it tends to identify European- and Indian-origin animals as species or subspecies of the genus *Bos*, though typically they are referred to as different breeds. Today, the European-origin animal is classified as *Bos taurus* and is acknowledged to have developed independently in southern Europe, although its initial origin was probably west Asia or Africa. The Indian subcontinent beast, *Bos indicus*, also developed separately, perhaps from the same obscure origins. The name "zebu" is derived from the French, which in turn perhaps was adapted from Tibetan. A similar breed in the United States is the American Brahman.[2]

TRADITIONAL CATTLE BREEDS
IN BRAZIL AND MATO GROSSO

The breeds of cattle entering Mato Grosso with the first neo-European settlers were those familiar to ranchers throughout the Americas. Descendants of Iberian stock that populated the hemisphere during the colonial era from southern Chile to northern New Spain, they were muscular, heavily boned animals with short legs, long, curved horns, and powerful front quarters (including the head), especially suited to draft labor. In Brazil, the animals were often called *alemtejanos*, or *minhotos*, after regions in Portugal, but they were scarcely distinguishable from the correntino, which dominated in the interior regions of Argentina and Paraguay. These were similar to the famous longhorn that roamed the North American range in the nineteenth and early twentieth centuries. With time, a series of regional breeds began to develop, especially with the infusion of genes from the odd animal from Africa or Goa. Often called chinos, such occasional crosses were

Young creole cattle, 1928, at Fazenda Margarida. Courtesy of Elza Dória Passos.

unplanned and had only a slight effect on the breeds then emerging in the Americas. As the centuries passed, cattle adapted to local conditions and developed characteristics unknown to their ancestors. Most noticeable were the larger horns, necessary in an extensive ranching system for defense against predators like wildcats and wolves. As a result of this type of ranching, cattle became semiferal, shied from humans, and were more inclined to put up a fight at roundup. The now American animals were the products of natural selection; they became adapted to regional environments, particularly in the tropics, where they developed wider hooves, thicker hides, and stronger constitutions. These were the animals that entered and populated Mato Grosso from the eighteenth into the twentieth centuries.[3]

The most common breeds in most of Brazil from the mid-nineteenth century had various names. Although all were similar, they received different names depending on the region where they were prevalent. The term crioulo (creole) was used generically. In the Pantanal, the most common breeds after the Paraguayan War seem to have been the pantaneiro or cuiabano. In the local vernacular, the pantaneiro was called tucura, and until ranchers in the Pantanal began purchasing zebu in significant numbers, in the 1930s and 1940s, it dominated ranching in the region. The reason was simple: the tucura was at home in the specific conditions of the Pantanal. It was small and agile, sported large narrow horns, had a thick hide, and was adapted to regular flooding of its habitat. Its rate of survival under arduous conditions was remarkably high, with close to 20 percent annual herd growth, a characteristic that guaranteed increases with little effort on the part of the rancher. The conditions under which the pantaneiro were raised were hardly conducive to weight gain, however. The animal's slow maturation, light weight, and allegedly weak hindquarters were mentioned by observers as the principal results of the breed's so-called degeneration in the Pantanal. Its meat was also judged to be tough and sinewy, and it could be an aggressive animal, occasionally known to attack even men on horseback, a temperament hardly suited for the raising of cattle on a commercial scale.[4]

Ranching in the Pantanal remained poorly developed until the entry of the zebu. The pantaneiro was raised for its hide, bouillon, or, after the beginning of the twentieth century, for jerky. Animals were sometimes rounded up *a bala* (by the bullet) because of their undomesticated nature. Whether by direct influence or as a part of overall changes in the ranching industry, once the zebu entered the Pantanal and were crossed with the pantaneiro, the structure of ranching in the region was transformed. What was once a subsistence existence developed into a successful business, and the pantaneiro became almost extinct. Today only one ranching family in the northern Pantanal still raises purebred tucura, though the Brazilian government runs a program to maintain its gene pool.[5]

In the cerrado and campo limpo, the traditional breeds included the caracu, franqueiro, and chino. Caracu were rare in Mato Grosso throughout the period, although they were sometimes crossed with other breeds, usually more by chance than by design. There is some doubt as to the origins of the franqueiro. Some suggest it had a French lineage, others that it originated in the municipality of Franca in São Paulo. Like correntino cattle, the franqueiro had exceptionally long horns (when removed, each horn was said to be capable of holding as much as 5 to 6 liters of liquid), a powerful head, and strong front quarters, making the breed well suited as draft oxen (once dehorned). It also carried a highly prized thick hide. On the down side, it had a low meat-to-bone ratio and was a slow breeder, lowering its value in the national market, especially with the increase in the demand for beef during World War I. Thereafter, the franqueiro was gradually crossbred with zebu and no longer exists in its original form.[6]

More widespread and useful was the chino, which was found not only in Mato Grosso but also in Minas Gerais, Goiás, and São Paulo. The origins of this breed are uncertain. Some argue that it was a cross between Iberian and Indian cattle, others that it carried franqueiro and correntino blood. The name probably came not from China or Asia but from the practice in Argentina and elsewhere of calling mestizos "chino." Allegedly brought to Mato Grosso from Minas Gerais by settlers after the Paraguayan War (as was likely the case with the franqueiro), the chino was prized for its rapid growth, good proportion of meat to bone (up to 50 percent), well-developed hooves, and high fertility. The chino had short horns and was skittish if not kept near humans, but it could become docile once it was accustomed to their presence. Chino cows were also reputed to be excellent mothers, a boon under the harsh conditions of extensive ranching. They were hardy enough to survive the long drives to Minas and São Paulo. The major drawbacks to the breed were its relatively small size, with correspondingly low quantity of meat, and its tendency to become commercially less desirable if subsequent generations were not regularly crossbred with more corpulent and docile animals. That was particularly the case once crossbreeding with zebu cattle began in the early years of the twentieth century, requiring greater care on the part of ranchers than normally was possible. Like most of the other so-called native breeds of Brazil, the chino virtually disappeared into zebu crosses, and today it is little more than a memory.[7]

The rustic character of these original breeds indicates the uncertain ability of the local ranching industry to produce cattle that could be marketed at a profit outside the state. For the most part animals were raised for their hides, local consumption, the making of jerky, or, in the northern Pantanal, for beef bouillon. There were regular calls for breed improvement over the decades, particularly to promote the importation of European breeds, which reflected an eagerness to emulate Argentina's phenomenal success in cattle raising. Government officials, ranchers, veterinarians, even presidents of the state, suggested the introduction of breeds like Hereford, Durham, and Polled Angus to improve local stock. The campaign was a clear recognition that the Mato Grosso ranching industry required larger, more productive animals if it was to provide greater wealth for ranchers and the state, but there was less understanding of the character of tropical ranching and the suitability of certain breeds to the region. The result was a debate that swirled around yet another breed introduction, the zebu, which became key to Brazil's cattle history. For Mato Grosso, it created a ranching revolution in the state.[8]

THE ZEBU REVOLUTION

It is unclear when the first zebu entered Brazil, but animals probably arrived irregularly during the colonial era. Some have speculated
that the first were imported during the transatlantic slave trade, a reasonable assumption considering the extensive importation from Africa of not only humans but also grasses and other domestic animals
like goats and guinea fowl. Brazil also had irregular contact with Portuguese colonies in India (Goa) and Asia (Macao), where zebu reigned
supreme. Nevertheless, most cattle were imported from Portugal itself,
and the zebu that did arrive in Brazil were soon absorbed into the national herd through crossbreeding, losing their identifying characteristics, particularly the distinctive hump. Only in the late nineteenth century did the importation of zebu become an organized effort.[9]

Alberto Santiago recorded imports of zebu through the nineteenth
and into the twentieth centuries, noting that until the 1870s there was
no organization; isolated animals in breeder pairs or small groups came
to the country more by chance than by design (such as through the mu-
tiny on a British ship at Recife in 1873). Brazilian Emperor Dom Pedro I was said to have established a small herd of African zebu on his
royal ranch in the province of Rio de Janeiro in 1826. His experiment,
if it was one, appears to have had little immediate influence in stimulating other imports. A few decades later such animals, sometimes called
nilos (suggesting an African, or at least a perceived African, origin),
were noticeable for their rapid adaptability to the Brazilian environment. Apparently few were purebreds, eliciting the opinion of one author that they should be crossbred in order to preserve their characteristics. The situation changed in the 1870s, however, as a few ranchers in
Rio de Janeiro began to import zebu commercially from India through
the European import houses then prominent in Brazil. Over the next
two decades, several lots of animals were imported into Brazil, largely
by three Fluminense ranching families, the Azevedos, Lutterbachs, and
Lemgrubers. Exact numbers are unknown, but probably no more than
a few dozen animals actually entered the country during this time. The
effect on ranching was minimal, at least initially, but by the 1890s the
adaptability of the breed had been noticed by ranchers in an unlikely
region of the country.[10]

Over the course of the nineteenth century, ranching had spread
into the far western panhandle of Minas Gerais, the Triângulo Mineiro. In this cerrado of natural grasses, the hot and dry climate was

hard on European breeds, especially in drought years when forage selection was limited. As explained in chapter 2, in the 1830s and 1840s a number of ranchers from the region had emigrated to Mato Grosso in search of better ranching conditions, but with the establishment of a military headquarters in Uberaba during the Paraguayan War, the region itself attracted a wave of settlers. Reliant on slow-developing Brazilian breeds limited by local environmental conditions, and in response to a growing demand for improved cattle, wealthier ranchers in the 1870s began to buy zebu breeder animals from Rio de Janeiro. Their efforts were quite successful, but the Mineiros then found themselves reliant on the supplies and prices dictated by either Fluminense ranchers or the import houses, particularly the German firm of Hagenbeck. That led one rancher, Teófilo de Godoy, to make the expensive and risky voyage to India himself in 1893–1894.[11]

Stimulated by an expanding cattle market in São Paulo and the arrival of the Mogiana railroad at Uberaba in 1895, Godoy began the first of a series of direct imports of zebu by ranchers in the Triângulo. Between 1893 and 1914, more than 2,000 breeder zebu were imported into Brazil from India, half of them destined directly for the Triângulo. The animals traveled in specially constructed pens on the upper decks of ships carrying jute for the coffee trade. The ships could accommodate 60 to 90 animals, usually accompanied by Indian cow herders, who were then repatriated after arrival in Santos. The Triângulo soon became the focus of zebu raising in Brazil, and even the advent of World War I did not discourage importers. From 1914 to the end of extensive imports in 1921, more than 3,300 zebu were brought into the country, most heading to the Triângulo.[12]

Mato Grosso and Zebu

The first and only major effort to import European breeds into Mato Grosso was by Brazil Land, which introduced 1,000 purebred Durham and Shorthorn cattle to one of its ranches in the Vacaria just before World War I. This experiment was copied by a few ranchers in the region, particularly other foreign interests, but it was an abysmal failure. European breeds of the time were not appropriate to the Mato Grosso environment, since they had been bred to digest grasses rapidly and efficiently, a distinct disadvantage in hot tropical regions. At the same time, they suffered from the intense sun, insect plagues, lower parturition, and slower growth. In this case, they became susceptible to the extremely harsh winters of 1917 and 1918, when nutritive forage was scarce. Considering their relatively delicate nature, the animals may not have received the care they required, but the experience convinced Brazil Land of the need to raise more rustic breeds, and other ranchers decided that importing more European animals was inadvisable. The inability of the railroad to provide regular livestock transport also contributed to the decision. The solution chosen by Brazil Land, and already followed by some ranchers in the state, was to rely on zebu and zebu crosses. It was a practical approach, reflected by observations from ranch manager John Mackenzie:

> With few exceptions the native Brazilian prefers cattle of the Brahma type, chiefly because they are hardy and capable of withstanding the numerous pests the country is infested with, such as ticks and flies, and because they recover quickly from foot and mouth which is prevalent in the country.

In addition, the market of World War I made minimal demands on the quality of beef exported to armies in the field. The much-disparaged Indian breed was thus given a boost from foreigners that helped to consolidate its penetration into Mato Grosso, in the process stimulating a ranching revolution in the state.[13]

Considering the geographic proximity of Mato Grosso to the Triângulo, it is no surprise that zebu soon saw their way into that state. Though it took some decades, their increasing presence established Mato Grosso as a significant source of cattle for the national market. The first zebu was said to have arrived in 1880, in what appears to have been an isolated case. Occasional introductions of the breed into Mato Grosso probably began after 1895, but only after the turn of the century did the use of the breed expand. Several ranchers, particularly in the Campo Grande area, bought zebu in Minas for resale in Mato Grosso or for their own ranches. Paulo Coelho Machado reports that his grandfather, Antônio Rodrigues Coelho, drove a herd of local cattle to Uberaba for sale in 1906, but because of a national economic recession he could not sell the animals and was forced to trade his herd for 400 zebu. He brought the zebu back to his ranch near Nioác, keeping half for himself and selling the rest. Coelho later declared that it was the best investment he had ever made, since the quality of his stock improved significantly when the zebu were crossed with "degenerated" local cattle. They were also admired for their endurance on the

Zebu imported by Brazil Land in the 1920s. Courtesy of Elza Dória Passos.

long drives to fattening pastures in São Paulo. The importation of zebu into Mato Grosso extended across the border to Paraguay, where by the early twentieth century its northern regions had a considerable number of zebu crosses, and some ranchers importing breeder zebu from Brazil.[14]

Ranchers in Mato Grosso did not always acquire zebu as a matter of choice. Until the boom of World War I, the most important route for the export of Mato Grosso cattle was through Minas, which was also the major source for breeder animals (see map 3, p. xiv). Drovers from Minas often arrived in Mato Grosso for the annual cattle drives trailing small herds of breeder zebu as part or full payment. In many cases ranchers in Mato Grosso had little option but to accept zebu blood into their herds, since other breeds were almost impossible to find or at the very least prohibitively expensive. The zebu were not exactly cheap either. Around the turn of the century, a purebred bull in Mato Grosso could fetch 5,000–6,000 milreis (US$1,000–$1,200), while a half-breed cross demanded 2,000–3,000 milreis (US$400–$600). Those were exorbitant prices, to say the least, but as more animals became available, prices declined. By 1903, a purebred zebu bull commanded around 1,000 milreis (US$240) in Uberaba, and by 1910, oversupply had depressed prices by half. The situation prompted ranchers to agree to suspend most imports into the Triângulo for five years, on penalty of

5,000 milreis (US$1,650). By 1923, a purebred bull commanded between 500 and 1,000 milreis (US$50 to $100) in Campo Grande, a significant drop from previous decades.[15]

Still, few Mato Grosso ranchers could afford purebred cattle or even the periodic purchase of half-breed animals. The result was a gradual decline in animal precocity, average weight, and resistance to the extensive ranching conditions under which they lived. As early as 1907, Lisboa noticed that after four or five generations the zebu's initial hardiness had disappeared, and the cattle were degenerating in all respects, especially in terms of body weight. They were still considered ideal as traction animals and for their ability to tolerate the long drives, but no longer were they producing meat as they had. This observation fit well into a growing national debate over the suitability of zebu to Brazilian conditions.[16]

Nationalist Sentiments and Asiatic Plagues

The introduction of the zebu generated considerable controversy. Part of the argument questioned the viability of such an animal in a technological climate that promoted European (or Texan) breeds. Also, significant nationalism was involved, especially regarding attempts in the state of São Paulo to advance a regional breed as the best candidate for the development of an exclusively Brazilian breed. The caracu became the focus of the nationalist approach after 1900. Those who touted it as the only truly national breed in the country argued that it should be promoted as such, particularly in the face of an alleged zebu invasion. Claimed as a direct descendant of the original Portuguese cattle or of animals brought to the northeast of the country during the Dutch occupation of the seventeenth century, the promoters of the caracu praised the breed for its rusticity, mild temperament, and quality (if not quantity) of meat. According to Joaquim Carlos Travassos, the breed had become virtually extinct by the end of the nineteenth century but was given a new lease on life thanks to a campaign by a number of Paulista cattlemen. Dr. Eduardo Cotrim, a nationally respected civil engineer and cattle expert, suggested the establishment of zootechnical posts to improve the breed and to develop a breeding program that could supply the national market. His advice was followed, and in 1916 a caracu studbook was created. Breeding was conducted in the São Paulo state agricultural research station at Campinas, and improvements over the years in the breed's fertility, weight gain, and docility were cited as proof that it was the breed of the future. But the controversy went further than the competition between caracu and zebu.[17]

The debate over the suitability of the zebu was heated, and given its nationwide importance, it deserves extended discussion here. The first salvo was launched in 1904 by Dr. Luis Pereira Barreto, a respected São Paulo agricultural scientist and physician. There had been some concern among ranchers that the zebu did not live up to their expectations after the first or second generation. Pereira Barreto admitted that the first generation was "truly splendid," but succeeding generations, he argued, declined rapidly:

> the second is already much inferior; the third very bad; the fourth is a juvenile goat herd; the fifth a herd of long-eared hares; the sixth, finally is of debilitated rats, wretched, infertile. . . . Zebu meat has the rankness of capivara; cows don't have enough milk to raise their young; males and females are wild beasts.[18]

Such strong words touched off a bitter argument that pitted supporters of Pereira Barreto, most of whom were Paulistas, against defenders of the zebu, who, not surprisingly, were largely Mineiros. The debate went on for decades and was not fully resolved in the minds of some participants even into the 1940s.

An even harsher opinion was expressed by Dr. Assis Brasil, a renowned ranching specialist from Rio Grande do Sul. Assis Brasil admitted no positive characteristics for the zebu whatsoever, calling it the "Asiatic plague" and characterizing its champions as having succumbed to "collective hysteria." Somewhat less strident in his criticism was Cotrim, whose experience with cattle was extensive but largely confined to Rio Grande do Sul and Rio de Janeiro. In his highly regarded manual on cattle raising published in 1913, Cotrim opined that the immense popularity of this "hindu idol" in recent years had produced a painful experience for Brazilian ranchers, who did not understand that in the course of countless centuries in India, the zebu had "proven" it was incapable of improvement. He had argued earlier that the zebu was good only for traction, although somewhat hard to handle, and that as times changed the need for traction animals was rapidly diminishing. Cotrim disparaged the animal's meat, saying that it was of secondary quality because the breed did not adapt well to "luxurious" pasture areas. He warned that zebu milk production was exceptionally low, calf care by cows minimal, and the procreation rate well below that of the more traditional breeds, such as the caracu. The addition of Cotrim's voice to the discussion gave greater legitimacy to the anti-zebu lobby.[19]

The zebu's defenders were not idle in their responses. Carlos Fortes, a Mineiro animal expert, noted in 1903 that the state ranching industry was poorly developed, little cattle care was practiced, government support was negligible, and exports were low as a result. He explained that cattle breeds should be chosen depending on the location, with zebu being the best selection for remote regions, either as purebreds or as scientifically controlled crosses. His point was that the animal's rusticity and productivity under less than ideal conditions made it a natural choice for ranchers in regions like western Minas, Goiás, and Mato Grosso. Others, including some federal government technicians, argued that the zebu was ideally adapted to the tropical climate of Brazil, produced better than its detractors made out, and was especially resistant to diseases that frequently incapacitated European breeds.[20]

Probably the most important ally of the zebu was Joaquim Carlos Travassos, a powerful member of the National Agricultural Society in Rio. Travassos made a study of the zebu in Brazil and in India even before the anti-zebu forces launched their attack. Noting that zebu meat production was markedly less than that of European breeds in ideal conditions, he described a number of ways in which the Indian animal was best adapted to the local climate: longer ears to sweep away flies; usually light hair color, reflecting the sun; darker skin pigmentation, making resistance to ultraviolet light much greater; resistance to tropical parasites; less sweating, therefore greater absorption of heat by droplets. Travassos relied heavily on work by the English scientist Robert Wallace in India in the 1870s. Wallace, who made the first scientific study of zebu, also recounted failed attempts to improve local cattle by importing English breeds into the British colony. He and later scientists concluded that European cattle were unsuited to the tropics and that breeder stations and veterinary schools should be established to improve the productive quality of the zebu. That is apparently what occurred, and Travassos argued that the same should be done in Brazil. He lamented a lack of vision among Brazilian legislators and scientists, pointing out that several other Latin American nations had veterinary schools, India had several, and even distant Siam was so equipped.[21]

The conclusions of Travassos and Wallace were consistent with those of modern zoological scientists (except for the role of sweating, which functions opposite to the manner described by Travassos). Today, dark skin pigmentation and light hair color are considered important advantages for survival in tropical environments. Compared with European animals, the zebu breed is better adapted to the tropics for numerous other reasons: its longer legs and slower metabolism, ensuring greater conservation of energy; a smaller digestive tract and more efficient absorption of food; lower requirements of protein and calories, making it better able to survive in less than ideal pastures; greater herd sociability, ensuring increased safety from predators; large body surface in relation to weight, lower metabolic rate, and larger sweat glands, leading to improved regulation of body temperature in response to heat; short hairs and oily skin secretions that discourage ectoparasites (as do their independent hypodermal muscles, like horses); acquired resistance to some diseases, such as foot-and-mouth, Texas fever, even rinderpest; greater blood irrigation, permitting increased oxygen absorption into the lungs and brain; and longer eyelashes, which protect against intense light and dust.[22]

Government was not entirely absent, however. Importation was aided by support from both the federal and the Minas Gerais state governments beginning in 1907. Federal laws in that year and in 1912 loosened import regulations, and the Minas government also entered the business. The president of Minas, João Pinheiro, encouraged the display of zebu at state cattle exhibitions in 1908, despite objections by some veterinarians and ranchers, and in 1910 Minas subsidized the import of 242 animals. President Pinheiro took an entirely utilitarian attitude when he declared:

> I do not want to know if the science of zoo-technicians recommends the zebu or not; what I know is that the ranchers of Uberaba and other regions of Minas Gerais are getting rich with zebu, and that's enough for me.[23]

Nevertheless, many scientists and ranchers rejected the work of Travassos. The debate was kept alive particularly by the untiring lobbying of Pereira Barreto and other Paulistas, obsessed as they were with the political agenda of promoting caracu as the national breed. In Mato Grosso, however, zebu continued to grow in importance, particularly in crosses with local breeds of the franqueiro and chino. In some cases where ranchers provided little care, these animals became so unruly that they posed a risk to humans and other, more docile cattle, forcing owners to destroy them. In general, however, the zebu was the breed of choice by World War I, notably because of its ability to withstand the long and arduous cattle drives to São Paulo, which became even more difficult during the wartime cattle boom. Indeed, it was the war that persuaded most ranchers throughout Central Brazil to accept zebu as the breed best adapted to the Brazilian milieu. Extensive exports of meat to warring Europe also convinced many doubters of zebu meat's marketability and its place in the national economy. Zebu was on the way to predominating in Campo Grande and the Vacaria, and ranchers in the Pantanal, which was much slower in adopting the breed, first began to import zebu breeders during and after the war. The controversy died down at that point, and even Eduardo Cotrim accepted the inevitability of zebu for Central Brazil's ranching industry, although he emphasized the need for selection and breeding care.[24]

That was the crux of the matter, as several less emotional observers pointed out during the course of the debate. In itself, the zebu was not prone to inevitable degeneracy as its detractors claimed, but inadequate care led to a decline in quality over several generations. Fernando Ruffier argued during the war that crosses between *Bos taurus* and *B. indicus* were more akin to hybridization than to crossbreeding. Although, unlike mules, sterility did not occur, it was still necessary to crossbreed regularly in order to avoid degeneration.[25] He and others also argued that the best way to improve cattle quality in Brazil (as always, compared with the success of Argentina) was to create conditions under which the animals could prosper, including better pastures, closer attention by ranchers, and the establishment of zootechnical educational facilities. Allegedly this was not understood by Brazilian ranchers, who believed that by simply injecting some zebu blood into their herds they would produce some miraculous breed requiring no further care. That attitude did exist, but much of the reluctance of remote ranchers to practice such crossbreeding, particularly in regions like Mato Grosso, was more a matter of expense. Until World War I, the financial reward from such care was nonexistent, and as such only the richest ranchers could afford the investment.[26]

The debate continued, as in 1918 and 1921 the anti-zebu lobby received a tremendous boost from external sources, influences that would keep the controversy prominent for most of the 1920s. At issue was the quality of zebu meat. Zebu detractors argued that the breed had less fat than European cattle and was thus unpalatable to the European consumer, compared with beef exported by Argentina. That was part of the argument used by the London Board of Trade in 1918, when it banned

Mixed cattle, 1928, at Fazenda Margarida. Courtesy of Elza Dória Passos.

the import of Brazilian beef. It is true that zebu carry their flesh quite differently from other varieties of beef cattle. Animals of European origin have a thicker layer of subcutaneous fat that protects them against the cold. Zebu, on the other hand, generally show less marbling since their fat is found between the muscles, making zebu meat not only leaner but also drier and potentially tougher when cooked. In today's world of health consciousness, leaner meat might be considered a benefit, but it was a definite disadvantage at that time, particularly since Brazilian beef competed with prized Argentine beef. Fresh beef was seldom consumed in Brazil, except by the wealthiest, and even that group preferred lean cuts, so zebu meat found a ready market among them. Nevertheless, since the future of the industry depended on exports, especially to Britain, the ban imposed by London caused concern.[27]

A furor developed over the ban. Some extremists called for an end to the import of zebu breeder stock and the slaughter of all zebu, insisting that the country concentrate on raising only European breeds, above all English. Ruffier responded by making a few salient points. He explained that the English had rejected Brazilian meat based on its poor quality, which, despite the rhetoric of the London decree, had nothing to do with its zebu origin. The Frenchman blamed poor preparation by the frigoríficos. In response to the feverish demand for meat to supply the war market, scrawny animals were often slaughtered immediately on arrival from grueling three-month drives, followed by an excessively rapid freezing process that damaged the meat. As a result, the meat reaching England was tough, discolored, and suffered from freezer burn. He concluded that such a poor product was the consequence of slaughterhouse incompetence, since zebu meat from other areas of the world, including Australia, was not part of the ban. Ruffier also pointed out that along with the ban, the Board of Trade had recommended that Brazil import purebred English bulls to rejuvenate its herds:

> [T]o neutralize those defects [of zebu] it is necessary to import purebred English cattle, like Hereford, Devon, Shorthorn, etc. . . . England is the country best equipped to provide Brazil with the necessary breeds.[28]

Ruffier likened the ban to a previous British strategy taken against Argentina. London had banned Argentine meat imports in the 1880s because of an outbreak of foot-and-mouth in that southern nation, with the result that British influence over Argentina increased and eventually the British had almost exclusive access to Argentine production. Although there may have been an element of bias in Ruffier's assessment regarding the historical precedents of the Board of Trade action, there is no masking an obvious British attempt to promote the interests of that nation's breeders. It was not the only time London used a seemingly minor issue to manipulate the market in its favor.[29]

While part of the Brazilian agricultural sector went into a frenzy, others did not take the ban too seriously. After all, there was still a market in the rest of Europe, especially France and Italy. Many officials and cattle raisers had come to the conclusion that the zebu was indeed the beast of the future, and imports from India were resumed in 1919.[30] The São Paulo outbreak of rinderpest in 1921, however, significantly complicated the widespread adoption of the zebu.

In Brazil, only one outbreak of cattle plague has ever been recorded, but its damage to the cattle industry extended far beyond its place of incidence. Rinderpest is a viral disease that in its acute form lasts from four to ten days. It was a constant problem in Europe until it was eradicated in 1877. In the past it had devastated African cattle, and it is endemic in parts of Africa and Asia to this day. Since mortality frequently reaches 90 percent, an epidemic can destroy a country's food supplies, resulting in famine. Today vaccination is common, but in the past, rinderpest was commonly controlled through quarantine and slaughter.[31]

In 1921, the disease was detected in several animals in a zebu herd imported from India on arrival in São Paulo. The imports had either picked it up on a stopover in a Belgian port, where other cases were found, or had passed it on there. Once diagnosed, the outbreak was quickly controlled by Brazilian authorities. The herd was quarantined, while the infected animals were destroyed, and government agents disinfected the rail cars, stock, and packing areas and temporarily closed the frigoríficos. As a precaution, even dogs and ravens in the vicinity were killed. Nevertheless, many cattle in the surrounding area came down with the disease. In total the epidemic killed some 855 animals and another 2,500 were destroyed as a preventive measure. Though rinderpest never spread beyond the stockyards of São Paulo, the danger was considered so acute that the federal government prohibited further imports of zebu, European nations suspended imports of Brazilian beef, Argentina and Uruguay shut their borders to Brazil, and slaughterhouses temporarily shut down. Trade was halted between Mato Grosso and São Paulo for three months, effectively terminating exports for that year, at a time when the market was already declining. Simultaneously, neighboring Paraguay suspended imports of Mato Grosso cattle, and Argentina took advantage of the situation to ban imports of Paraguayan animals for the rest of the year, thus protecting its own ranchers from the cheaper Paraguayan cattle. Such drastic measures were criticized by some at the time as overkill, but in hindsight they can be considered an example of quick response, today duplicated in cases of other diseases, and were highly successful in protecting the region from a potentially devastating epidemic. No cases of rinderpest have since been registered in the Americas.[32]

Within Brazil, however, the plague provided ammunition to ranchers and agricultural bureaucrats who opposed the further introduction of zebu. To assertions that zebu was a degenerate breed was added another alarm, that its introduction risked the spread of a devastating disease, and thus the breed should be eradicated from the country. The immediate result was a collapse in zebu prices and the ruin of many breeders. Minas Gerais, where zebu breeding had begun, was most strongly affected. The distress was only peripherally felt in Mato Grosso, since at the time few ranchers had fully converted to the Indian breed.[33]

The rinderpest outbreak depressed further an already stagnant postwar ranching sector. After the boom in imports between 1914 and 1919, only two more shipments were transported from India during the period, in 1921 and 1930, involving 360 animals. Meat exports had picked up by 1923, and expansion in the later 1920s served to mute zebu disparagers. The change was aided by the improved treatment of cattle and meat by the frigoríficos. Other support came from an unexpected source, a representative of the British Ministry of Agriculture who traveled to the country in the late 1920s. The official, John Lamb Frood, visited the Uberaba area and was impressed by the quality of zebu found there. He suggested that greater crossbreeding with breeds like Hereford or Polled Angus might be beneficial, but he assured Brazilian ranchers that he was returning to England convinced of the value of zebu in providing beef for the European market. He also judged that Brazilian slaughterhouses were processing the product even better than those in Argentina.[34]

Frood's opinion was echoed by Sir Edmund Vestey, head of the British conglomerate that controlled a number of Brazil's frigoríficos. Vestey visited Brazil in 1927 and confirmed Ruffier's earlier assessment that the low quality of the meat, not the breed of cattle, was to blame for Brazil's loss of clients for its meat in the postwar period. He too suggested crosses with European breeds but believed that zebu was the best animal for remote interior regions like Mato Grosso. He also urged completion of the bridge over the Paraná River and recommended that rail freight rates should be based on carload and not per head. The visits by foreigners legitimated the zebu as a viable animal in the production of beef for export, albeit with the caveat that more scientific breeding methods be employed.[35]

These opinions were elaborated on in replies to questionnaires sent to the major slaughterhouses in the country. In 1929 or 1930 representatives were interviewed from the three largest packing houses in Brazil, owned by Continental Products, Armour, and Anglo, a Vestey company. They were asked their opinion of the quality of zebu steers sold to the frigoríficos and the quality and marketability of zebu beef in Europe. All three reported that between 70 and 80 percent of all steers processed were zebu crosses, which should be no surprise, considering that most of the animals came from Minas, Goiás, and Mato Grosso. The representatives went on to praise the quality of most of the animals received and the favorable quantity and quality of the meat rendered. They emphasized that their production was based on zebu and suggested that ranchers should breed purebred animals or at least use purebred bulls for crosses. It was admitted that zebu meat was considered inferior in Britain and that the price of zebu beef ranged from three-quarters to one-half as much as beef from Argentina and Uruguay. The

◀ "David e Golias" ▶
Um impressionante paralelo entre um nelore e um curraleiro

"David and Goliath: An impressive parallel between a Nelore and a Curraleiro."
Reprinted from Revista Zebu, *August 1943, 12. (Also published in Robert W. Wilcox,*
"Zebu's Elbows: Cattle Breeding and the Environment in Central Brazil, 1890–1960,"
in Christian Brannstrom, ed., Territories, Commodities and Knowledges: Latin
American Environmental History in the Nineteenth and Twentieth Centuries
[London: Institute for the Study of the Americas, 2004], 235. Used with permission.)

interviewees, however, emphasized their belief that the breed was most
suitable for Brazilian environmental conditions.[36]

In the end, between 1921 and the 1940s, zebu came to dominate as
the major breed in Central Brazil. Most of the opposition to the breed ·
had been overcome by the 1940s, and the bulk of ranchers, buyers, and
veterinarians were believers. In Mato Grosso, the zebu had become the
main breed by the 1930s, particularly in Campo Grande and the Vacaria. Andrade explained that the reasons were simple: hardiness, precoc-

ity of calves, resistance to parasites, and ability to swim, essential in the fording of rivers and streams during drives. The capacity to endure the long drives to market and to regain weight quickly was central in the decisions of Mato Grosso ranchers to opt for zebu, despite the controversy surrounding its value. In fact, by 1940 Mato Grosso reportedly had a higher proportion of zebu in its herd than any other region of Brazil, including Minas and Goiás. Most of the zebu were not purebred, however, as they were the products of both deliberate and uncontrolled crosses with other stock. That began to change only in the 1940s and 1950s, when more purebred animals were introduced.[37]

The region of Mato Grosso where ranching continued to resist the zebu until the 1940s was the Pantanal. Part of the reason was a lack of resources to buy breeders, plus ranchers seemed to have the erroneous perception that zebu could not adapt to regional annual flooding. That was largely a result of the propaganda campaign emanating from São Paulo, which for undetermined reasons was disproportionately influential in the region. Rancher Renato Ribeiro, for example, explained that his father was slow in adopting the breed because he was a reader of *O Estado de S. Paulo*, the São Paulo daily newspaper through which Pereira Barreto publicized his campaign. There also was a negative attitude toward zebu because those left to wander in a semiferal state, as was common in the Pantanal, turned quite feral and were extremely difficult to round up. That was more a reflection of the character of ranching in the region than the quality of the zebu, and it was not long before ranchers began to pay greater attention to animal care than before. The result was that by the mid- to late 1940s, when access to higher quality animals increased and prices declined, the zebu also came to predominate in the Pantanal.[38]

Finally, it is important to point out that the zebu is not genetically uniform but rather includes a number of distinct varieties. The breed's diversity played an important role in shaping the cattle industry and the debate. Apparently, the first regular imports of zebu were the Nelore variety, called Ongole in India. The animal, from the region around Madras, has a distinctive predominantly white hair color. It is the most prominent zebu breed in South America today, particularly Central Brazil. Other important breeds were Gir, from the west of India near Ahmadabad, and Guzerá (Kankrej), from the same region, more specifically the modern Indian state of Guzerat. Other breeds were imported in small numbers and have made only slight genetic contributions to Brazilian herds.[39]

Why these three breeds were chosen is not entirely clear, but it seems to have had something to do with breeding in India, their dominance in that country, and their ability to satisfy the fundamental criterion of Brazilian buyers: durability, docility, and adaptability to a variety of purposes. Brazilian inexperience with Indian cattle led to the choice of animals on the basis of external features, particularly color, extent of dewlap, and, in later years, length of ears. At the height of Brazilian purchases, Indian cattle traders jocularly referred to the Brazilians as "buyers of cattle ears." These nonscientific arbitrary criteria influenced the marketability of animals throughout the 1920s and into the 1930s, and they became part of a debate over the development of a distinctly Brazilian zebu breed, the Indubrasil.[40]

Up to World War I, purebred animals were crossed with local animals, producing mixed results. The degeneracy that occurred because of a lack of control over breeding fueled the zebu controversy and, coupled with continued high import costs, influenced the decision by a number of Mineiro ranchers to develop a distinctly Brazilian zebu breed. Independent of any government assistance, they developed the Indubrasil in response to the wartime market, and it became the dominant zebu breed during the 1920s. A pure zebu breed, with no input from *Bos taurus*, it was mainly a cross between Gir and Guzerá, although other zebu stock was not entirely absent. The creators of the Indubrasil succeeded in obtaining a much greater quantity of meat per animal than provided by zebu crossed with European breeds, and by 1930 it was being touted as the savior of the Brazilian cattle industry.[41]

As much as possible, Indubrasil cattle were bred with long ears, one of the measures by which Brazilians judged animals for purity. Most ranchers had no experience with zebu, and like the buyers in India, they relied on its unusual external features to determine their individual purchases. In many cases, animals with the longest ears fetched the highest prices in the market, whether they were carefully bred or not. Ranchers often believed they were breeding with pure stock, when in fact most animals were no more than three-quarters or seven-eighths zebu. That led to uncontrolled crossbreeding, which eventually caused herds to degenerate, further fueling the debate over zebu quality. Along with their colleagues elsewhere in the country, some ranchers in Mato Grosso suffered unsatisfactory results, which helped to discourage greater expansion of zebu in the state.[42]

The boom in Indubrasil inspired another advance. By the late 1930s it had become clear that the stocks of purebred Indian cattle were precariously low in Brazil. That prompted the federal government to stimulate the breeding of pure-blooded stock and to guarantee genealogical lines through herd and studbooks. Between 1934 and the early 1940s, federally funded experimental ranches were set up in the Triângulo Mineiro and in São Paulo. The 1930s saw the development of the national Genealogical Registry Service through the Rome Agricultural Conference of 1936, and a zebu registry was authorized two years later. Those measures were instrumental in guaranteeing the zebu a permanent place in Brazilian ranching circles and in preserving the small numbers of purebred stock still in existence. They also ensured a place for Indubrasil, which became a source of national pride for Brazilian stock raisers and was officially recognized as a distinct breed in 1936.[43]

From 1940 until the 1950s, Indubrasil was the dominant breed among registered zebu purebreds, although there was some concern over the extent of necessary human intervention, particularly in order to ensure that newborn calves would suckle. The concerns contributed to the perceived need to import new blood into the industry, although a federal government ban on imports had been in effect, with only the occasional exception, since 1930. Official refusal to sanction increased imports led in 1955 to the illegal import from Bolivia through Corumbá of more than 100 high-quality Gir. Through artificial insemination, those imports became instrumental in guaranteeing the predominance of that breed for a decade, particularly in Mato Grosso. Under pressure from the nation's ranching sector, which sought to expand herds to reach broader markets, particularly overseas, in 1962 the legal importation of 318 Nelore began to inject that blood back into Brazilian herds, and the industry came full circle.

No significant numbers of zebu have been imported since then, as the industry became sophisticated enough to engage in its own breeding program, leading to the overwhelming predominance of Nelore in the national herd, with the exception of Rio Grande do Sul. In fact, most zebu in southern South America are Nelore, and many are the descendants of Brazilian-bred animals. Nelore animals are most prized because the breed is not only highly resistant to climatic extremes, particularly under conditions of extensive ranching as in Mato Grosso, but is also highly fertile and gains weight easily. Zootechnicians from other countries, such as South Africa and Australia, also have affirmed the Nelore's significance for large-scale tropical ranching. The beast's light hair color plays a role in its adaptability to conditions of hot sun and little shade, as is common in Brazilian pastures. It should be pointed out that although pure white is the normal color, other tints do exist among the Nelore. In Brazil, color clearly played a part in selection for breeders, since most zebu throughout the country are white Nelore. Unlike the arbitrary prejudice in favor of superficial characteristics seen in the past, the choice of light hair color today has scientific approbation behind it.[44]

As uncertain and prone to error as the Brazilian breeding program was in the past, it has now become a major player in the industry, not only consolidating the dominance of the zebu in Brazil but also serving as a repository for possible export to India. It may seem paradoxical that Brazil could export to the region of the world where zebu are most numerous, but in recent years the purity of Indian herds has declined considerably. The cause has been the lack of genealogical controls, coupled with an emphasis on utilitarian needs in the face of staggering human poverty, which has led to the introduction of European milk cows to increase milk production. As Brazilian experience shows, such experiments have been less than successful in the long term, particularly since follow-up control has been minimal. In the mid-1980s, Alberto Santiago predicted that by the end of the twentieth century Brazil would be exporting purebred zebu semen to India, though that apparently did not come to pass.[45]

The advent of the zebu guaranteed Mato Grosso a significant place in the national cattle market, a position already developing before the entry of the breed but not assured at the time. The process was slow, and the local ranching sector was a good example of how traditional methods will coexist with innovative techniques for some time, as economic growth stimulates the transition from one to the other. Mato Grosso ranchers eventually accepted the lead of the Triângulo Mineiro and their willingness to press on, despite several setbacks and arguments to the contrary, ensuring that they would be in the forefront of the developing technology along with western Minas and Goiás. In fact, Mato Grosso became one of the most important of the several zebu breeding regions. As a result, the state benefited first from breakthroughs and then consolidated its place in national ranching strategies, establishing abundant good-quality grazing land as well as a stock of excellent animals for the future.

By the 1940s, Mato Grosso and Minas had surpassed São Paulo in cattle breeding technology. Some observers have suggested that in the long term the anti-zebu campaign was a great benefit to the zebu-producing states because it permitted them to develop a technology that São Paulo would otherwise have dominated. It was even estimated that the campaign may have cost São Paulo as much as US$100 million in lost sales and production. Be that as it may, the opportunity that Mato Grosso found in the zebu was golden. The region adopted a technology developed in Brazil itself, through both the private and public sectors, which overcame obstacles of meat quality and limited opportunities offered by jerky production. Indeed, the conditions were created for Mato Grosso's present status as a major beef cattle producing zone. At the same time, the story of zebu in the state and the country affirms that ranching in the tropics was a business highly dependent on sometimes controversial experimentation. Knowledge was still limited, and as one wag had it, the introduction of zebu was like a *jogo de bicho*, a game of chance, though a lucrative one in the end.[46]

Transformation and Continuity

*[The Pantanal is] a problem area of great potential, whose
abandonment to ordinary [rotineira] and obdurate [secular]
economic exploitation, without doubt constitutes a true national
waste of the potential of the land and the capacity of the Brazilian
resident [homem] of the interior.*

MATO GROSSO PRESIDENT JOSÉ FRAGELLI, 1974

These words, written almost 100 years after Vice President Ramos Ferreira expressed the same concerns, reveal how the recurrent theme of unfulfilled development lived on in the state of Mato Grosso. Beginning in the late nineteenth century, the transformation of ranching in Mato Grosso was almost as dramatic as that of the United States during the same period. Compared with North America, however, growth in Mato Grosso was gradual and erratic, exhibiting little of the dynamism that characterized ranching in the north, or even in Argentina. That was the crux of the problem, given that all eyes looked to the spectacular successes of other regions as the criteria by which Mato Grosso ranching should be judged.

The principal argument running through this study is that even where the will existed for the expansion of ranching, the opportunities and successes were limited by geographical, ecological, and economic restraints. Husbandry methods common in other biomes frequently are not successful in the tropics, and since Mato Grosso is made up of several ecosystems that require a range of ranching practices, even routines in one region were inapplicable to bordering zones. Traditional ranching had adapted to specific environmental conditions, permitting ranchers to make a modest living supplying the local market, with some minor exports to other regions of Brazil. At the same time, government attention was limited, despite repeated expressions of concern, leaving the sector to rely largely on its own endeavors.

Certainly efforts were made to expand the region's cattle industry, and over the long run it grew. After the Paraguayan War, Mato Grosso was recognized for its potential for settlement, and migration increased significantly, though infrastructure remained precarious, land tenure ambiguous, and distances great. Aside from the lure of extractive boom products like rubber and mate, that left ranching as the best option for making a living. Coupled with the constraints of geography, the problems of property legalization and limited financial options forced many to operate at a subsistence level. As in the past, open-range ranching predominated. Consequently, investors with capital (primarily foreign concerns) had a tremendous advantage. Entrepreneurs from Argentina and Uruguay were the first to take advantage of the opportunities, but soon others, especially from Europe and the United States, came to dominate specific sectors of the local economy.

Cattle products, primarily jerky and beef bouillon, became major export articles, though export markets were limited compared with other regions of the continent. Still, it is remarkable that a significant business could be built from beef extract in a region as remote as the northern Pantanal, an example that speaks to the irrevocable influence of commoditization on all parts of Latin America. Although the story of Descalvados is fascinating, it is more notable as an exception than as a rule. Circumstances dictated that most ranching remained open range and supplied local consumption, though some ranchers were able to produce enough to contribute small amounts of cattle and meat to the national market. No truly robust cattle production developed until the twentieth century, leaving the enduring impression, especially among elites, that the sector was not being exploited to the extent it should.

That judgment stimulated tentative fiscal interest in the early years of the twentieth century. A growing belief that the region could be integrated into the nation through transportation led to hesitant investments by Rio in shipping and support for a railroad across the state. The interest became urgent with the advent of World War I, as feverish war demand attracted serious attention to ranching in Mato Grosso and in Brazil in general. Meat and leather exports skyrocketed, placing the region and nation on the map of notable global cattle zones. The attention did not continue on the same scale after the war, but enough investment, Brazilian and foreign, had been made to consolidate Mato Grosso's role as a significant player in Brazilian ranching, a position that gradually expanded over the following decades.

Until the 1960s, though, Mato Grosso persisted as a secondary player in national and international markets, relying primarily on jerky and leather production, sent downriver to Uruguay and beyond, and live cattle exports to São Paulo frigoríficos. The expansion of national and global consumption of beef after World War II afforded opportunities for the sector, but until the 1970s, other regions of Brazil and Latin America supplied most national and international markets.

The twentieth century ushered modern ideas of animal husbandry into the region, many of them inapplicable to the tropical conditions of the region. That characteristic runs through the study, since it is crucial to understanding the irregular evolution of the ranching sector, especially in the context of backward versus developed. The failure of Mato Grosso to grow as expected led to a greater push to apply the latest husbandry techniques to improve ranching's productive capacity. The approaches vital to Mato Grosso's evolution in that regard came from both outside and within Brazil, though there were lessons to be learned from local experience as well. As practitioners and outside observers sluggishly came to understand the ecological specificities of Mato Grosso, the sector began to be transformed by experimentation with forage species, the introduction of exotic breeds, and more intensive management of cattle, including expansion of fencing, regular provision of salt, and disease and parasite control. The process was anything but smooth, for it was not uncommon for modern applications to result in spectacular failure, as in the case of Brazil Land. Often traditional practices persisted well into the twentieth century, including land tenure and use, access to credit, labor relations, and the handling of cattle.

As expected in a relatively remote frontier territory like Mato Grosso, land was always in contention. For the most part, large landholding prevailed, at least up to the 1930s and 1940s. The limited availability of land for smallholders, whether agriculturalists or small ranchers, caused criticism from certain sectors of the country. By the 1920s, much land was held by large foreign concerns, but there were also many huge private ranches in the state, particularly in the Pantanal. With direct state intervention during the Vargas years and internal population pressures, properties gradually fragmented and opportunities opened up for family operations. Still, environmental conditions, relative geographic isolation, and the expense of running a ranch with uncertain market prospects guaranteed the disproportionate influence of large ranches.

Large establishments also had greater access to credit to maintain or expand their operations. Thanks largely to the perceived lack of easily collected collateral in the state, private banking was virtually nonexistent for much of the period. In its absence, fiscal attempts were made to stimulate credit accessibility; as with transportation, these efforts tended to be lethargic and inconsistent. The result was a system of informal credit relations that linked large property owners to other significant operators, or with export agents, but denied opportunities to smaller entrepreneurs. Such limiting conditions changed only over the course of subsequent decades, particularly in the latter years of the twentieth century.

Conflict was a part of the process, of course, but as I have outlined, the greater influences came from the day-to-day operations of the ranches, the application or not of imported technologies, and the persistence of traditional work relations. For the most part, relations between employers and employees were amicable, if hierarchical. There were also rather distinct divisions of labor among those directly working the cattle, creating the unique circumstance in which cowboys and ranch hands performed specific tasks based on status. Undoubtedly, a certain discrimination existed, but for the most part it did not cause discord among workers. Labor relations had the added dimension of ethnic specificities. Brazilian-born workers dominated, but Paraguayans, Uruguayans, and Argentines were common in Mato Grosso, working in various jobs on the ranches and in the bouillon and jerky factories. Besides their labor, the immigrants' customs and traditions significantly influenced the cultural makeup of Mato Grosso, including food ways and language patterns. To a lesser extent, aboriginal customs also had an influence.

As in all of the Americas, indigenous peoples were imposed on by the expansion of nonnative settlement, with ranching often in the forefront in dispossessing Indian territories. This history is the most egregious example of conflict in Mato Grosso. Native peoples were stripped of their lands, identities, families, and even their lives; they also adapted, and frequently negotiation was necessary to avoid complete disruption for tribes and settlers. Indian cowboys and ranch hands were common, especially in the Pantanal, and active resistance by some groups in alliance with thoughtful Brazilians permitted them to retain territories they may not have secured otherwise. That alliance gave rise to the Serviço de Proteção aos Índios, an agency that preceded today's FUNAI and was instrumental in guaranteeing permanent native presence in the region. That is not to say that the narrow cultural perceptions of SPI leaders did not negatively impact aboriginal communities, but in several cases Indians became cowboys and both communities and individuals raised cattle, often successfully.

Viable ranch operations require considerable expertise, knowledge often gained with practice. Certainly the long-term extensive nature of ranching in Mato Grosso required generations of trial and error to sustain a precarious industry through uncertain ecological and economic conditions. Even with modern concepts, the difficult conditions did not disappear; in fact they have persisted into the twenty-first century. In part, this is the nature of raising cattle, since the practicality of some tasks is undeniable. For example, the latest equipment can help in constructing fences and vaccinating animals, but roundup still requires human power and expertise, as does the taming of horses and branding of cattle. At the same time, historical knowledge of local environmental conditions was essential in planning most ranching activities.

A crucial component of this story is the complex role of ecosystems in defining how ranching developed. Routine tasks such as determining appropriate pasturage, times to engage in the roundup, branding, and so on, were dependent on climate, terrain, and availability of new husbandry technologies. Central was the role of breeding in Mato Grosso ranching. The effect of the zebu breed on ranching in Brazil and elsewhere has been spectacular. Viewed from the air, the white specks of Nelore cattle spread across the landscape demonstrate the magnitude of this Indian-origin breed in revolutionizing modern tropical cattle ranching. Mato Grosso's position in the expansion and the experience of ranchers in adopting or rejecting the breed speaks volumes about both tradition and modernization. The story of the zebu explains in large part why the region eventually was able to develop as it did. As the numbers of animals grew and the industry expanded, environmental consequences became more pronounced. With the zebu's rise to prominence in the last decades of the twentieth century, the previous limited impact of cattle on the land changed dramatically. That transformation leads to my other main argument—that much of what characterizes recent endeavors in ranching modernization, particularly in the Amazon Basin, drew on the experiences of ranching development in Mato Grosso prior to the mid-twentieth century. To clarify that point and bring the discussion full circle, my final reflection addresses pertinent developments in Mato Grosso and beyond since 1950.

LATE TWENTIETH-CENTURY MODERNIZATIONS

Though poor communications and official and economic inertia restricted pre-1950s access to resources and technical inputs throughout Brazil, by mid-century concerted efforts to integrate the interior into the nation began to bear fruit. Starting in the 1930s, national government interest in the interior gradually came to fruition, particularly through the planning and construction of a new national capital at Brasília, completed in 1960. The intention of the Brazilian State was to occupy its remote regions finally and definitively, the success of which depended on economic viability. Fiscal subsidies were the vehicle through which the military government, which ruled from 1964 to 1985, sought to achieve that end, and ranching, above all in the Mato Grossos, was instrumental.

As noted in the Introduction, I appreciate that my contention that ranching in Mato Grosso from the nineteenth through the twentieth centuries influenced future endeavors in Brazil runs the risk of regarding subsequent experiences in other regions, particularly in the Amazon, as a form of reading backward into early Mato Grosso. Indeed, the Amazon's particular set of biomes only partially parallel those of the Center West, and the economic and social imperatives that stimulated ranching in Pará, for example, do not necessarily duplicate those some decades earlier in southern Mato Grosso. Yet economic commitments and the application of new technologies in raising cattle, such as the introduction of exotic grasses, development of hardier cattle breeds, approaches to land acquisition and control, myriad forms of labor relations, and most especially the ecological restrictions and opportunities that informed struggles for land and capital, all saw their application first and most heavily in Mato Grosso. Although not fully developed until late into the twentieth century, these advances and impacts of Brazilian tropical ranching expanded rapidly from the 1960s, in both Mato Grosso and Amazonia.[1]

The cattle sector was relatively stable by the 1950s, as interest in its expansion strengthened, and fiscal and private investment began to filter, then flow, into the interior of Brazil. By the second decade of the twenty-first century, cattle had become a major economic component of the nation's prosperity. The national herd increased from some 50 million head in the 1950s to well over 200 million by 2011.² Those population increases sprang from official policy and private opportunity, occurring most dramatically between the 1970s and 1990s, when the beef cattle population across the nation doubled in just 25 years (1970–1995).³ The most heavily affected regions were the Center West and Amazonia, both now focal points of socio-environmental conflict. Stimulated by government development programs such as the Programa de Desenvolvimento dos Cerrados (POLOCENTRO), Programa de Desenvolvimento do Pantanal (PRODEPAN), and Superintendência do Desenvolvimento da Amazônia (SUDAM), human and cattle populations in both regions surged. The human population almost tripled in the Center West and quadrupled in Amazonia, as rich and poor southerners flocked into the interior seeking opportunities not offered in their home areas. Cattle populations increased most spectacularly, roughly 300 percent in the Center West, over 1000 percent in Amazonia.⁴ The military dictatorship deliberately sought to expand on previous intentions and stimulate occupation of the vast interior, and that policy did not change significantly with the coming of democracy. The hope remained to develop the region for the nation, an endeavor that led to the promotion of economic activities that could generate increasing amounts of foreign exchange through trade. Cattle were seen as the easiest and cheapest way to achieve that dream.

Part of the cost of expansion was social. A component that has attracted considerable international attention is the role of land occupation and accompanying violence in newly settled regions. A general perception of ranching history across the Americas is that it has been a region of violent dispute among ranchers, smallholders, and native peoples.⁵ That is still the assessment today. In recent decades, multiple news reports and studies have revealed how acute some of the conflict over land has become, even when the state attempts to address it. The well-told story of the assassination of rubber tapper, trade unionist, and environmentalist Chico Mendes by a ranching family in 1988 in Acre is one example.⁶ There are many others, most notably in Amazonia, but also in the Mato Grossos, that indicate how seriously small settlers, as well as native communities, have struggled to protect their access to land, address environmental concerns, protect traditional culture, and make a living.⁷ As expected, those with the most resources are the ones who usually triumph, though we should be circumspect when it comes to the goals of such players, if not their sometimes nefarious methods.

Although in the past some clashes in Mato Grosso were linked to struggles over land, the actions of bandits and hired killers were governed in large part by politics and individuals who could take advantage of official connections for personal benefit. State attempts to mitigate violence were irregular and largely insufficient, but ironically, hostility also was tempered by the vastness of the territory and the availability of land. That changed as more settlers entered the region and as the best land was occupied.[8] More recently in Amazonia and the Mato Grossos, even with national and international attention and plentiful resources, there seems to be a reluctance of the state to provide regular monitoring of local actors, many of whom are well connected to regional politicians. Though most cattle regions are guardedly peaceful, extrajudicial actions over access to land have attracted headlines across the country and world, such as the recent conflict with the Guarani-Kaiowá peoples near Dourados. Native land claims that include properties belonging to ranching families date back almost to the original settlement of the Guarani peoples, though disputes have become most intense in the last few years, involving owners of cattle, soy, and sugarcane properties. Reflecting the situation nationwide, indigenous groups are asserting their rights while state and federal governments have made some attempts to address inequalities, though they have been frustratingly slow in resolving land claims.[9]

Other disputes pit small settlers against large landholders, or environmental activists against medium and large cattlemen. In Mato Grosso do Sul, most are underreported, though the overwhelming control of most large landholders ensures that such conflict is less than in other regions of the country. Nonetheless, it seems unlikely that acts of violence in the cattle regions of tropical Brazil will end soon. If the experience of Mato Grosso is any guide, that would require a concerted effort by multiple players, not an easy sell in a country with such an exaggerated gap between rich and poor.[10]

The human side of the story is compelling, but coupled with it are the technical and environmental characteristics that heavily informed the dramatic transformation. Until the 1970s, the Center West attracted the most attention from the State, which supported a number of studies to determine its suitability for agricultural expansion. As a result, the Mato Grossos before and after their division in 1979 received a considerable number of agronomists and entrepreneurs. The internationally respected Brazilian agricultural research corporation, EMBRAPA, was created specifically for this purpose, with a number of specialized units set up around the country to stimulate the modernization of agricultural and animal production.[11] The units were inspired by the notion that both the Cerrado and the Amazon Rainforest were underexploited and simply needed modern science to render them economically viable and thereby attractive regions for settlement. Not by chance did EMBRAPA establish its cattle research headquarters in Campo Grande, which in many ways vindicated the struggles of previous decades, as outsiders came to appreciate not only the potential but also the complexities of the region. With the creation of Mato Grosso do Sul, the effort only expanded, though by the 1980s its northern brother and Amazonia were seen to offer even more opportunities for development. In all cases, the science concentrated on several of the technical issues discussed in this book, such as forage development, breeding, nutrition, and disease control. EMBRAPA–Gado de Corte and EMBRAPA-Pantanal, as well as other units around the country, fund and publish a tremendous number of studies on those and other elements of cattle care and improvement. Agronomists, zootechnicians, ranchers, and other researchers regularly visit from around the world, and there are several exchange programs between the units and similar entities worldwide.

The effect of EMBRAPA efforts (though until the 1980s not the direct result) was to shift attitudes regarding the science of cattle ranching, which again incorporated some of the initiatives seen in previous decades. The most important programs for Brazilian tropical ranching have been in forage improvement, stock breeding, and disease control, and most were applied first in the Mato Grossos. For the majority, the economic rewards of expanded cattle production and beef exports fully justified the scientific and fiscal investment.

Thanks to government incentives and facilitated access to cheap credit in the 1970s, establishing a ranch in southern Mato Grosso became much easier. With no requirements to employ current husbandry technology, personal investment was minimal, and little care was taken to rotate pasture, to seek grasses suited to the breed or age of animals, or to preserve soil fertility. Part of the motivation was the same as in the Amazon Basin during the same period—speculation. The result for the environment was predictable. It also disrupted ranch social structure. As in Argentina many decades earlier, land was turned over to renters or sharecroppers, with the stipulation that a section had to be sown in artificial pasture, usually jaraguá, after two or three years. Hence, initial investment was minimal for the owner, and there appeared to be little need to dedicate much attention to the ranch. In this way, artificial pasture increased 170 percent in Mato Grosso do Sul between 1970 and 1980, and 103 percent for Brazil as a whole.[12]

As an integral part of this economic imperative, roads were constructed to access more remote regions of the state. The intention was to facilitate the export of cattle, and a major result was increased activity and pressure on the land. For example, exotic grasses were eagerly adopted with no concern for the environment, or even for economic rationality. In fact, there is probably no region of the world more invested in planting artificial pasture than Brazil. In Europe and the United States, for example, Ana Primavesi reported that in the 1980s only 6 percent of pasture was planted; the rest was native or spontaneous, observed and managed if necessary, but not sown. She revealed no data, but in Brazil the percentage was much higher, since in the cerrado of Mato Grosso do Sul in 1980 close to 20 percent of pasture was planted. This expanded dramatically in succeeding decades.[13]

What truly changed the landscape was the widespread introduction of brachiaria (signalgrass). In the early 1980s several varieties of brachiaria, including *Brachiaria brizantha* (marandu), *B. decumbens*, and *B. humidicola*, were introduced into ecologically diverse regions because they did well in soils of varied fertility, were resistant to insects, and fattened cattle well. By the early 2000s, estimates indicated that in the Cerrado alone over 50 million hectares were in pasture, roughly 26 percent of the total area, while more than 60 million hectares of the Amazon, over 10 percent of the entire basin, had been cleared, primarily for pasture. Of that total, 80 percent was in brachiaria. Virtually all plantings in Mato Grosso do Sul have been brachiaria species, even including the Pantanal, where the amount of land in planted pasture totaled over 16 million hectares by 2002, between 8 and 11 percent of the total area.[14]

The environmental consequences of this expansion have been extensive: deforestation leading to erosion and silting, consequent flooding and permanent transformation of river courses, soil impoverishment, native species elimination, and local climate change, including increased drought. For example, reports indicate that deforestation for pasture expansion in the elevated regions of the Pantanal has caused regular flooding to become more severe in recent years. By contrast, in some regions ranchers faced greater forest encroachment as soils deteriorated. Initially, defoliants (including Agent Orange) were used to combat the spread of forest, but the practice was abandoned after public outcry, and many ranchers turned to mechanical and manual stripping of the young forest. Those who could not keep up were soon forced to sell out. At the same time, the practice of burning has continued. With so many more ranching units, however, the ecological effects of fire were extensive. Besides the substantial release of greenhouse gases and harm to native species, the success of introduced species allowed for overstocking of pastures, which resulted in rapid soil degradation and the proliferation of termite mounds. For Mato Grosso, Allem and Valls have documented that by the 1980s in the upper Pantanal, soils had deteriorated significantly and invasive vegetation had become common. The same was the case in the campo limpo, turning campos limpos into *campos grossos* (coarse fields). The invaders were commonly carona and barba de bode, the same as seen in the cerrado to the east. In some areas where as little as two hectares of original grasses had supported one head of cattle, it now required five hectares of degraded pastures to carry one head, and the cattle took longer to reach optimal weight for slaughter. Those obstacles led EMBRAPA and others to seek new varieties of grasses, including improved brachiaria, and more diversified methods of pasture maintenance. To highlight the irony in those efforts, other imported species, such as Pensacolo (*Paspalum notatum*) and the legume Townsville lucerne (*Stylosanthes gracilis*), are eagerly sown by ranchers unaware that they are enhanced varieties of species native to the Americas.[15]

A similar revolution occurred with the zebu. Following on earlier achievements, more recent developments highlight how important this breed is to Brazilian tropical ranching. By the late twentieth century, zebu accounted for over 80 percent of the national cattle herd, and 85 percent of the zebu were Nelore.[16] Reflecting the rapid expansion of modern technology in global ranching, today the Mato Grosso breeding sector is dominated by artificial insemination and, most recently, embryo transplantation, cutting-edge technologies that clearly represent the degree of scientific sophistication that overlays the Brazilian cattle sector in the twenty-first century. The environmental stresses have only intensified, exacerbated by the dramatic increase in cattle populations, while the number of regions involved are much more numerous. Unlike the earlier years in Mato Grosso, in many places ranching has extended into forests, contributing heavily to the dramatic ecological degradation so decried globally.

While the world's attention largely has been attracted to the Amazon Basin, we have overlooked the damage that ranching and agriculture produced on the Cerrado, arguably even more severe. The biome played host to much of early Mato Grosso ranching, with mixed environmental effects, but today it is considered to be a biodiversity hot spot, as over 50 percent of its 2 million square kilometers has been converted to pasture and commercial agriculture since 1980. More than 500,000 square kilometers are in pasture alone, half already degraded by fire, species invasion, and the proliferation of termite mounds. Since the 1970s, ranchers in Mato Grosso have played a key role in this degradation.[17]

Another aspect of this explosive expansion has been the evolution of soybean agriculture. By the twenty-first century, Brazil was the second largest producer of soybeans in the world. It is argued that cattle occupation of the land helped pave the way for soy planting on soils that previously were considered infertile for commercial agriculture, although occupation of the land by soy often bypassed cattle altogether. The expansion was impressive, from minuscule plantings in the 1960s to over 13 million hectares by 2000, half of that in the Cerrado of Mato Grosso and Goiás.[18] Once again, much of the change was stimulated by science developed at EMBRAPA and elsewhere. Since soybeans are also used for animal feed around the globe, the region contributes both directly and indirectly to animal husbandry worldwide.

Coupled with ranching, huge corporate soybean operations have led to significant environmental consequences. Large soy farmers have come under considerable criticism for their permanent transformation of the tropical ecosystems in which they operate, often in association with cattle ranching, and scientific observations seem unequivocal in their conclusions.[19] Yet those judgments have been contested or simply ignored by producers, primarily because official pressure has been irregular at best, including even enforcement of national laws, allowing many producers to develop the land in what they determined to be their best interests. There were signs early in the twenty-first century that some ecological and social concerns were beginning to be addressed, but today observations, official and independent statistics, and simple anecdotal evidence indicate that such successes may only be temporary.[20]

A recent phenomenon that speaks directly to the transformation of Mato Grosso do Sul ranching has been a shift from ranching to sugarcane production for ethanol. Biofuels have been an integral part of the Brazilian energy sector since the 1970s, and the nation's sugarcane production is overwhelmingly dominated by São Paulo, but biofuel production is new to Mato Grosso. Although the data are still premature, it is clear that Mato Grosso do Sul has seen a significant increase in lands dedicated to cane, to almost 500,000 hectares in 2011, which translated into nearly 5 percent of total national production.[21] As in São Paulo, and similar to soy farming, most of the production has been on land that formerly supported cattle, especially in regions where ranchers did not consider the long-term health of their land. Cattle still dominate most sectors of land use in the state, but the conversion of ranch land to commercial cane and soy production has shifted the dynamic of development. Given what we know, it is not difficult to imagine the long-term effect this is likely to have on the region's economic, social, and environmental structures.

Tropical ranching has a long history in the Americas that did not begin in Mato Grosso. Yet, in the modern period there is no better place to observe the development of this important sector of the Latin American economy. It was in Mato Grosso that innovation met tradition, feeding a series of conjunctures that many observers of the time saw as messy and backward and thus in need of correction. This caused some conflict, but mostly censure of the protagonists of ranching, who in turn often rejected outside advice as inappropriate to conditions on the ground, above all ecological.

The ambitions and innovations that drove cattle ranching in Mato Grosso in the early years eventually coalesced in a form that could be applied to the region and well beyond in great intensity today. The tenacity of the human drive for economic improvement comes in all shapes and travels many roads, and by examining the historical trajectory of ranching we can more easily appreciate the importance of cattle production to the expansion and economic growth of Brazil, as well as the consequences that both Mato Grosso and nation have had to face. Ranching is a formidable sector of the Brazilian economy today. Throughout the period under study and well beyond, market forces largely beyond the range influenced how quickly ranching developed, regionally and nationally. In turn, the many daily inputs of raising cattle reveal how firmly ranching is governed by a series of factors closely linked with land, labor, environment, and technology. Indeed, environmental and social concerns in recent years illustrate how much more influential ranching is to national well-being than simple economics. Most notably, there is the complication of a tropical-semitropical landscape, distinct from that of temperate zones. External and internal factors combined to sculpt the character of ranching in Mato Grosso, through a process that was complex, erratic, and sometimes violent. Ultimately, ranching expansion in Mato Grosso shaped the structure of a significant sector of the Brazilian tropical frontier, establishing a precedent for the rest of the country and outlining a complex course of development. In keeping with that complexity, the consequences have been decidedly mixed, and the future is yet uncertain.

Notes

INTRODUCTION. THE PARADOX OF TROPICAL RANCHING

1. Some extended discussions of this perspective are found in: Arturo Escobar, *Encountering Development: The Making and Unmaking of the Third World* (Princeton: Princeton University Press, 1995); Gilbert Rist, *The History of Development from Western Origins to Global Faith*, 2nd ed. (London: Zed Books, 2002); Robert N. Gwynne and Cristóbal Kay, *Latin America Transformed: Globalization and Modernity*, 2nd ed. (New York: Routledge, 2014); Andréa Zhouri, "'Adverse Forces' in the Brazilian Amazon: Developmentalism Versus Environmentalism and Indigenous Rights" *Journal of Environment and Development* 19 (September 2010): 252–273.

2. Other important cattle states include Goiás (21 million), Minas Gerais (24 million), Rio Grande do Sul (14 million), and São Paulo (11 million). Brazil, IBGE, "Efetivo dos rebanhos de grande porte em 31.12, segundo as Grandes Regiões e as Unidades da Federação—2011," table 3 in *Produção da Pecuária Municipal*, 2011, vol. 39 (Rio de Janeiro: IBGE, 2011): 1–63. ftp://ftp.ibge.gov.br/Producao_Pecuaria/Producao_da_Pecuaria_Municipal/2011/tabelas_pdf/tabo3.pdf (accessed January 2013). A significant percentage of the total for Minas Gerais includes dairy cattle, and although dairy cattle are included in all figures, with the exception of that state beef cattle overwhelmingly dominate in all jurisdictions. According to the 2010 census, Brazil's human population in that year had reached 190 million. Brazil, IBGE, *Synopse do Censo Demografico, 2010* (Rio de Janeiro: IBGE, 2011). http://www.censo2010.ibge.gov.br/ (accessed March 6, 2016). Until 1979, Mato Grosso was one state. In that year, the state was divided into two, with the south forming the state of Mato Grosso do Sul, while the north retained the original name. For much of the period, raising cattle was almost exclusively the domain of the southern sector of the state, since until recent years the northern region offered less optimal conditions for large-scale ranching. As a consequence, this study concentrates primarily on what is today the state of Mato Grosso do Sul, with occasional reference to other regions of Mato Grosso. Unless

otherwise noted, my use of "Mato Grosso" refers to the region as it fits into the historical timeline of this study.

3. US Department of Agriculture, Foreign Agricultural Service, *Livestock and Poultry: World Markets and Trade*, April 2015. http://apps.fas.usda.gov /psdonline/circulars/livestock_poultry.PDF. India's exports of beef are almost exclusively buffalo meat, referred to as carabeef. See Rishi Iyengar, "India Stays World's Top Beef Exporter Despite New Bans on Slaughtering Cows," *Time* online, April 23, 2015, http://time.com/3833931/india-beef-exports-rise-ban-buffalo -meat/, and Sena Desai Gopal, "Selling the Sacred Cow: India's Contentious Beef Industry," *Atlantic*, February 12, 2015, http://www.theatlantic.com/business /archive/2015/02/selling-the-sacred-cow-indias-contentious-beef-industry /385359/. All sources accessed June 2015.

4. Many studies of contemporary ranching in the American tropics have been published over the past decades, particularly from an economic and environmental perspective. Some of the most notable in English are: Susanna B. Hecht, "Cattle Ranching in Amazonia: Political and Ecological Considerations," in *Frontier Expansion in Amazonia*, ed. Marianne Schmink and Charles H. Wood (Gainesville: University of Florida Press, 1984), 366–398; John Hemming, ed., *Changes in the Amazon Basin* (Manchester: Manchester University Press, 1985); Douglas R. Shane, *Hoofprints on the Forest: Cattle Ranching and the Destruction of Latin America's Tropical Forests* (Philadelphia: Institute for the Study of Human Issues, 1986); Marc Edelman, *The Logic of the Latifundio: The Large Estates of Northwestern Costa Rica Since the Late Nineteenth Century* (Stanford: Stanford University Press, 1992); David Kaimowitz, *Livestock and Deforestation in Central America in the 1980s and 1990s: A Policy Perspective* (Jakarta: Center for International Forestry Research, 1996); Charles H. Wood and Roberto Porro, eds., *Deforestation and Land Use in the Amazon* (Gainesville: University of Florida Press, 2002); Mark London and Brian Kelly, *The Amazon in the Age of Globalization* (New York: Random House, 2007); Jeffrey Hoelle, *Rainforest Cowboys: The Rise of Ranching and Cattle Culture in Western Amazonia* (Austin: University of Texas Press, 2015). Two significant historical examinations of the greater Amazon include far northern Roraima on the border with Venezuela and northern Goiás (now Tocantins), though neither addresses environmental considerations explicitly. See, respectively, Peter Rivière, *The Forgotten Frontier: Ranchers of North Brazil* (New York: Holt, Rinehart and Winston, 1972), and Júlio César Melatti, *Índios e Criadores: a Situação dos Krahô na Área Pastoril do Tocantins* (Rio de Janeiro: Instituto de Ciências Sociais, Universidade Federal do Rio de Janeiro, 1967).

5. Studies of historical frontiers are numerous, but ones that most directly address ranching include: Silvio Zavala, "The Frontiers of Hispanic America," in *The Frontier in Perspective*, ed. W. D. Wyman and C. B. Kroeber (Madison: University of Wisconsin Press, 1965); Alistair Hennessy, *The Frontier in Latin American History* (London: Edward Arnold, 1978); Silvio R. Duncan Baretta and John Markoff, "Civilization and Barbarism: Cattle Frontiers in Latin America," *Compara-*

tive Studies in Society and History 20 (October 1978): 587–620; David J. Weber and Jane M. Rausch, eds., *Where Cultures Meet: Frontiers in Latin American History* (Wilmington: Scholarly Resources, 1984). Studies of Brazilian frontiers in English include: Mary Lombardi, "The Frontier in Brazilian History: An Historiographical Essay," *Pacific Historical Review* 44, no. 4 (Nov. 1975): 437–457; E. Bradford Burns, "Brazil: Frontier and Ideology," *Pacific Historical Review* 64, no. 1 (Feb. 1995): 1–18; Jane M. Rausch, "Frontier History as an Explanatory Tool for Brazilian History: A Viable Construct?" *Latin American Research Review* 43, no. 1 (2008): 201–207. For Brazil and specifically Mato Grosso, see Sérgio Buarque de Holanda, *Caminhos e Fronteiras* (Rio de Janeiro: Editora José Olympio, 1957), and Buarque de Holanda, *O Extremo Oeste* (São Paulo: Brasiliense, 1986); Lúcia Salsa Corrêa, *História e Fronteira: O Sul de Mato Grosso, 1870–1920* (Campo Grande: Editora UCDB, 1999); Maria de Fátima Costa, *História de um país inexistente: O Pantanal entre os séculos XVI e XVIII* (São Paulo: Estação Liberdade, Kosmos, 1999); Valmir Batista Corrêa, *Fronteira Oeste* 2nd ed. (Campo Grande: Editora UNIDERP, 2005). Two works that have briefly addressed ranching somewhat along the lines I propose here are: Arnold Strickon, "The Euro-American Ranching Complex," in *Man, Culture, and Animals: The Role of Animals in Human Ecological Adjustments*, ed. Anthony Leeds and Andrew P. Vayda (Washington, D.C.: American Association for the Advancement of Science, 1965), 229–258; and David McCreery, *Frontier Goiás, 1822–1889* (Stanford: Stanford University Press, 2006), chapter 5.

6. Mato Grosso was a province until the republican coup of 1889 that ended the empire and led to the reformulation of provinces into states throughout the country.

7. Since this is a study of beef cattle ranching, I only mention sheep in Argentina and Uruguay to underline the importance of livestock to their economies. Sheep were always an insignificant proportion of the Brazilian livestock sector, and pigs were important only in Minas Gerais. As such, all comparative figures refer only to beef cattle.

8. Rollie E. Poppino, "Cattle Industry in Colonial Brazil," *Mid-America* 31 (October 1949): 219–247; João António Andreoni (André João Antonil), *Cultura e Opulência do Brasil* (1711), (São Paulo: Companhia Editora Nacional, 1967), 307–316; Celso Furtado, *The Economic Growth of Brazil: A Survey from Colonial to Modern Times* (Berkeley: University of California Press, 1965), 62–71, 83–85. The literature on ranching spans the continent, of course. For our purposes I offer here only the most significant studies of Argentina, Uruguay, and Brazil. Argentina: Richard W. Slatta, *Gauchos and the Vanishing Frontier* (Lincoln: University of Nebraska Press, 1983); Horacio C. Giberti, *Historia económica de la ganadería argentina* (Buenos Aires: Raigal, 1954); Alfredo J. Montoya, *Historia de los saladeros argentinos* (Buenos Aires: Raigal, 1956); Samuel Amaral, *The Rise of Capitalism on the Pampas: The Estancias of Buenos Aires, 1785–1870* (New York: Cambridge University Press, 1998); Osvaldo Barsky and Julio Djenderendjian, *Historia del capitalismo agrario pampeano, tomo 1: La Expansión ganadera hasta 1895* (Buenos Aires: Siglo Veinteuno editores,

2003); Carmen Sesto, *Historia del capitalismo agrario pampeano, tomo 2: La Vanguardia ganadera bonaerense, 1856–1900* (Buenos Aires: Siglo Veinteuno editores, 2005). Uruguay: Agustín Ruano Fournier, *Estudio economico de la producción de las carnes del Río de la Plata* (Montevideo: Peña y Cía., 1936); José Pedro Barrán, *Apogeo y crisis del Uruguay pastoril y caudillesco, 1839–1875*, vol. 4: *Historia uruguaya*, 2nd ed. (Montevideo: Banda Oriental, 1975). Brazil: Spencer Leitman, "Slave Cowboys in the Cattle Lands of Southern Brazil, 1800-1850," *Revista de Historia (São Paulo)* 51 (January–March 1975): 167-177; Sandra Jatahy Pesavento, *República velha gaúcha: charqueadas, frigoríficos, criadores* (Porto Alegre: Movimento/IEL, 1980); Stephen Bell, *Campanha Gaúcha: A Brazilian Ranching System, 1850–1920* (Stanford: Stanford University Press, 1998); David McCreery, *Frontier Goiás, 1822–1889* (Stanford: Stanford University Press, 2006), 130–154; Nelson Werneck Sodré, *Oeste: Ensaio sobre a grande propriedade pastoril* (Rio de Janeiro: Livraria José Olympio Editora, 1941).

9. Brazil, Ministério da Agricultura, Indústria e Commercio, Directoria do Serviço de Estatística, *Synopse do censo pecuário da república pelo processo indirecto de avaliações em 1912–1913 resultados provisorios* (Rio de Janeiro: Typographia do Ministério da Agricultura, 1914), 36; US Department of Agriculture, "South American Cattle Census," Records of the Bureau of Agricultural Economics, General Correspondence of the Bureau of Markets and Bureau of Agricultural Economics, 1912–1952: Brazil, no. 39 [1918], Record Group 83, United States National Archives, College Park, Maryland; Brazil, Fundação Instituto Brasileiro de Geografia e Estatística (hereafter IBGE), Séries Estatísticas Retrospectivas, vol. 3, *Estatísticas históricas do Brasil: Séries econômicas, demográficas e sociais de 1550 a 1988*, 2nd ed. (Rio de Janeiro: IBGE, 1990), 300; Antonio Carlos de Oliveira, "Economia pecuária do Brasil Central; Bovinos (Estudo gêo-estatístico e político-econômico) II Parte," *Boletim do Departamento Estadual de Estatística* (São Paulo) 1 (January 1942): 15, 21–22, 24; Brazil, IBGE, Séries Estatísticas Retrospectivas, vol. 3, Section 6.24, http://seculoxx.ibge.gov.br/economicas/tabelas-setoriais/agropecuaria (accessed September 2015); Dwight R. Bishop, *Argentina's Livestock and Meat Industry* (Washington, D.C.: US Department of Agriculture, Foreign Agricultural Service, 1963), 3. Accessible at https://archive.org/stream/argentinaslivest149bish#page/n3/mode/2up (accessed September 2015); Dwight R. Bishop, *Uruguay's Livestock and Meat Industry* (Washington, D.C.: US Department of Agriculture, Foreign Agricultural Service, 1963), 7. Accessible at https://archive.org/stream/uruguayslivestoc150bish#page/6/mode/2up (accessed September 2015). It should be emphasized that early cattle censuses in all these countries often were unreliable. Nonetheless, the figures indicate trends that continued over subsequent years.

10. Stephen Bell, *Campanha Gaúcha: A Brazilian Ranching System, 1850–1920* (Stanford: Stanford University Press, 1998), chapters 5–7.

11. Agustín Ruano Fournier, *Estudio economico de la producción de las carnes del Río de la Plata* (Montevideo: Peña y Cía., 1936), 67. European meat consumption involved breed types such as Shorthorn, Hereford, Devon, Angus, and others that originated in Britain and were raised in large numbers in the United States,

Canada, Australia, and Argentina. In comparison, most consumers considered the meat of animals originating in Iberia and raised for centuries in the Americas to be tough, stringy, and rank in flavor, and therefore suited only to the production of canned meat, salt beef, or beef bouillon. Fascinating discussions of the contemporaneous debate over beef quality can be found in Jeffrey M. Pilcher, "Empire of the 'Jungle': The Rise of an Atlantic Refrigerated Beef Industry, 1890–1920," *Food, Culture and Society* 7, no. 2 (fall 2004): 63–78; Roger Horowitz, Jeffrey M. Pilcher, and Sydney Watts, "Meat for the Multitudes: Market Culture in Paris, New York City, and Mexico City over the Long Nineteenth Century," *American Historical Review* 109, no. 4 (October 2004): 1055–1083; and Roger Horowitz, *Putting Meat on the American Table: Taste, Technology, Transformation* (Baltimore: Johns Hopkins University Press, 2006).

12. This has been reinforced by the tourist industry, particularly ecotourism. Of course, closer study has revealed a more complicated interrelationship between man and nature in the Pantanal. See chapter 1 and Frederick A. Swarts, ed., *The Pantanal of Brazil, Bolivia, and Paraguay: Selected Discourses on the World's Largest Remaining Wetland System* (Gouldsboro, PA: Hudson MacArthur Publishers, 2000).

13. Though not a focus of this study, one approach to the process has been through the concept of commodity chains, defined as: "the production of tradable goods from their inception through their elaboration and transport to their final destination in the hands of consumers." Steven Topik, Carlos Marichal, and Zephyr L. Frank, eds., *From Silver to Cocaine: Latin American Commodity Chains and the Building of the World Economy, 1500–2000* (Durham, NC: Duke University Press, 2006), 14. The introduction to this volume explains the authors' approach in detail.

14. It is difficult to find reliable statistics on beef consumption in Brazil during the period. A report from the beginning of the twentieth century suggests that nationwide as many as 12 million head were sacrificed annually. The same report indicated that residents of Rio de Janeiro consumed 22 kilos of beef a year: J. C. Oakenfull, *Brazil in 1911* (London: Butler and Tanner, 1912), 240. Maria-Aparecida Lopes notes that urban consumption was higher than rural and suggests that annual national consumption of all meat (though mostly beef) was roughly 15 kilos per capita. She also notes that the *availability* of salt beef or jerky (*charque*) in the Rio market indicates potential consumption at 60 kilos per capita, per annum. Maria-Aparecida Lopes, "Struggles over an 'Old, Nasty, and Inconvenient Monopoly': Municipal Slaughterhouses and the Meat Industry in Rio de Janeiro, 1880–1920s," *Journal of Latin American Studies* 47, no. 2 (May 2015): 362–363. Until World War I there were few exports of beef except salt beef to Cuba. More on this in chapter 2.

15. See Valmir Batista Corrêa, *Fronteira Oeste* 2nd ed. (Campo Grande: Editora UNIDERP, 2005), and *Coronéis e bandidos em Mato Grosso (1889–1943)* 2nd ed. (Campo Grande: Editora UFMS, 2006).

16. This also was the case in US Western lands. See Patricia Nelson Limerick, *The Legacy of Conquest: The Unbroken Past of the American West* (New York:

W. W. Norton, 1987); Donald Worster, "New West, True West: Interpreting the Region's History," *Western Historical Quarterly* 18 (1987): 141–156; Donald Worster, "Cowboy Ecology," in *Major Problems in American Environmental History: Documents and Essays*, ed. Carolyn Merchant and Thomas G. Paterson, 2nd ed. (Boston: Houghton Mifflin, 2005), 295–301. The same was true for Colombia, as outlined by Shawn Van Ausdal, "Productivity Gains and the Limits of Tropical Ranching in Colombia, 1850–1950," *Agricultural History* 86, no. 3 (summer 2012): 1–32.

17. While I discuss the introduction of the zebu and the controversies that accompanied its arrival, I have not done so in the context of the promising new field of Animal Studies. This is in part a product of the limited information available in Mato Grosso on cattle behavior during the period. Animal Studies offers a clearer understanding of animal-human relations and the importance of the individual nonhuman animal in those relationships, and with some innovative approaches I hope my study might encourage an adventurous scholar to undertake a journey down this auspicious path of investigation. Recent relevant publications in the field include: Harriet Ritvo, *Noble Cows and Hybrid Zebras: Essays on Animals and History* (Charlottesville: University of Virginia Press, 2010); Margo DeMello, *Animals and Society: An Introduction to Human-Animal Studies* (New York: Columbia University Press, 2012); Martha Few and Zeb Tortorici, eds., *Centering Animals in Latin American History* (Durham, NC: Duke University Press, 2013).

CHAPTER 1. MIRROR OF THE LAND

1. Antonio Carlos Simoens da Silva, *Cartas Mattogrossenses* (Rio de Janeiro: n.p., 1927), 51.

2. Mato Grosso, Instituto Nacional de Estatística, Secretaria da Agricultura, Indústria, Comercio, Viação e Obras Públicas, Junta Executiva Regional de Estatística, *Sinopse estatística do Estado, no. 1*, Separata, com acréscimos, do Anuário Estatístico do Brasil, Ano II, 1936 (Cuiabá: Tipografia A. Calhão, 1937), 12; Luiza Rios Ricci Volpato, *A Conquista da terra no universo da pobreza: Formação da fronteira oeste do Brasil, 1719–1819* (São Paulo: Editora HUCITEC, 1987), 34; Brazil, Ministério das Minas e Energia, Secretaria-Geral, *Projeto Radambrasil: Levantamento de Recursos Naturais, vol. 27, Corumbá* (Rio de Janeiro: IBGE, 1982), 23. The state of Amazonas has always been the largest in the federation. In 1943, the federal territories of Ponta Porã and Guaporé were carved out of Mato Grosso. Ponta Porã was reinstated in 1946, but Guaporé continued to be a federal territory until it became the state of Rondônia in 1983.

3. Afonso Simões Corrêa, "Pecuária de corte em Mato Grosso do Sul," unpublished report delivered at Encontros Regionais de Pecuária de Corte, 27 November 1984, Brasília, 1. I am grateful to the author, retired from EMBRAPA-Gado de Corte, Campo Grande, for making this report available to me.

4. The advent of global climate change appears to be altering climate pat-

terns in the region today, but the information here is relevant through the period of this study.

5. Orlando Valverde, "Fundamentos Geográficos do Planejamento Rural do Município de Corumbá," *Revista Brasileira de Geografia* 34 (January–March 1972), 71–79; Brazil, *Projeto Radambrasil, vol. 27,* 203. In the 1970s and 1980s the Brazilian government conducted a series of studies of natural resources, geology, soils, climate, vegetation, etc., of the nation's interior through aerial radar photography called Projeto Radambrasil. Even though it was not able to complete its survey for all of the nation, until recently the project was probably the most comprehensive information on Brazil's natural resources. It is used here as a guide to understanding the geographical and environmental makeup of Mato Grosso in a more recent period compared to the past. Today, Landsat images from the early 2000s are available through the Brazilian Instituto Nacional de Pesquisas Espaciais (INPE): http://www.dgi.inpe.br/CDSR/ (accessed August 2015).

6. Antonio C. Allem and José F. M. Valls, *Recursos forrageiros nativos do Pantanal mato-grossense* (Brasília: EMBRAPA, 1987), 3–4; Valverde, "Fundamentos," 59; Brazil, *Projeto Radambrasil, vol. 27,* 110–111; Brazil, Ministério dos Transportes, *Síntese do estudo do sistema rodoviária coletor do Pantanal Matogrossense* (Brasília: n.p., 1974), 7. There is some disagreement about the total area of the Pantanal. I have followed the more conservative estimate, but for more see Frederick A. Swarts, ed., *The Pantanal of Brazil, Bolivia, and Paraguay: Selected Discourses on the World's Largest Remaining Wetland System* (Gouldsboro, PA: Hudson MacArthur Publishers, 2000), 2–3.

7. Valverde, "Fundamentos," 80, 82; Brazil, Ministério das Minas e Energia, Secretaria-Geral, *Projeto Radambrasil, Levantamento de Recursos Naturais, vol. 28, Campo Grande* (Rio de Janeiro: IBGE, 1982), 139.

8. Manoel de Barros, quoted in Sociedade de Defesa do Pantanal (SODEPAN), "Apresentação," *O Pantanal e o Pantaneiro* (Campo Grande: SODEPAN, 1990).

9. Allem and Valls, *Recursos forrageiros,* ix; Brazil, *Projeto Radambrasil, vol. 27,* 190; Valverde, "Fundamentos," 55, 71, 85.

10. Valverde, "Fundamentos," 60; Brazil, *Projeto Radambrasil, vol. 27,* 190–191.

11. Allem and Valls, *Recursos forrageiros,* 17–18, 31–60.

12. Jorge Adámoli, "A dinamica das inundações no Pantanal," in *Anais do Primeiro Simpósio sobre Recursos Naturais e Socio-Econômicos do Pantanal,* Corumbá, 28 November to 4 December 1984 (Brasília: Ministério da Agricultura, EMBRAPA, 1986), 53, 56; Valverde, "Fundamentos," 84–87.

13. Allem and Valls, *Recursos forrageiros,* 19–21; Valverde, "Fundamentos," 95–98; Brazil, *Projeto Radambrasil, vol. 27,* 272–279.

14. Allem and Valls, *Recursos forrageiros,* 183–187.

15. Allem and Valls, *Recursos forrageiros,* 180–187; M. Cavalcanti Proença, *No têrmo de Cuiabá* (Rio de Janeiro: Biblioteca de Divulgação Cultural, Série A-XVI, 1958), 21: "O que é que viaja comendo a casa; quando a casa afunda êle vai embora?"

16. EMBRAPA, Centro de Pesquisa Agropecuária do Pantanal, *Aspec-*

tos Hidrológicos do Pantanal de Mato Grosso (1900–1978), Foleto A 570 (Corumbá: EMBRAPA, 1978), 8–9; Miguel Arrojado Ribeiro Lisboa, *Oeste de São Paulo, Sul de Mato-Grosso: Geologia, Indústria Mineral, Clima, Vegetação, Solo Agrícola, Indústria Pastoril* (Rio de Janeiro: Typografia do Jornal do Commercio, 1909), 149; Esther de Viveiros, *Rondon conta sua vida* (Rio de Janeiro: n.p., 1958), 195; Econ J. Monserrat and Carlos A. Gonçalves, *Observações sobre a pecuária no Brasil Central: Relatório de viagem apresentado ao Ilmo. Sr. Dr. Manoel Corrêa Soares, D.D., Presidente do Instituto Sul Rio Grandense de Carnes, em 8 de agosto de 1953* ([Porto Alegre]: n.p., 1954), 53; Brazil, Ministério da Agricultura, Indústria e Commercio, Serviço de Inspecção e Fomento Agrícola, *Estudo dos Factores da Producção no Municípios Brasileiros e Condições economicos de cada um: Estado de Matto Grosso, Município de Corumbá* (Rio de Janeiro: Ministério da Agricultura, 1924), 15; José de Barros, *Lembranças para os meus filhos e descendentes* (São Paulo: n.p., 1987), 65; Allem and Valls, *Recursos forrageiros*, 184; Cleber J. R. Alho and João S. V. Silva, "Effects of Severe Floods and Droughts on Wildlife of the Pantanal Wetland (Brazil): A Review," *Animals* 2 (2012): 597. Regarding such regularity, recent deforestation upriver of the region has led to more frequent severe flooding. Climate change will likely be a significant factor as well.

17. Allem and Valls, *Recursos forrageiros*, 12, 175–187.

18. Carlos Vandoni de Barros, *Nhecolândia* (Mato Grosso: n.p., 1934), 13; Renato Alves Ribeiro, *Taboco—150 anos: Balaio de Recordações* (Campo Grande: n.p., 1984), 28, 30–31; José de Barros, *Lembranças para os meus filhos e descendentes* (São Paulo: n.p., 1987), 87–89.

19. Ribeiro, *Taboco*, 28, 228; interview with Dr. Cassio Leite de Barros, Corumbá, September 1990; Allem and Valls, *Recursos forrageiros*, 183.

20. Valverde, "Fundamentos," 90–98; Octavio Domingues and Jorge de Abreu, "A pecuária nos pantanaes de Matto Grosso," these apresentada ao Terceiro Congresso de Agricultura e Pecuária, 1922 (São Paulo: n.p., 1922), 12–13.

21. Valverde, "Fundamentos," 98; Domingues and Abreu, "A pecuária," 12–13.

22. Allem and Valls, *Recursos forrageiros*, 34–38, 48–60.

23. Allem and Valls, *Recursos forrageiros*, 78–82; S. Cardoso Ayala and Feliciano Simon, *Album Graphico do Estado de Matto-Grosso* (Corumbá/Hamburg, n.p., 1914), 286–288; João Leite de Barros, "A Pecuária em Corumbá," in *Anuário de Corumbá, 1939* (Corumbá: n.p., 1939), 8–11. The name "mimoso" apparently originates in the Portuguese term for tender or delicate, hence palatable to cattle, and should not be confused with the mimosa legume, which is not present in the region.

24. Allem and Valls, *Recursos forrageiros*, 31–60, 78–82, 155, 158; Leite de Barros, "A Pecuária," 8–11; Brazil, Ministério da Agricultura, Indústria e Commercio, Serviço de Inspecção e Fomento Agrícola, *Estudo dos Factores . . . Corumbá*, 9; Arnildo Pott, A. K. M. Oliveira, G. A. Damasceno-Junior, and J. S. V. Silva, "Plant diversity of the Pantanal wetland," *Brazilian Journal of Biology* 71, no. 1 suppl. (2011): 265–273.

25. Allem and Valls, *Recursos forrageiros*, 78–82. Contemporary authors enumerated some grasses important in the Pantanal that have not been mentioned by

Allem and Valls or other modern scientists. It is hard to determine if the differences are the result of the disappearance of certain species over time, variations in naming, or the geographic restrictions of previous studies; however, since these are not constant in the world of botanical science anyway, they need not be of serious concern. For more, see Virgílio Alves Corrêa Filho, *Fazendas de gado no pantanal Mato-Grossense*, Documentario da vida rural, no. 10 (Rio de Janeiro: Ministério da Agricultura, 1955), 10; Rodolpho Endlich, "A criação do gado vaccum nas partes interiores da America do Sul," *Boletim da Agricultura* 4, no. 1 (1903): 28–32 (from a series of articles that appeared in 1902 and 1903, in seven continuous issues); Lisboa, *Oeste de São Paulo*, 141–150.

26. Allem and Valls, *Recursos forrageiros*, 78–82, 164; Brazil, Ministério dos Transportes, *Síntese do estudo do sistema rodoviária coletor do Pantanal Matogrossense* (Brasília: n.p., 1974), 2; James J. Parsons, "Spread of African Pasture Grasses to the American Tropics," *Journal of Range Management* 25 (January 1972): 12–14. The rest of the region was made up of forest, with insignificant agriculture. Parenthetically, it should be mentioned that native legumes do exist in the region; however, their role in the diet of local cattle has not been studied fully even today and historically ranchers paid them little attention. The most common tend to be shrubs, usually grazed by cattle when tender or semicoarse grasses are unavailable, an uncommon occurrence in an area with an abundance of water and native grasses. See Allem and Valls, *Recursos forrageiros*, 115.

27. Parsons, "Spread of African Pasture Grasses," 12–14; Sandro Dutra Silva, Rosemeire Aparecida Mateus, Vivian da Silva Braz, and Josana de Castro Peixoto, "A Fronteira do Gado e a *Melinis minutiflora* P. Beauv. (Poaceae): A História Ambiental e as Paisagens Campestres do Cerrado Goiano no Século XIX," *Sustentabilidade em Debate* (Brasília) 6, no. 2 (mai/ago 2015): 17–32.

28. The most common brachiarias in Brazil are *Brachiaria decumbens*, *B. brizantha*, and *B. humidicola*.

29. Joaquim Carlos Travassos, *Monographias agrícolas* (Rio de Janeiro: Typographia Altina, 1903), 23–26; Campo Grande, Matto Grosso, *O município de Campo Grande em 1922*, Relatório do anno de 1922, Apresentado á Camara Municipal pelo Intendente Geral Dr. Arlindo de Andrade Gomes (São Paulo: Cia. Melhoramentos, 1923), 48; Parsons, "Spread of African Pasture Grasses," 12–14; Monserrat and Gonçalves, *Observações*, 53; Arnildo Pott, *Pastagens no Pantanal* (Corumbá: EMBRAPA-CPAP, 1988), 49.

30. Frederik A. B. Mayerson, *O aproveitamento e proteção do Pantanal*, análise crítica do "Estudo de desenvolvimento integrado da Bacia do Alto Paraguai" (Brasília: SEMA/MINTER, 1981), 13; Allem and Valls, *Recursos forrageiros*, 287–288. "Disclimax" refers to alteration of a stable ecosystem community ("climax"), as defined by the *Free Dictionary*: "A climax community that has been disturbed by various influences, especially by humans and domestic animals, such as a grassland community that has been altered to desert by overgrazing" (http://www .thefreedictionary.com/disclimax, accessed June 3, 2015; American Heritage Dic-

tionary of the English Language, 5th edition, copyright © 2011 by Houghton Mifflin Harcourt Publishing Company, published by Houghton Mifflin Harcourt Publishing Company, all rights reserved). By 2002 it was estimated that close to 11 percent of the Pantanal was cultivated pasture: João dos Santos Vila da Silva, Myrian de Moura Abdon, Arnildo Pott, "Cobertura vegetal do bioma Pantanal em 2002," in *XXIII Congresso Brasileiro de Cartografia*, Rio de Janeiro, Brasil, 21 a 24 de outubro de 2007 (Rio de Janeiro: Sociedade Brasileira de Cartografia, 2007), 1037 (1030–1038).

31. Lisboa, *Oeste de São Paulo*, 145–150; Endlich, "A criação," *Boletim da Agricultura* 4, no. 2 (1903): 82–84.

32. Allem and Valls, *Recursos forrageiros*, 226–227; Mayerson, *O aproveitamento*, 13.

33. Virgílio Alves Corrêa Filho, *Pantanais Matogrossenses (Devassamento e Ocupação)* (Rio de Janeiro: IBGE, 1946), 128; Virgílio Alves Corrêa Filho, *Fazendas de Gado*, 9–10; Endlich, "A criação," *Boletim da Agricultura* 4, no. 1 (1903): 24–27.

34. G. Sarmiento, "The Savannas of Tropical America," in *Tropical Savannas: Ecosystems of the World, 13*, ed. François Bourlière (Amsterdam: Elsevier Scientific Publishing Co., 1983), 282; Jean Koechlin, "Végétation et mise en valeur dans le sud du mato Grosso," in *Le Bassin Moyen du Paraná Brésilien: L'Homme et son Milieu*, ed. Raymond Pebayle et al., 114, 118–119 (Bordeaux: Centre d'Études de Géographie Tropicale, CNRS, 1978); Fernando Flávio Marques de Almeida and Miguel Alves de Lima, *Excursion Handbook 1: The West Central Plateau and Mato-Grosso "Pantanal,"* trans. Richard P. Momsen, Jr., Eighteenth International Geographical Congress (Rio de Janeiro: International Geographical Union, Brazilian National Committee, 1956), 90–93.

35. Marília Velloso Galvão, "Clima," in *Grande Região Centro-oeste*, Vol. 2 of *Geografia do Brasil*, org. Marília Velloso Galvão, 71, Publ. 16, Conselho Nacional de Geografia (Rio de Janeiro: IBGE, 1960); Ana Primavesi, *Manejo Ecologico de Pastagens Em Regiões Tropicais e Subtropicais* (São Paulo: Nobel, 1986), 63; Simões Corrêa, "Pecuária," 1; Edgar Kuhlmann, "A Vegetação de Mato Grosso: Seus Reflexos na Economia do Estado," *Revista Brasileira de Geografia* 16 (January/March 1954): 82–90; Valverde, "Fundamentos," 85.

36. Brazil, *Projeto Radambrasil, vol. 28*, 141–147, 311–312; Brazil, Mato Grosso, Instituto Nacional de Estatística, Diretoria de Estatística e Publicidade, *Sinopse estatística do Estado, no. 2*, Ano III, 1937 (Cuiabá: Imprensa official, 1938), 6; Velloso Galvão, "Clima," in *Grande Região Centro-oeste*, 91–94, 98, 108–109, 114.

37. Letter from Francisco de Borja Sabeuma Garção, Superintendente da linha fluvial, Cia. Nacional de Navegação a Vapor, to company president João Antonio Mendes Motta, October 15, 1888, Montevideo, *Documentos avulsos* lata 1888-F, Arquivo Público do Estado de Mato Grosso, Cuiabá; US Department of State, Despatches from United States Ministers to Paraguay and Uruguay, 1859–1906, "United States Minister Finch to Department of State, report 126, 19 January 1899," microcopy M-128, roll 10, Record Group 59, United States National Ar-

chives; José de Barros, *Lembranças*, 40, 60, 69; Brazil, Ministério da Agricultura, Indústria e Commercio, Serviço de Inspecção e Fomento Agrícola, *Estudo dos Factores da Producção nos Municípios Brasileiros e condições economicas de cada um: Estado de Matto Grosso, Município de Campo Grande* (Rio de Janeiro: Imprensa Nacional, 1929), 16, 47–48.

38. José de Barros Maciel, "A pecuária nos pantanaes de Matto Grosso," these apresentado ao 3º Congresso de Agricultura e Pecuária, 1922 (São Paulo: Imp. Metodista, 1922), 7–8; John C. Tothill, *Report on the native pasture research program, 1982, EMBRAPA-CPAC, Planaltina, Brasília*, mimeo, fol. 750 (Campo Grande: EMBRAPA, 1982), 4; Almeida and Lima, *Excursion Handbook 1*, 88–89; Brazil, Ministério da Agricultura, Indústria e Commercio, Serviço de Inspecção e Fomento Agrícola, *Estudo dos factores . . . Campo Grande*, 14; Allem and Valls, *Recursos forrageiros*, 90, 99.

39. William Sanford and Elizabeth Wangari, "Tropical grasslands: Dynamics and utilization," *Nature and Resources* 21 (July–September 1985): 16; Brazil, *Projeto Radambrasil, vol. 28*, 245; Antonio Carlos Simoens da Silva, *Cartas Mattogrossenses* (Rio de Janeiro: n.p., 1927), 17–18.

40. Edgar Kuhlmann, "Os tipos de vegetação," in *Grande Região Centro-oeste*, Vol. 2 of *Geografia do Brasil*, org. Marília Velloso Galvão, 129–135, Publ. 16, Conselho Nacional de Geografia (Rio de Janeiro: IBGE, 1960).

41. Maurício Coelho Vieira, "A pecuária," in *Grande Região Centro-oeste*, Vol. 2 of *Geografia do Brasil*, org. Marília Velloso Galvão, 196–199; Eduardo Cotrim, *A Fazenda Moderna: Guia do Criador de Gado Bovino no Brasil* (Brussels: Typ. V. Verteneuil et L. Desmet, 1913), 90–94; Carlos Alberto Dos Santos et al., "Aproveitamento da pastagem nativa no cerrado," in *Cerrado: Uso e manejo*, V Simposio sobre o cerrado (Brasília: Ministério da Agricultura, 1979), 428; E. V. Komarek, "Fire Ecology: Grasslands and Man," *Proceedings of the Fourth Annual Tall Timbers Fire Ecology Conference*, Tallahassee, Florida (1965): 181, 193, 212. It should be understood that the common names of grasses vary between regions and as such are not absolute.

42. Primavesi, *Manejo*, 54; Brazil, *Projeto Radambrasil, vol. 28*, 297; Simões Corrêa, "Pecuária," 8.

43. Roy Nash, *The Conquest of Brazil* (New York: AMS Press, 1926), 262–264; Mato Grosso, *O Município de Campo Grande, Estado de Matto-Grosso* (n.p.: Publicação official, 1919), 57; Dolor F. Andrade, *Mato Grosso e a sua pecuária* (São Paulo: Universidade de São Paulo, 1936), 9; Parsons, "Spread of African Pasture Grasses": 12–14; Endlich, "A criação," *Boletim da Agricultura* 4, no. 2 (1903): 74–81; Nicolao Athanassoff, "Forragens," *Brasil Agrícola* 1 (March 1916): 79–81; Tarciso de Souza Filgueiras, "Africanos no Brasil: Gramíneas introduzidas da África," *Cadernos de Geociências* 5 (July 1990): 59; Santos et al., "Aproveitamento," 434; Primavesi, *Manejo*, 54; Gustavo D'Utra, "Capim de Jaraguá," *Boletim da Agricultura* 1 (1900): 467; Cotrim, *A Fazenda Moderna*, 90–109. In general terms, the advantage of planted over native grasses is confirmed by scientific study, although it also depends on the

makeup of the soil and climatic conditions: Primavesi, *Manejo*, 14–16; Monserrat and Gonçalves, *Observações*, 53. Also, it should be noticed that the carrying capacity on natural cerrado pastures is not overly different from the Pantanal, though this depends on the region and regular precipitation.

44. Virgílio Corrêa Filho, *A propósito do Boi pantaneiro*, Monografias Cuiabanas, vol. 6 (Rio de Janeiro: Pongetti e Cia., 1926), 50–52; Andrade, *Mato Grosso e a sua pecuária*, 10–11; Primavesi, *Manejo*, 43, 124–127, 140.

45. Primavesi, *Manejo*, 54; Vieira, "A pecuária," in *Grande Região Centro-oeste*, 196–197; Brazil, *Projeto Radambrasil, vol. 28*, 297; Santos et al., "Aproveitamento," 428. Recently EMBRAPA began a program of consortium planting for forage, though little information is yet forthcoming. For a teaser, see http://agrolink .com.br/viagenstecnicas/noticia/rede-virtual-de-forrageiras-tropicais-entra -em-acao_137118.html (accessed June 2015).

46. Marília Velloso Galvão, "Clima," in *Grande Região Centro-oeste*, 120 (facing page); Simões Corrêa, "Pecuária," 1.

47. Brazil, *Projeto Radambrasil, vol. 28*, 165–166.

48. Brazil, *Projeto Radambrasil, vol. 28*, 310–313.

49. Brazil, *Projeto Radambrasil, vol. 28*, 297, 300–301.

50. Kuhlmann, "Os tipos de vegetação," 139–140; Komarek, "Fire Ecology," 212.

51. In southern Brazil one square league equals 4,356 hectares, though in Mato Grosso until the 1930s it was generally measured at 3,600 hectares.

52. Vieira, "A pecuária," in *Grande Região Centro-oeste*, 193–196; Brazil, *Projeto Radambrasil, vol. 28*, 314; Estevão Mendonça, *Quadro chorografico de Matto-Grosso* (Cuiabá: n.p., 1906), 8; *Commercial Almanach "Matto-Grossense"* (São Paulo: 1916), 179–180; Kuhlmann, "A Vegetação de Mato Grosso," 103–104; Simões Corrêa, "Pecuária," 7–8.

CHAPTER 2. ESTABLISHING ROOTS

1. The Paraguayan War (1864–1870) was the most destructive nineteenth-century war in South America, killing over half the Paraguayan population and most adult males. The impact on Paraguay resonates loudly in that country today, and the war had a transformative effect on Mato Grosso, as we will see. Among many studies, perhaps the most useful in English are Thomas Whigham, *The Paraguayan War*, vol. 1 (Lincoln: University of Nebraska Press, 2002); Hendrik Kraay and Thomas Whigham, eds., *I Die with My Country: Perspectives on the Paraguayan War, 1864–1870* (Lincoln: University of Nebraska Press, 2004); and Bridget M. Chesterton, *The Grandchildren of Solano López: Frontier and Nation in Paraguay, 1904–1936* (Albuquerque: University of New Mexico Press, 2013).

2. Uacury Ribeiro de Assis Bastos, "Expansão territorial do Brasil colonia no Vale do Paraguai (1767–1801)," *Boletím No. 4, Departamento de História No. 3, Curso*

de História da América Colonial No. 1 (São Paulo: Universidade de São Paulo, 1978), 124–135, 132–133; David M. Davidson, "How the Brazilian West Was Won: Freelance and State on the Mato Grosso Frontier, 1737–1752," in *Colonial Roots of Modern Brazil: Papers of the Newberry Library Conference*, ed. Dauril Alden, 61–106 (Berkeley: University of California Press, 1973); John Hoyt Williams, "The Undrawn Line: Three Centuries of Strife on the Paraguayan-Mato Grosso Frontier," *Luso-Brazilian Review* 17 (1980): 19–20; Paulo Marcos Esselin, *A Pecuária Bovina: No Processo de Ocupação e Desenvolvimento Econômico do Pantanal Sul-Mato-Grossense, 1830–1910* (Dourados, MS: Editoria da UFGD, 2011), chapter 1. The Spaniards referred to the Guaikuru as Mbayá; the Portuguese called them Guaicuru and later Caduveo (today Kadiwéu).

3. Esselin, *A Pecuária Bovina*, 102–103; J. Lucídio N. Rondon, *Tipos e aspectos do Pantanal* (São Paulo: n.p., 1972), 58; João Batista de Souza, *Evolução Histórica sul Mato Grosso* (São Paulo: Organização Simões, [1961]), 108; Dolor F. Andrade, *Mato Grosso e a sua pecuária* (São Paulo: Universidade de São Paulo, 1936), 5; Lécio Gomes de Souza, "Retrospectiva histórica do Pantanal," in *Anais do Primeiro Simpósio sobre Recursos Naturais e Socio-Econômicos do Pantanal*, Corumbá, 28 November to 4 December 1984 (Brasília: Ministério da Agricultura, EMBRAPA, 1986), 200–201.

4. Virgílio Alves Corrêa Filho, *Pantanais Matogrossenses (Devassamento e Ocupação)* (Rio de Janeiro: IBGE, 1946), 68–70. As an illustration, in the 1840s the ranch served as a welcome exile for Francisco Sabino Alves da Rocha Vieira, the leader of the Sabinada revolt of 1837–1838 in Bahia. It was later renamed Descalvados.

5. Raul Silveira de Mello, *Corumbá, Albuquerque e Ladário* (Rio de Janeiro: n.p., 1966), 81–82; Luiz D'Alincourt, "Resultado dos trabalhos e indagações statisticas da Província de Matto-Grosso," *Anais da Biblioteca Nacional* 3 (1877–1878): 99–100, 238, 254; Estevão de Mendonça, *Datas Mato-grossenses*, vol. 1 (Niterói: Esc. Typ. Salesiana, 1919), 293.

6. Fazendas reais became *fazendas nacionaes* with independence. They were established in a number of remote regions, most notably Mato Grosso, northern Amazonia along what is today the Guyana border, and Piauí.

7. Luiz D'Alincourt, "Resultado," 256; Joaquim Ferreira Moutinho, *Notícia sobre a província de Matto Grosso seguida d'um roteiro da viagem da sua capital á S. Paulo* (São Paulo: Typographia de Henrique Schroeder, 1869), 272; Armen Mamigonian, "Inserção de Mato Grosso ao mercado nacional e a gênese de Corumbá," *Géosul* 1 (May 1986): 45, 57 n. 8; Luiz D'Alincourt, *Memória sobre a viagem do Porto de Santos à Cidade de Cuiabá* (1825; reprint, São Paulo: Livraria Itatiaia Editora, 1975), 169–175; interrogation of Brazilian deserter Lourenço Ribeyros by the Paraguayan Army, Villa de Concepción, 1825–1826, Colección Rio Branco I-29, 34, 22, cat. 221, doc. 12, Archivo Nacional de Asunción (hereafter ANA). Ribeyros said he had been foreman (*capataz*) of one of the Miranda ranches from 1814 to 1819; Matto Grosso, *Relatorio do Exm. Snr. Coronel Barão de Maracajú, Presidente da Provincia de Matto Grosso, 5 de Dezembro de 1879* (Cuiabá: Typ. Official, 1879), 42–43.

8. Nelson Werneck Sodre, *Oeste: Ensaio sobre a grande propriedade pastoril* (Rio de Janeiro: Livraria José Olympio Editora, 1941), 82 n. 15.

9. David Wood, "An Artificial Frontier: Brazilian Military Colonies in Southern Mato Grosso, 1850–1867," *Proceedings of the Pacific Coast Council on Latin American Studies* 3 (1974): 95–96.

10. David Wood, "An Artificial Frontier," 95–108.

11. Letters from the commandants of Dourados and Miranda colonies, and between the Brazilian Minister in Asunción and the Paraguayan Minister of Foreign Relations, 1862, nos. 1–4, Colección Rio Branco I-30, 8, 52, cat. 1577, ANA.

12. J. Barbosa Rodrigues, *História de Mato Grosso do Sul* (São Paulo: n.p., 1985), 55–57, 107; letter from Alfredo D'Escragnolle Taunay to his family, Coxim, 23 December 1865, in *Cartas da campanha de Matto Grosso (1865 a 1866) por o Visconde de Taunay*, ed. Afonso D'Escragnolle Taunay (São Paulo: n.p., [1942]), 142–143.

13. J. Barbosa Rodrigues, *História de Campo Grande* (São Paulo: n.p., 1980), 16–17; Emilio G. Barbosa, *Os Barbosas em Mato Grosso: Estudo histórico* (Campo Grande: n.p., 1961), 33.

14. Virgílio Alves Corrêa Filho, "Terras devolutas: Evolução do processo de adquiril-as em Matto-Grosso," *Revista do Instituto Histórico de Matto-Grosso* 3 (1921): 65–69; J. R. de Sá Carvalho, "Memorias de sertanista Joaquim Francisco Lopes; O Povoamento do Sul de Matto Grosso, Centenario en Matto Grosso dos Barbosas—dos Lopes—dos Garcias, 1829–1929," *Diario do Sul* (Campo Grande) 29 December 1929, 1–3; Sodre, *Oeste*, 65 n. 8, 79–85; Astolpho Rezende, *O Estado de Matto-Grosso e as suppostas terras do Barão de Antonina* (Rio de Janeiro: Papeleria Sta. Helena, S. Monteiro & Cia., 1924).

15. Epifânio Candido de Sousa Pitanga, "Da viagem do porto do Jatahi à villa de Miranda," *Revista Trimensal do Instituto Historico, Geographico e Ethnographico do Brazil* (Rio de Janeiro) 27 (1864): 188–190; Sodre, *Oeste*, 81 n. 13; report from commander of Paraguayan forces in Santa Gertrudis, Martin Urbieta, 5 January 1865, Colección Rio Branco I-30, 11, 74, cat. 3441, ANA; report from commander of Paraguayan forces, José Alvarenga, Miranda, 9 February 1865, Colección Rio Branco I-30, 25, 22 cat. 3603, ANA; Afonso Taunay, ed., *Cartas*, 162 (letter from Alfredo Taunay, Morros, "close to the Aquidauana," 4 June 1866).

16. Corrêa Filho, *Pantanais Matogrossenses*, 69–70; Renato Alves Ribeiro, *Taboco, 150 anos: Balaio de Recordações* (Campo Grande: n.p., 1984), 26.

17. Corrêa Filho, *Fazendas de gado no pantanal Matogrossense*, Documentário da vida rural, no. 10 (Rio de Janeiro: Ministério da Agricultura, 1955), 22; João Leite de Barros, "A Pecuária em Corumbá," in *Anuário de Corumbá, 1939*, 8–11; Corrêa Filho, *Pantanais Matogrossenses*, 69, n. 150, 70; Moutinho, *Noticia*, 247; declaration of prisoner André Alves to the Paraguayan military on the condition of route and Brazilian troops between Miranda and S. Lourenço, n.d., Colección Rio Branco I-29, 27, 4, cat. 3392, ANA; letter from Venancio López, Minister of War, to commander of Paraguayan forces, Asunción, 2 March 1865, Colección Rio Branco I-30, 6, 25, cat. 3681, ANA; various reports, Colección Rio Branco I-30, 17, 1865–1866,

ANA; Visconde de Taunay, *Em Matto Grosso invadido, 1866–1867* (São Paulo: Cia. Melhoramentos, [1929]), 121. Mal das cadeiras, sometimes called peste de cadeiras, was first seen in Mato Grosso in the early 1850s and was soon endemic to the region. The impact of the disease on the raising of horses in Mato Grosso is discussed at length in chapter 6.

18. Virgílio Corrêa Filho, *A propósito do Boi pantaneiro*, Monografias Cuiabanas, vol. 6 (Rio de Janeiro: Pongetti e Cia., 1926), 20–21; Luiz d'Alincourt, "Resultado," *Anais da Biblioteca Nacional* 3 (1877–1878): 254–258, and 8 (1880–1881): 67.

19. José Jorge da Silva, "Communicado: O commercio de gado," *Correio da Tarde* (Rio de Janeiro) 4 (December 15, 1858): 1–2, Doc. Arm. 5, Gav. 3, No. 42, Instituto Histórico e Geográfico Brasileiro, Rio de Janeiro (hereafter IHGB). The réal, the standard Brazilian currency until the 1940s, was divided into 1,000 reis (1 milreis, US$1,000) and 1,000 milreis (1 conto). The milreis was worth approximately one US dollar at the time. Throughout this study, exchange rates are calculated using: Brazil, IBGE, Séries estatísticas retrospectivas, vol. 1, *Reportório Estatístico do Brasil (Separata do Anuário Estatístico do Brasil, Ano V, 1939–1940* (Rio de Janeiro: IBGE, 1986), 63; and United States, Federal Reserve, *Banking and Monetary Statistics* (Washington, D.C.: Federal Reserve System, 1943), 681. David McCreery reports that for Goiás the local price for cattle was lower, around 10 to 12 milreis. Given the relatively easier proximity to market in Minas for Goiás cattle, prices cited for Mato Grosso seem high and probably reflect a particular point in time. David McCreery, *Frontier Goiás, 1822–1889* (Stanford: Stanford University Press, 2006), 148, table 5.3. A valuable study of slaughterhouses in Rio during a slightly later period is found in Maria-Aparecida Lopes, "Struggles over an 'Old, Nasty, and Inconvenient Monopoly': Municipal Slaughterhouses and the Meat Industry in Rio de Janeiro, 1880–1920s," *Journal of Latin American Studies* 47:2 (May 2015): 349–376.

20. David Wood, "An Artificial Frontier," 98; Moutinho, *Noticia*, 33–35.

21. Various letters, Colección Rio Branco: October 1859, I-29, 25, 22, cat. 1348; 5 November 1859, I-29, 32, 9, cat. 1378, 1379; 17 October 1859, I-30, 26, 3, cat. 1370, ANA.

22. J. Bach, "Datos sobre los indios Terenas de Miranda," *Anales de la Sociedad Científica Argentina* (Buenos Aires) 82 (1916): 87; Virgílio Corrêa Filho, *A propósito do Boi pantaneiro*, 35–37; Moutinho, *Noticia*, 273.

23. See several military reports from the field, Colección Rio Branco: 9 April 1865, I-30, 15, 176, cat. 3794; 27 April 1865, I-30, 17, 43, cat. 3750 (no. 6); 7 September 1865, I-30, 17, 54, cat. 3976; 18 April 1866, I-30, 11, 56, cat. 4225, ANA.

24. Visconde de Taunay, *Em Matto Grosso invadido, 1866–1867*, 60–63, 120–121; letter from Caetano da Silva Albuquerque to the president of Mato Grosso, 1 March 1870, Miranda, Documentos avulsos lata 1870-C, Arquivo Público do Estado de Mato Grosso, Cuiabá (hereafter cited as APMT). It is a notable coincidence that this letter was dated the same day the Paraguayan dictator, Mariscal López, was killed at Cerro Corá in northern Paraguay near the present international frontier. His death ended the bloody five-year conflict.

25. Letter from Commander André da Costa Leite d'Almeida to the Paraguayan Commander of the Alto Paraguay, Colonel Vicente Barrios, 3 March 1865, Corumbá, Colección Rio Branco I-30, 17, 61, cat. 4640, ANA.

26. Ribeiro, *Taboco, 150 anos*, 46–49.

27. For more on the contribution of commerce and the production of erva and rubber to the Mato Grosso economy during the period, see Zephyr Lake Frank, "The Brazilian Far West: Frontier Development in Mato Grosso, 1870–1937," Ph.D. diss., University of Illinois, 1999, chapters 3 and 4.

28. Moutinho, *Noticia*, 9, 95–99, 113–117, 118–119; João Severiano da Fonseca, *Viagem ao Redor do Brasil, 1875–1878* (1880; Rio de Janeiro: Biblioteca do Exército Editora, 1986), vol. 1, 13; Brazil, IBGE, Séries estatísticas retrospectivas, vol. 1, 14.

29. Mato Grosso, *Relatório apresentado à Assemblêa Legislativa da Provincia de Matto-Grosso, 4 de outubro de 1872, pelo Presidente, Tenente Coronel Dr. Francisco José Cardozo Junior* (Rio de Janeiro: Typ. Official, 1873), 86–90. All presidential reports are found on microfilm in the Núcleo de Documentação e Informação Histórica Regional, Universidade Federal de Mato Grosso, Cuiabá (NDIHR); reports between 1830 and 1930 are available online at the Center for Research Libraries (http://www-apps.crl.edu/brazil/provincial/mato_grosso). Virgílio Alves Corrêa Filho, *História de Mato Grosso* (Rio de Janeiro: Ministério da Educação e Cultura, 1969), 553; Lúcia Salsa Corrêa, "Corumbá: O Comércio e o Casario do Porto (1870–1920)," in *Casario do Porto de Corumbá*, ed. Valmir Batista Corrêa, Lúcia Salsa Corrêa, and Luiz Gilberto Alves (Campo Grande: Fundação de Cultura de Mato Grosso do Sul; Brasília: Gráfica do Senado, 1985), 34.

30. The total civilian population of the town rose to 8,600 persons by 1876, with another 3,000 in the neighboring shipyard town of Ladário, fully one-half Paraguayans. Ironically, in order to avoid social disruption after the 1878 recession, free passages were again offered to Paraguayans who lost their jobs, this time downriver to Asunción. Michael G. Mulhall, *Journey to Matto Grosso* (Buenos Aires, n.p., 1876), 15–16; Fonseca, *Viagem ao Redor do Brasil, 1875–1878*, 9–10, 13, 317; *A Opinião* (Corumbá) 1 (Nov. 17, 1878): 1–2; J. J. Ramos Ferreira, Juiz de Direito de Corumbá, "Report," January 26, 1879, Corumbá, Documentos avulsos lata 1879-B, APMT. For a more in-depth discussion of this immigration and the role of Paraguayans in Mato Grosso, see Robert Wilcox, "Paraguayans and the Making of the Brazilian Far West, 1870–1935," *The Americas* 49, no. 4 (April 1993): 479–512.

31. Mato Grosso, *Relatório com que o exm. snr. Dr. João José Pedrosa, pres. da provincia de Matto Grosso abrio a 2ᵃ legislatura da respectiva assemblea, em 1 de outubro, 1879* (Cuiabá: n.p., 1879), 93–97; report by João Caetano Teixeira Muzzi, director of Brilhante military colony, to the president of Mato Grosso, Brilhante, April 12, 1882, Documentos avulsos lata 1882-F, APMT.

32. Letter from Bernardo Vasques, Ministerio dos Negocios da Guerra, to the President of Mato Grosso, Rio de Janeiro, January 14, 1896, Documentos avulsos lata 1896-D, APMT.

33. *Relatório apresentado à assembléia da legislativa provincial de Matto Grosso no*

dia 3 de maio de 1874, pelo presidente da provincia, Snr. General Dr. José de Miranda da Silva Reis (Cuiabá: n.p., 1874), Annexo 4, A; "Industria pastoril," *Relatório do vice-presidente Dr. José Joaquim Ramos Ferreira devia apresentar à Assembléia Legislativa Provincia de Matto Grosso, 2ª sessão da 26ª legislatura de Setembro de 1887* (Cuiabá: n.p., 1887). These figures should not be accepted at face value but taken only as an indication, since accounting at the time was highly inefficient and local corruption commonplace. The milreis was worth about US$0.49 in 1871 and US$0.38 in 1885. Brazil, IBGE, Séries estatísticas retrospectivas, vol. 1, 63; and US, *Banking and Monetary Statistics*, 681.

34. *Colletoria Provincial de Corumbá, 1875*, Coletoria Corumbá, caixa 1848–1878, 134a-137b, APMT; *Colletoria de Corumbá: Impostos de exportação e consumo, 1879–80*, Coletoria Corumbá, caixa 1878 a 1880, 63a-72a, APMT; *Colletoria da Cidade de Corumbá, 1883*, Coletoria Corumbá, caixa 1883–1884, 1a-9b, APMT. These figures do not include jerky and hides exported from operations farther upriver. Certainly cattle were slaughtered for local consumption as well. Unfortunately, no consistent runs of figures for these cattle have been found for any municipality, making speculation difficult. One can reasonably assume that the slaughter of animals for their hides contributed to local consumption, but it is not clear how much meat the average Matogrossense ate during the period. At the same time, tax collection was uncertain over these years, making it difficult to document definite trends. The general political and economic upheaval of the 1880s and the republican coup of 1889 and subsequent transfer of many fiscal powers to the regions (political designation also changed, from provinces to states), contributed to the fallibility of tax figures. More reliable collection only began in the mid-1890s.

35. Brazil, IBGE, Séries estatísticas retrospectivas, vol. 1, 14.

36. Corrêa Filho, *História de Mato Grosso*, 687; Brazil, IBGE, Séries estatísticas retrospectivas, vol. 1, 120, 124; Virgílio Corrêa Filho, *Á sombra dos hervaes matogrossenses*, Monografias Cuiabanas, vol. 4 (São Paulo: São Paulo Editora, 1925), 15 n. 8, 18, 19–20 n. 14, 15. Brazil went through a period of economic and financial recession after a speculative boom went bust in 1891; the overproduction of coffee caused a serious decline in the international price of the bean by the turn of the century. The result was a retraction of earnings and subsequent economic activity, and the milreis was devalued from US$0.46 in 1890 to US$0.19 in 1900.

37. Valmir Batista Corrêa, *Coronéis e bandidos em Mato Grosso (1889–1943)*, 2nd ed. (Campo Grande: Editora UFMS, 2006), 39–40, 57–72; report from Antonio Canale, mayor of Miranda, Miranda, December 31, 1895, Documentos avulsos lata 1895-D, APMT; Mato Grosso, *Mensagem apresentado à Assemblea Legislativa em 1º de Fevereiro de 1897 pelo Dr. Antonio Corrêa da Costa, Presidente do Estado* (Cuiabá: Typ. do Estado, 1897), 4–6.

38. Corrêa Filho, *História de Mato Grosso*, 592–598; Mato Grosso, *Mensagem do Presidente do Estado, Cel. Antônio Pedro Paes de Barros, 14 de novembro de 1901* (Cuiabá: n.p., 1901), 10–11; *Mensagem . . . Paes de Barros . . . 1904*, 15; *Mensagem . . . Paes de Barros . . . 1905*, 5–8, 25.

39. Brazil, IBGE, Séries estatísticas retrospectivas, vol. 1, 124; Corrêa Filho, *História de Mato Grosso*, 691–693; Brazil, IBGE, "Tomo 1, Indústria Extrativa: Secção 1, Reino Vegetal," in Séries estatísticas retrospectivas, vol. 2, *O Brasil, suas riquezas naturais, suas indústrias* (1907), 21–22, 47–48 (Rio de Janeiro: IBGE, 1986); Mato Grosso, *Mensagem do Presidente de Mato Grosso, Dom Francisco de Aquino Corrêa, 7 de setembro de 1919* (Cuiabá: n.p., 1919), 112–114; S. Cardoso Ayala and Feliciano Simon, *Album Graphico do Estado de Matto Grosso*, EEUU do Brasil (Corumbá/Hamburg: n.p., 1914), 101–102.

40. Mato Grosso, "Recenseamento feito no ano de 1872," Documentos avulsos lata 1872-F, APMT; Brazil, Ministério da Indústria, Viação e Obras Publicas, Directoria Geral de Estatística, *Synopse do recenseamento de 31 de dezembro de 1890* (Rio de Janeiro: Oficina da Estatistica, 1898); Cardoso Ayala and Simon, *Album Graphico*, 414–420.

41. "Industria pastoril," *Relatório . . . Ramos Ferreira . . . 1887* (Cuiabá: n.p., 1887). For these early years estimates are unavoidably speculative, but at least they offer some indication of numbers and production. The 20-percent growth rate and 10-percent export and consumption rates (10-percent growth overall) are based on sources that use these figures throughout the period. As confirmation, export and consumption figures for 1875 from the provincial tax collector in Corumbá reveal the export of more than 32,000 hides, some from animals consumed in the city. At a 10 percent rate, this would indicate a total of 320,000 head at the time. Extrapolating back five years, this rate suggests some 200,000 animals in 1870 and is consistent with the same rate up to the 1885 estimate of 800,000 head. *Colletoria Provincial de Corumbá, 1875*, Coletoria Corumbá, caixa 1848–1878, 111a–120b, 134a–137b, APMT.

42. Mato Grosso, *Relatório . . . Barão de Maracajú . . . 1879*, 42–43; report by José [illegible], Contador da Thesouraria da Fazenda, to the President of Mato Grosso, Cuiabá, July 14, 1883, Documentos avulsos lata 1883-D, APMT.

43. Armen Mamigonian, "Inserção de Mato Grosso ao mercado nacional e a gênese de Corumbá," *Geosul* (UFSC, Florianópolis) 1 (May 1986): 48–49. Export taxes in 1883 were 400 reis per hide (US$0.17) and 20 reis per kilo of dried beef or beef bouillon (US$0.01); *Colletoria da Cidade de Corumbá, 1883*, 1a–9b, APMT.

44. Francisco Antônio Pimenta Bueno, *Memória justificativa dos trabalhos de que foi encarregado à Provincia de Matto Grosso, segundo as instrucções do Ministério da Agricultura de 27 de maio de 1879* (Rio de Janeiro: Typ. Nacional, 1880), 85–86; Corrêa Filho, *A propósito do Boi pantaneiro*, 38 n. 48. Jacobina was said to have 60,000 head in 1828. Allowing for a population increase of 10 percent per annum, this would have meant there were 600,000 head by 1850 and well over 5 million in 1879! This example underlines the difficulty of making population estimates for the early years. I stick by the more modest count of the total Mato Grosso herd in 1870, though not including Jacobina, which likely supported a more believable (but still substantial) 150,000 head in 1880, as reported by Jaime Cibils: "Jaime Cibils y Bu-

xareo," n.p., 10. This source is an unpublished pamphlet produced by the Cibils family.

45. Pimenta Bueno, *Memória*, 85–86; Virgílio Corrêa Filho, *A propósito do Boi pantaneiro*, 39; Rafael del Sar, *Persecución de Florencio Garay a Rafael del Sar de 1870 a 1878. En Asunción del Paraguay, Villa Concepción, Corumbá, y Cuiabá* (Corumbá: Typographia do Iniciador, 1879), 1–2, 4–5, 10 (No. 1739, Colección Enrique Solano López, Biblioteca Nacional de Asunción [hereafter BNA]); Domingos Sávio da Cunha Garcia, *Território e Negócios na 'Era dos Impérios': Os Belgas na Fronteira Oeste do Brasil* (Brasília: Fundação Alexandre de Gusmão, 2009), 88. Del Sar, whose business required that he frequently travel between his plant and Montevideo, suffered constant harassment from a former partner in Asunción. The conflict, described in a pamphlet published by del Sar, went on until at least 1878. His departure may have been because he was simply worn down by the litigation.

46. "Jaime Cibils y Buxareo," 10; Hilgard O'Reilly Sternberg, "Presencia catalana en Matogrosso, Brasil," paper presented to the International Conference on Demographic History, July 1988, Barcelona, Spain, 2; Hilgard O'Reilly Sternberg, "Tentativas expansionistas Belgas no Brasil: O Caso 'Descalvados,'" *Revista do Instituto Histórico e Geográfico de Mato Grosso* (Cuiabá) 119–120 (1983): 49, 54 n. 14, 15; Eddy Stols, "O Brasil se defende da Europa: Suas relações com a Bélgica (1830–1914)," *Boletín de Estudios Latinoamericanos y del Caribe* 18 (June 1975): 69–70.

47. "Jaime Cibils Buxareo," 11–12; *Colletoria da Cidade de Corumbá, 1883*, Coletoria Corumbá, caixa 1883 à 1884, 1a, 9b, APMT; Mato Grosso, *Relatório... Ramos Ferreira... 1887*, n.p.; Tesouraria da Fazenda Nacional am Mato Grosso, Alfândega de Corumbá, Capatazia, Guias de Exportação, 1870–1894, ano 1884, microfilm roll 01, NDIHR; Mato Grosso, *Relatório apresentado à Assembléia Legislativa da Provincia de Matto Grosso no 1ª sessão da 26ª legislatura, 12 de julho de 1886, pelo Presidente da Provincia Dr. Joaquim Galdino Pimentel* (Cuiabá: Typ. da Situação, 1886), 37–38.

48. *Relatório... Galdino Pimentel... 1886*, 38.

49. "Indústria fabril," *Relatório... Ramos Ferreira... 1887*, n.p.; *Alfândega de Corumbá... 1870–1894*, anos 1890, 1895, NDIHR.

50. Cardoso Ayala and Simon, *Album Graphico*, 293–294; report by Luiz Felippe Cuiabano, captain of a commission to provide food aid to inhabitants of the Military District of Mato Grosso, to the President of Mato Grosso, June 8, 1888, Cuiabá, Documentos avulsos lata 1888-A, APMT; "Indústria pastoril," *Relatório... Ramos Ferreira... 1887*, n.p.

51. Fonseca, *Viagem ao Redor do Brasil, 1875–1878*, 369; "Collectoria das rendas do Estado na Cidade de Santa Cruz de Corumbá, Diversos Impostos, 1891," 1a–9b, book 2, *Coletorias avulsas*, APMT. For more on Cibils's complex involvement in Mato Grosso political intrigue of the early 1890s and the title transfer of Descalvados, see Cunha Garcia, *Território e negócios*, 91–117, 121–132. Garcia's overall study of Belgian interests in Mato Grosso is a fascinating examination of extraordinary imperial pretensions.

52. Stols, "O Brasil se defende," 69–71; Eddy Stols, "Les Belges au Mato Grosso et en Amazonie, ou la Récidive de l'Aventure Congolaise (1895–1910)," in *La Belgique et l'Étranger aux XIXe et XXe Siècles*, ed. Michel Dumoulin and Eddy Stols (Louvain-la-Neuve: n.p., 1987), 82–84, 87, 89; letter from José Duarte da Cunha Pontes, mayor of Cáceres, Cáceres, January 3, 1896, Documentos avulsos lata 1896-D, APMT. Throughout the period, the Belgian franc maintained a level exchange rate of 5.19 to 5.22 francs to the dollar. *The Times* (London), July 1, 1895: 13; *The Times* (London), July 1, 1899: 6; *New York Daily Tribune* (New York), July 1, 1911: 14.

53. Franz Van Dionant, *Le Rio Paraguay et l'État Brésilien de Matto-Grosso* (Brussels: L'Imprimerie Nouvelle, 1907), 126; Stols, "Les Belges," 82–83, 94–95, 101–103; "Collectoria das Rendas estadoaes na Cidade de Corumbá, escripturação do imposto de exportação, 1895," *Coletorias avulsas*, APMT; "Collectoria do Estado na Cidade de Corumbá, Impostos de exportação, 1900," 1a-28a, *Coletorias avulsas*, APMT; "Mesa das Rendas de Corumbá, 1905," 2a-34a, *Coletorias avulsas*, APMT; Alfândega de Corumbá, Capatazia, Guias de Exportação, Importação, e Reexportação, 1895–1909, anno 1905, microfilm roll 01, NDIHR; Sternberg, "Tentativas," 50, 55 n. 21.

54. Stols, "Les Belges," 82–83, 107; Stols, "O Brasil se defende," 71–73; Sternberg, "Tentativas," 49, 52. Cunha Garcia, *Território e Negócios*, 133–135. In his monograph on Mato Grosso, cited previously, Van Dionant made no reference to his relations or problems while at Descalvados, restricting himself to observations on conditions in Mato Grosso.

55. Letter from T. G. Chittenden, General Manager of Brazil Land, Cattle and Packing Company, through the British Embassy in Rio de Janeiro, to the President of Brazil, São Paulo, August 9, 1920, Documentos avulsos lata 1920-C, APMT; J. F. de Mello Nogueira, *Excursão a Matto Grosso: Artigos publicados no Commercio de S. Paulo* (São Paulo: Typ. do Commercio de S. Paulo, 1915), 13–15; Mato Grosso, *O Município de Campo Grande, Estado de Matto Grosso* (n.p.: Publicação official, 1919), 57–59; Campo Grande, Matto Grosso, *O município de Campo Grande em 1922, Relatório do anno de 1922, Apresentado á Camara Municipal pelo Intendente Geral Dr. Arlindo de Andrade Gomes* (São Paulo: Cia. Melhoramentos, 1923), 46, 48; Brazil, Ministério da Agricultura, Indústria e Commercio, Serviço de inspecção e fomento agrícolas, *Estudo dos factores da Producção nos Municípios Brasileiros e condições economicas de cada um: Estado de Matto Grosso, Município de Campo Grande* (Rio de Janeiro: Imprensa Nacional, 1929), 32.

56. Charles A. Gauld, *The Last Titan: Percival Farquhar, American Entrepreneur in Latin America* (Stanford: Institute of Hispanic American and Luso-Brazilian Studies, 1964), 209, 220; note on letter from Storey, Thorndike, Palmer and Dodge attorneys, to US Secretary of State Robert Lansing, Boston, Massachusetts, Dec. 16, 1916, in Box 15, "Notes of Charles Gauld relative to Farquhar," Percival Farquhar Papers, Manuscripts and Archives, Yale University Library, No. 205; Mato Grosso, *Mensagem apresentado pelo Dr. Joaquim A. da Costa Marques,*

Presidente do Estado de Mato Grosso à Assemblea Legislativa, 13 de maio de 1914 (Cuiabá: Typ. Official, 1914), 76–77; "Meza de Rendas de Corumbá, Despachos de exportação, 1910," book 1: 1a-26b, book 2: 1a-5b, *Coletorias avulsas,* APMT. For more detail of the Farquhar operations and the environment, see Robert W. Wilcox, "Ranching Modernization in Tropical Brazil: Foreign Investment and Environment in Mato Grosso, 1900–1950," *Agricultural History* 82, no. 3 (summer 2008): 366–392.

57. Miguel Arrojado Ribeiro Lisboa, *Oeste de S. Paulo, Sul de Mato-Grosso: Geologia, Indústria, Mineral, Clima, Vegetação, Solo Agrícola, Indústria Pastoril* (Rio de Janeiro: Typ. do Jornal do Commercio, 1909), 145–150; Corrêa Filho, *Pantanais Matogrossenses,* 69 n. 150; Lécio Gomes de Souza, "Retrospectiva histórica do Pantanal," in *Anais do Primeiro Simpósio sobre Recursos Naturais e Socio-Econômicos do Pantanal,* Corumbá, 28 November to 4 December 1984 (Brasília: Ministério da Agricultura, EMBRAPA, 1986), 201.

58. "Indústria pastoril," *Relatório . . . Ramos Ferreira . . . 1887,* n.p.; Rodolpho Endlich, "A criação do gado vaccum nas partes interiores da America do Sul," *Boletim da Agricultura* 4, no. 3 (1903): 128–129; Matto Grosso, *O Estado de Matto-Grosso (Brazil): Notas e apontamentos uteis aos immigrantes e industrias europeus* (Cuiabá: Typ. Official, 1898), 20; Marshall Saldanha Moreira, *De Juiz de Fora a Cuiabá e sul de Matto Grosso (Impressões do viagem)* (Juiz de Fora: n.p., 1914), 90; Brazil, Ministério da Agricultura, Commercio e Obras Públicas, *Directoria da Agricultura, 1890–1892,* letters of November 3, December 6, 11, 12, 28, 1888, GIFI, 5F, caixa 602, Arquivo Nacional de Rio de Janeiro (hereafter ANRJ); Virgílio Alves Corrêa Filho, "Terras devolutas," 65–69; "Destino Para O Nabileque," *Terra e gente* 2 (January 1957): 78–80.

59. Emílio G. Barbosa, *Esbôço Histórico e Divagações sôbre Campo Grande* (Campo Grande: Tipografia Pindorama, 1964), 5–13; report from Captain Rogaciano Monteiro de Lima, director of Colonia Dourados, to the president of Mato Grosso, Dourados, February 5, 1882, Documentos avulsos lata 1882-F, APMT; João Batista de Souza, *Evolução Histórica sul Mato Grosso,* 89–144.

60. *Relatório apresentado á Assembléa Geral dos Accionistas do Banco Rio e Matto Grosso, 1895, pelo Presidente Dr. Joaquim Duarte Murtinho* (Rio de Janeiro: n.p., 1895), 11, Folleto No. 607, BNA; *Relatório apresentado á Assembléa Geral dos Srs. Accionistas da Companhia Matte Larangeira, 1897, pelo Presidente Dr. Francisco Murtinho* (Rio de Janeiro: n.p., 1897), 6–7, 12–13, Folleto No. 611, BNA; *Relatório apresentado á Assembléa Geral dos Srs. Accionistas da Companhia Matte Larangeira, 1898* (Rio de Janeiro:, n.p., 1898), 10–11, Folleto No. 612, BNA; Banco Rio e Matto Grosso, *Relatório apresentado à Assembléa geral dos accionistas, na sua reunião ordinaria de 1897, pelo vice-presidente Dr. Manoel Martins Torres* (Rio de Janeiro: Typ. Leuzinger, 1897), 7–8; report from Olympio Lima, customs agent in Ponta Porã, to the president of Mato Grosso, Ponta Porã, December 15, 1897, Documentos avulsos lata 1897-C, APMT; report from João Fernandes de Lima, customs agent in Ponta Porã, to the president of Mato Grosso, Ponta Porã, August 15, 1904, Documentos avulsos lata 1904-B, APMT; Mario Monteiro de Almeida, *Episódios históricos da formação geográfica do*

Brasil: Fixação das raias com o Uruguai e o Paraguai (Rio de Janeiro: Pongetti, 1951), 420 n. 142; Arsenio López Decoud, *Album Gráfico de la República del Paraguay* (Buenos Aires: n.p., 1911), XLV-XLVI, XCVII. For those readers familiar with yerba mate, Larangeira continued in the Argentine Cruz de Malta brand until a recent takeover by Buenos Aires food giant Molinos.

61. Barbosa Rodrigues, *História de Mato Grosso do Sul*, 135–146; Virgílio Corrêa Filho, *Ervais do Brasil e ervateiros*, Documentário da vida rural no. 12 (Rio de Janeiro: Ministério da Agricultura, Serviço de Informação Agrícola, 1957), 50–53; report by Hugo Heyn, Mate Larangeira superintendent, to the president of Mato Grosso, Concepción, November 11, 1897, Documentos avulsos lata 1897-C, APMT.

62. Letter from Hugo Heyn, Mate Larangeira superintendent, to the vice-president of Mato Grosso, Asunción, July 15, 1898, Documentos avulsos lata 1898-B, APMT; letter from João Fernandes de Lima, erva customs agent in Ponta Porã, to the president of Mato Grosso, Ponta Porã, December 19, 1903, Documentos avulsos lata 1903-B, APMT; Corrêa Filho, *Á sombra dos hervaes matogrossenses*, 44–49 n. 47–50. For more on Larangeira, see Gilmar Arruda, *Frutos da terra: Os trabalhadores da Matte-Larangeira* (Londrina, Brazil: Ed. da Universidade Estadual de Londrina, 1997).

63. Lisboa, *Oeste de S. Paulo*, 141–145.

64. Letter from the Santana do Paranaíba municipal council to the president of Mato Grosso, S. Anna do Paranahyba, October 10, 1881, Documentos avulsos lata 1881-D, APMT. Another letter revealed that mail had not been received from Cuiabá for six months. Letter from the local vicar of Santana do Paranaíba to the secretary of the government in Cuiabá, Sant'Anna do Paranahyba, January 1, 1884, Documentos avulsos lata 1884-B, APMT.

65. Cattle census for Santana do Paranaíba, sent to the president of Mato Grosso, signed by J. A. de Salles Fleury, Cuiabá, December 23, 1895, Documentos avulsos lata 1895-D, APMT; Brazil, Estado de Mato-Grosso, Instituto Nacional de Estatística, Diretoria de Estatística e Publicidade, *Sinopse estatística do estado, no. 2*, Separata, com acréscimos, do Anuário Estatística do Brasil, Ano III, 1937 (Cuiabá: Imprensa Oficial, 1938), 14; A. Marques, *Matto Grosso: Seus recursos naturaes, seu futuro economico* (Rio de Janeiro: Papelaria Americana, 1923), 138–139, 190–191.

66. José Jorge da Silva, "Communicado," *Correio da Tarde* (Rio de Janeiro) 4 (December 15, 1858): 1–2, Instituto Histórico e Geográfico Brasileiro (hereafter IHGB); Lopes, "Struggles over an 'Old, Nasty, and Inconvenient Monopoly'," 349–376.

67. Corrêa Filho, *Pantanais Matogrossenses*, 117; Jesuino da Silva Mello, *A pecuária no Brazil (S. Paulo e Minas Gerais)* (Rio de Janeiro: n.p., 1903), 47–49; São Paulo, Secretaria de Estado dos Negocios da Agricultura, Commercio e Obras Públicas, *Almanach para o anno de 1917* (São Paulo: Typ. do Estado, 1917), 273; Antonio Carlos de Oliveira, "Economia pecuária do Brasil Central, Bovinos (Estudo gêo-estatístico e político-econômico) I Parte," *Boletim do Departamento Estadual de Estatística* (São Paulo) 10 (October 1941): 27.

68. Mato Grosso, *Mensagem apresentado à Assembléa Legislativa em 1° de Fev. de 1896 pelo Dr. Antonio Corrêa da Costa, Presidente do Estado* (Cuiabá: Typ. do Estado, 1896), 26–27; *Mensagem do 2° vice-presidente do Estado Cel. Antonio Cesario de Figueiredo à Assembléa Legislativa em 1° de Fev. de 1899* (Cuiabá: Typ. do Estado, 1899), 24; *Mensagem . . . Alves de Barros . . . 1903*, 25; *Mensagem do Presidente de Matto Grosso, Cel. Antonio Paes de Barros à Assembléa Legislativa, 4 de março de 1905* (Cuiabá: Typ. do Estado, 1905), n.p.; Dióres Santos Abreu, "Communicações entre o sul de Mato Grosso de o sudoeste de São Paulo. O comércio de gado," *Revista de História* (São Paulo) 53 (1976): 197, 199–211; Paulo Coelho Machado, *Pelas ruas de Campo Grande, Vol. 1: A Rua Velha* (Campo Grande: Tribunal de Justiça de Mato Grosso do Sul, 1990), 95–99.

69. Mato Grosso, *Mensagem . . . Paes de Barros . . . 1905*, 18; Abreu, "Communicações," 191–214.

70. "Agencia Fiscal do Porto XV de Novembro, 1911," 1a-4b, *Coletorias avulsos*, APMT; US Department of State, Division of Latin American Affairs, Report "Matto Grosso and Its Finances," by C. R. Cameron, US consul in São Paulo, São Paulo, 24 December 1927, No. 832.51 M 43/2, 12–13, Reports of the US Consuls in Brazil, 1910–1929, Record Group 59, Microfilm M-519, roll 27, United States National Archives (hereafter USNA).

CHAPTER 3. A BOOM OF SORTS

1. Warren Dean, "Economy," in *Brazil, Empire and Republic, 1822–1930*, ed. Leslie Bethell (New York: Cambridge University Press, 1989), 232; Brazil, Fundação Instituto Brasileiro de Geografia e Estatística, Séries Estatísticas Retrospectivas, vol. 3, *Estatísticas históricas do Brasil: Séries econômicas, demográficas e sociais de 1550 a 1988* (Rio de Janeiro: IBGE, 1990), 307–312, 524.

2. Brazil, Fundação Instituto Brasileiro de Geografia e Estatística, Séries Estatísticas Retrospectivas, vol. 1, *Repertório estatística do Brasil: Quadros retrospectivos*, Separata do Anuário Estatístico do Brasil, Ano V, 1939-1940 (Rio de Janeiro: IBGE, 1986), 124; Virgílio Alves Corrêa Filho, *História de Mato Grosso* (Rio de Janeiro: Ministério da Educação e Cultura, 1969), 691–693; S. Cardoso Ayala and Feliciano Simon, *Album Graphico do Estado de Matto Grosso*, EEUU do Brasil (Corumbá/Hamburg: n.p., 1914), 102; Mato Grosso, *Mensagem dirigido à Assembléa Legislativa, 7 de Setembro de 1921, por D. Francisco de Aquino Corrêa, Bispo de Prusiade, Presidente do Estado* (Cuiabá: Typ. Official, 1921), annexo; Virgílio Corrêa Filho, *A propósito do Boi pantaneiro*, Monografias Cuiabanas, vol. 6 (Rio de Janeiro: Pongetti e Cia., 1926), 59. In 1913, the milreis was worth US$0.32, but in 1919 it had declined to US$0.26.

3. Mato Grosso, *Mensagem dirigido pelo Dr. Joaquim A. da Costa Marques, Presidente do Estado de Mato Grosso, à Assembléa Legislativa, 13 de maio de 1915* (Cuiabá: Typ. Official, 1915), 31–32; Mato Grosso, *Mensagem à Assembléa Legislativa, 7 de Setembro de 1920 por D. Francisco de Aquino Corrêa, Bispo de Prusiade, Presidente do Es-*

tado (Cuiabá: Typ. Official, 1920), 88; Brazil, Directoria Geral de Estatística, *Anuário estatística do Brasil, 1908–1912* (Rio de Janeiro: Typ. Nacional, 1912), 327–328; Brazil, Ministério da Agricultura, Commercio e Obras Públicas, Directoria Geral de Estatística, *Recenseamento realizado em 1 de Setembro de 1920, vol. 4, parte 1 "População"* (Rio de Janeiro: Typ. da Estatística, 1926), 408–409. A detailed discussion of land acquisition is in chapter 4.

4. Brazil, IBGE, *Séries estatísticas retrospectivas*, vol. 1, 69, 85, 87; Jorge Chami, "O Setor Externo Brasileiro no Século XX," *Estatísticas do Século XX* (Rio de Janeiro: IBGE, 2006), 432. One factor that should not be forgotten is the devaluation of the Brazilian currency and currency controls imposed by the federal government in 1931–1934 and 1937, aiding the export sector as Brazilian goods became considerably cheaper in the international market. See Chami, 454 n. 124.

5. Brazil, IBGE, *Séries estatísticas retrospectivas*, vol. 1, 29–45, 95; Brazil, IBGE, Recenseamento Geral do Brasil (1º de Setembro de 1940), Série Regional, Parte XXII, *Mato Grosso: Censo Demográfico, Censos Econômicos* (Rio de Janeiro: Serviço Gráfico do IBGE, 1952), 51; Brazil, IBGE, Conselho Nacional de Estatística, Serviço Nacional de Recenseamento (1950), Série Regional, Volume XXIX, *Estado de Mato Grosso: Censos Demográficos e Econômicos* (Rio de Janeiro: IBGE, 1956), 64.

6. Mato Grosso, Instituto Nacional de Estatística, Diretoria de Estatística e Publicidade, *Sinopse estatística do Estado, no. 2*, Separata, com acréscimos, do Anuário Estatística do Brasil, Ano III, 1937 (Cuiabá: Imprensa Oficial, 1938), 128.

7. Corrêa Filho, *História de Mato Grosso*, 692; Mato Grosso, "Exportação de Matto-Grosso, Decennio de 1926–1935," *Relatório do Interventor Federal de Matto-Grosso, Manoel Ary Pires, 13 de junho de 1937* (Cuiabá: Imprensa Oficial, 1937), annexo.

8. Brazil, Ministério da Agricultura, Indústria e Commercio, Directoria do Serviço de Estatística, *Synopse do censo pecuário da república pelo processo indireto das avaliações em 1912–1913* (Rio de Janeiro: Typ. do Ministério da Agricultura, Indústria e Commercio, 1914), 36; US Department of Agriculture, "South American Cattle Census," Records of the Bureau of Agricultural Economics, General Correspondence of the Bureau of Markets and Bureau of Agricultural Economics, 1912–1952, Brazil No. 39 [1918], Record Group 83, USNA. It appears this was 1918, though there was no specific notation. Brazil, IBGE, *Séries Estatísticas Retrospectivas*, vol. 1, 87; Brazil, IBGE, *Séries Estatísticas Retrospectivas*, vol. 3, 307–312, 524; US Department of State, "Livestock, 1930–193-," Records of the Foreign Agricultural Service, Narrative Agricultural Reports, 1904–1954, Brazil, entry 5, box 64, 1–3, Record Group 166, USNA. This entry was misfiled and should be 1920–1925. Also, the final date in the range is missing.

9. Agustín Ruano Fournier, *Estudio economico de la producción de las carnes del Río de la Plata* (Montevideo: Peña y Cía., 1936), 35–37, 273; Wilson de Faria, "A pecuária de Barretos (SP) e os fatores de sua implantação," *Caderno de Ciências da Terra* (Universidade de São Paulo), 41 (1973): 9–10; Charles A. Gauld, *The Last Titan: Percival Farquhar, American Entrepreneur in Latin America* (Stanford: Institute of Hispanic American and Luso-Brazilian Studies, 1964), 219, 277 n. 12. The Rio

Grande do Sul operations did not really get under way until the war ended, but they expanded rapidly in the 1920s: Stephen Bell, *Campanha Gaúcha: A Brazilian Ranching System, 1850–1920* (Stanford: Stanford University Press, 1998), 153–155.

10. US Department of State, Division of Latin American Affairs, Report "Stock Raising and Meat Packing," by A. T. Haeberle, US Consul, São Paulo, 18 February 1924, no. 832.6222/15, Reports of US Consuls in Brazil, 1910–1929, Microfilm M-519, roll 44, Record Group 59, USNA; Sandra Jatahy Pesavento, *República velha gaúcha: Charqueadas, frigoríficos, criadores* (Porto Alegre: Movimento/IEL, 1980), 125; US Department of Agriculture, Preliminary report "Markets for Pure Bred Live Stock in South America," by David Harrel and H. P. Morgan, Brazil, No. 39, [1918], 12, Record Group 83, USNA; United States, Report "Packing Houses in Operation in Brazil," by A. Ogden Pierrot, Assistant US Commercial Attaché, Rio de Janeiro, 21 August 1931, Brazil, entry 5, box 65, Record Group 166, USNA.

11. Mato Grosso, *Mensagem do Presidente de Mato Grosso, Dom Francisco de Aquino Corrêa, 7 de setembro de 1919* (Cuiabá: Typ. Official, 1919), annexo; *Mensagem dirigido pelo Dr. Caetano Manoel de Faria e Albuquerque, Presidente de Matto Grosso, à Assembléa Legislativa, 15 de maio de 1916* (Cuiabá: Typ. Official, 1916), 26–31; *Mensagem à Assembléa Legislativa, 7 de Setembro de 1920 por D. Francisco de Aquino Corrêa, Bispo de Prusiade, Presidente do Estado* (Cuiabá: Typ. Official, 1920), annexo, quadro 5; Corrêa Filho, *A propósito do Boi pantaneiro*, 55–58.

12. Brazil, Ministério da Agricultura, Indústria e Commercio, Directoria Geral de Estatística, *Estimativa do Gado existente no Brazil em 1916* (Rio de Janeiro: Typ. da Estatística, 1917), 9, 18; Mato Grosso, *Mensagem pelo D. Francisco de Aquino Corrêa, Bispo de Prusiade, Presidente do Estado de Matto Grosso, dirigida à Assembléa Legislativa, 13 de maio de 1918* (Cuiabá: Typ. Official, 1918), 42–43; Corrêa Filho, *A propósito do Boi pantaneiro*, 55–58; Brazil, Ministério da Agricultura, Industria e Commercio, Directoria Geral de Estatística, *Synopse do recenseamento realizado em 1 de setembro de 1920; população pecuária* (Rio de Janeiro: Typographia da Estatística, 1922), 8.

13. Brazil, IBGE, *Séries estatísticas retrospectivas*, vol. 1, 87; Corrêa Filho, *A propósito do Boi pantaneiro*, 56–58; Pesavento, *República velha gaúcha*, 151–174, 183–184.

14. Rodolpho Endlich, "A criação do gado vaccum nas partes interiores da America do Sul," *Boletim da Agricultura* (São Paulo) 4, no. 6 (1903): 279–280. In 1913, a yearling steer fetched 40 to 45 milreis (US$13 to $14.50), and a cow sold for 26 milreis (US$8.30). Cardoso Ayala and Simon, *Album Graphico*, 293; Paulo Coelho Machado, *Pelas ruas de Campo Grande, Vol. 1: A Rua Velha* (Campo Grande: Tribunal de Justiça de Mato Grosso do Sul, 1990), 87–88; Mato Grosso, *Mensagem dirigido à Assembléa Legislativa em 13 de maio de 1924 pelo Cel. Pedro Celestino Corrêa da Costa, Pres. de Matto Grosso* (Cuiabá: Typ. Official, 1924), 75–76; letter from J. G Pereira Lins, Ministério da Agricultura, Indústria e Commercio, to the president of Mato Grosso, Rio de Janeiro, June 14, 1918, Documentos avulsos lata 1918-E, Arquivo Público do Estado de Mato Grosso, Cuiabá (hereafter APMT).

15. *Mensagem dirigido à Assembléa Legislativa, 13 de maio de 1922, pelo Coronel*

Pedro Celestino Corrêa da Costa, Presidente do Estado (Cuiabá: Typ. Official, 1922), 38–41.

16. Paulo de Moraes Barros, "A crisis da pecuária," *Revista da Sociedade Rural Brasileira* (São Paulo) 24 (Junho 1922): 326–327; US Department of State, Report "Government Efforts to Reduce Cost of Living," by R. R. Bradford, US Consul in Rio de Janeiro, February 12, 1925, Brazil, entry 5, box 65 n.p., Record Group 166, USNA.

17. Brazil, IBGE, *Séries estatísticas retrospectivas*, vol. 1, 25; Mato Grosso, Instituto Nacional de Estatística, Secretaria da Agricultura, Indústria, Comercio, Viação e Obras Públicas, Junta Executiva Regional de Estatística, *Sinopse estatística do Estado, no. 1*, Separata, com acréscimos, do Anuário Estatística do Brasil, Ano II, 1936 (Cuiabá: Tipografia A. Calhão, 1937), 35; Earl Richard Downes, "The Seeds of Influence: Brazil's 'Essentially Agricultural' Old Republic and the United States, 1910–1930" (Ph.D. diss. in History, University of Texas, 1986), 479–481; Luiz Miguel do Nascimento, "As Charqueadas em Mato Grosso: Subsídios para um estudo de história econômica" (dissertação de mestrado apresentada ao Departamento de História da Faculdade de Ciências e Letras da Universidade Estadual Paulista Júlio de Mesquita Filho, Assis, São Paulo, 1992), tabela II, 56–57; Arlindo Sampaio Jorge, "A pecuária em Mato Grosso," *Revista dos Criadores* 19 (Janeiro 1948): 73; Mato Grosso, *Mensagem apresentado à Assembléa Legislativa pelo Presidente de Mato Grosso, Dr. Annibal Toledo, 13 de maio de 1930* (Cuiabá: Imprensa Oficial, 1930), annexo. The destination for Mato Grosso hides commonly was listed as Uruguay, but in Montevideo cargoes were transferred to larger, usually non-Brazilian, vessels headed to Europe and the Caribbean.

18. "Exportação de Matto-Grosso, Decennio de 1926–1935," *Relatório . . . Manoel Ary Pires, 13 de junho de 1937*, annexo; Dolor F. Andrade, *Mato Grosso e a sua pecuária* (São Paulo: Universidade de São Paulo, 1936), 8–9.

19. Mato Grosso, *Sinopse estatística do estado, no. 2*, 30; Dolor F. Andrade, *Mato Grosso e a sua pecuária*, 9; Econ J. Monserrat and Carlos A. Gonçalves, *Observações sobre a pecuária no Brasil Central: Relatório de viagem apresentado ao Ilmo. Sr. Dr. Manoel Corrêa Soares, D.D., Presidente do Instituto Sul Rio Grandense de Carnes, em 8 de agosto de 1953* ([Porto Alegre]: n.p., 1954), 115; Brazil, Ministério da Agricultura, Indústria e Commercio, Serviço de Inspecção e Fomento Agrícola, *Estudo dos Factores da Producção nos Municípios Brasileiros e condições economicas de cada um: Estado de Matto Grosso, Município de Campo Grande* (Rio de Janeiro: Imprensa Nacional, 1929), 54; "Relatorio," *Correio do Sul* (Campo Grande) 13 (May 5, 1928), 1. According to OECD estimates, in Brazil the average annual per capita consumption of beef in 2014 was 27 kilos. Only Argentina and Uruguay consumed more beef. See Organization of Economic Cooperation and Development, *Meat consumption 2015* (indicator), OECD Data, Agricultural Output (doi: 10.1787/fa29ofdo-en), https://data.oecd.org/agroutput/meat-consumption.htm (accessed August 20, 2015).

20. Brazil, IBGE, Estatísticas do século XX, *Bovinos abatidos, peso das carca-*

ças e produção de carne (1936–1987) http://seculoxx.ibge.gov.br/economicas/tabelas
-setoriais/agropecuaria; Brazil, IBGE, Conselho Nacional de Estatística, Serviço Nacional de Recenseamento, Série Nacional, vol. 2, *Censo Agricola* (Rio de Janeiro: IBGE, 1956), 128; Eugênio da Silva Parão, "Formação, Estrutura e dinâmica
da Economia de Mato Grosso do Sul no context das transformações da Economia Brasileira" (dissertação de mestrado, Universidade Federal de Santa Catarina,
Florianópolis, 2005), 129–131. World War II encouraged increased Brazilian cattle production, but since the previous markets were the Axis powers and regions
of conflict like North Africa, most Brazilian supplies went to its own army, for
which statistics are elusive.

21. Brazil, *Estudo dos factores . . . Campo Grande*, 34; "A creação em Matto
Grosso," *Brasil Agrícola* (São Paulo) 1 (December 1916): 363; US Department of
State, reports by C. R. Cameron, US Consul in São Paulo, "Cattle prices," 7 February 1927, and "São Paulo Livestock Industry and Exposition," 26 June 1929, Brazil, entry 5, box 64, Record Group 166, USNA; US Department of State, Division
of Latin American Affairs, Report "Matto Grosso and Its Finances," by C. R. Cameron, US Consul, São Paulo, 24 December 1927, Reports of US Consuls in Brazil,
1910–1929, no. 832.51 M43/2, microfilm M-519, roll 27, Record Group 166, USNA.
The border region of Paraguay also absorbed these cattle, since prices of animals
from Mato Grosso were lower than those in Paraguay itself.

22. Andrew Sluyter, *Black Ranching Frontiers: African Cattle Herders of the Atlantic World, 1500–1900* (New Haven: Yale University Press, 2012), chapter 6.

23. José de Barros, *Lembranças para os meus filhos e descendentes* (São Paulo: n.p.,
1987), 40. For the most part jerky *(charque)* is less fatty than salt beef, though in
Brazil the term charque was commonly used to refer to salt beef. Dried beef is
popularly called carne de sol. Another Río de la Plata term for charque was tasajo.

24. Miguel Arrojado Ribeiro Lisboa, *Oeste de S. Paulo, Sul de Mato-Grosso:
Geologia, Indústria, Mineral, Clima, Vegetação, Solo Agrícola, Indústria Pastoril* (Rio de
Janeiro: Typografia do Jornal do Commercio, 1909), 157–158.

25. Cardoso Ayala and Simon, *Album Graphico*, 292–294; *Mensagem . . . D. Fran-
cisco de Aquino Corrêa . . . 1919*, 110–111; José de Barros Maciel, "A pecuária nos
pantanaes de Matto Grosso," these apresentada ao 3° Congresso de Agricultura e
Pecuária, 1922 (São Paulo: Imp. Metodista, 1922), 22; Corrêa Filho, *A propósito do
Boi pantaneiro*, 52–54; report "São Paulo Livestock Industry and Exposition," by
C. R. Cameron, US Consul, São Paulo, 26 June 1929, Brazil, entry 5, box 64, Record
Group 166, USNA; Monserrat and Gonçalves, *Observações*, 63–65; Abílio Leite de
Barros, *Gente pantaneira: Crônicas da sua História* (Rio de Janeiro: Lacerda Edito-
res, 1998), 238–240; interview with Sr. Clóvis de Barros, Corumbá, Mato Grosso
do Sul, June 12, 1990. Despite Mato Grosso's increased production of jerky, it still
paled in comparison with Rio Grande do Sul, where in 1927 almost 600,000 animals were slaughtered. See Cameron, "São Paulo Livestock," Record Group 166,
USNA, and Bell, *Campanha Gaúcha*, chapter 6.

26. Mato Grosso, *Mensagem . . . Annibal Toledo . . . 1930*, annexo; *Relatório . . . Manoel Ary Pires, 13 de junho de 1937*, annexo; Nascimento, "As Charqueadas em Mato Grosso," 55–67.

27. Herbert H. Smith, *Do Rio de Janeiro a Cuyabá: Notas de um naturalista* (1886; S. Paulo: n.p., 1922), 137–140.

28. Interview with Sr. Clóvis de Barros, Corumbá, June 12, 1990; Fortunato Pimentel, *Charqueadas e frigoríficos: aspectos gerais da indústria pastoril do Rio Grande do Sul* (Porto Alegre: n.p., [1947]), 56. Although the amount of jerky rendered by one animal and the weight of hides varied, for the most part a "thin" steer of 200 kilos provided no more than 95 kilos of meat, while the average hide weighed between 20 and 30 kilos. Some charqueada statistics suggest that the amount of meat rendered per animal could be as little as 60 kilos, while others mention 120 kilos. Part of the problem was that the salt and soaking process of making jerky itself caused a loss of up to 25 percent of the original carcass. See Pimentel, *Charqueadas e frigoríficos*, 56, 59–62.

29. Marion Katheryn Pinsdorf, *German-Speaking Entrepreneurs: Builders of Business in Brazil* (New York: Peter Lang, 1990), 206. The amount of salt imported by Pinsdorf was reported as 500,000 kilos, but this was clearly a misprint since at the time the factory produced roughly 200 head per day. Alberto Maranhão, *These no. 30: Sal*, Conferencia nacional de pecuária (Rio de Janeiro: n.p., 1917), 3–4; advertisement for "Sal Brasileiro," *Oasis* (Corumbá) 6 (December 18, 1892): 4; J. Sampaio Fernandes, *Indústria do sal: Relatório apresentado, em setembro de 1937, ao Sr. Ministro da Agricultura, Dr. Odilon Braga* (Rio de Janeiro: n.p., 1939), 57–61; Pesavento, *República velha gaúcha*, 237; Alvarino da Fontoura Marques, *Evolução das charqueadas rio-grandenses* (Porto Alegre: Martins Livreiro, 1990), 88.

30. Lisboa, *Oeste de S. Paulo*, 157–158; Dolor F. Andrade, *Mato Grosso e a sua pecuária*, 11; J. Resende Silva, *A Fronteira do sul* (Rio de Janeiro: n.p., 1922), 513–514; Pimentel, *Charqueadas e frigoríficos*, 59–62; Alexandre A. De Castro, ed. *Anuario de Matto Grosso 1930, 1º Anno* (Corumbá: n.p., 1930), 113; Paulo de Assis Ribeiro, "Conservação da carne pelo salga e dessecação," *Revista dos Criadores* (São Paulo) 19 (December 1948): 47; Bell, *Campanha Gaúcha*, 68–71; Eugênio Hoinacki and Nelson Carlos Gutheil, *Peles e couros: Origins, defeitos, industrialização* (Porto Alegre: n.p., 1978), 55–57, 59–60. Despite these descriptions, in Mato Grosso most hides exported were dried without salt. The reason was cost. On Pantanal ranches local conditions generated little capital for the import of salt, and the charqueadas preferred to use their salt for the meat.

31. Alvarino da Fontoura Marques, *Evolução*, 70–74. It was estimated that in the earlier years up to 14 kilos of waste came from each animal. In Río de la Plata operations, blood residue was used to feed pigs. For charqueada pollution in Argentina, see Sluyter, *Black Ranching Frontiers*, 183.

32. Alvarino da Fontoura Marques, *Evolução*, 58–59, 74; interview with Dr. Cassio Leite de Barros, Corumbá, 6 September 1990.

33. During the boom period of World War I, estimates suggest that at least

50,000 head were exported tax-free to São Paulo annually, about 20 percent of annual production. This continued well into the 1950s. In 1929, observers calculated that from the municipality of Campo Grande alone some 25,000 head left without paying any taxes, municipal or otherwise. Brazil, *Estudo dos factores . . . Campo Grande*, 48; Mato Grosso, *Mensagem apresentado pelo Presidente de Mato Grosso, Mario Corrêa, à Assembléa Legislativa, 5 de maio de 1928* (Cuiabá: Typ. Official, 1928), 73–75. Oscar Brito noted that into the 1940s cattle smuggling was endemic throughout Central Brazil, with contraband of 50,000 head a year from Goiás and up to 100,000 head per year from Minas Gerais. Oscar da Silva Brito, "A pecuária do Brasil Central e sua produção de bovinos de corte," *Boletim de Indústria Animal* (São Paulo) 13 (January–April 1944): 13.

34. For a plethora of complaints, see Mato Grosso presidential reports across the years, as well as letters from customs officials, found in Documentos avulsos, APMT.

35. Walther L. Bernecker, "Contrabando: Ilegalidad y corrupción en el México decimonónico," *Espacio, Tiempo y Forma, Serie V, Historia Contemporánea* 6 (1993), 418.

36. Pimentel, *Charqueadas e frigoríficos*, 57–58; Pesavento, *República velha gaúcha*, 154, 183, 235; Bell, *Campanha Gaúcha*, chapter 6.

37. Ruano Fournier, *Estudio economico*, 37, 140, 142; report "O xarque Uruguayo," by Decio Coimbra, Brazilian Commercial Attaché in Uruguay, to Minister of Foreign Relations, Montevideo, 1 September 1928, Documentos avulsos lata 1928-C, APMT.

38. Guilhermino Cesar, *O Contrabando no sul do Brasil* (Caxias do Sul, RS: Universidade de Caxias do Sul, 1978), 99; J. Resende Silva, *A Fronteira do sul*, 670–671; report "A industria do xarque no Paraguay," by J. T. Nabuco de Gouvêa, Brazilian Minister in Paraguay, to the Brazilian Minister of Foreign Relations, Asunción, October 1928, Documentos avulsos lata 1928-C, APMT; report "Xarque brasileiro em transito pelo Uruguay," by Brazilian Commercial Attaché in Uruguay, unidentified, to the Brazilian Minister of Foreign Relations, Montevideo, 12 December 1928, Documentos avulsos lata 1929-C, APMT. It should be noted that profits on Mato Grosso jerky, and other products such as hides, were still modest. Report "Xarque brasileiro no mercado cubano," by Ildefonso D'Abreu Albano, Interim Brazilian Commercial Attaché in Cuba and Central America, to the Brazilian Minister of Foreign Relations, Havana, 30 October 1928, Documentos avulsos lata 1929-C, APMT; reports by C. R. Cameron, US Consul in São Paulo, São Paulo, "Cattle Prices," 7 February 1927, and "São Paulo Livestock Industry and Exposition," 26 June 1929, Brazil, entry 5, box 64, Record Group 166, USNA. Uruguayan jerky tended to come from the flanks of the animal (*mantas*), where there was more fat content. Jerky from Mato Grosso and Rio Grande do Sul was primarily from the legs and other extremities (*postas*), which were much leaner. This distinction also explains the observation that Brazilian jerky was inferior to that of the Río de la Plata. Report "O xarque Uruguayo," by Decio Coimbra, Brazilian Commercial

Attaché in Uruguay to the Brazilian Minister of Foreign Relations, Montevideo, 1 September 1928, Documentos avulsos lata 1928-C, APMT. Andrew Sluyter argues that because leaner jerky was appreciated in Cuba, it was more expensive and generated greater profits for Uruguayan exporters. Sluyter, *Black Ranching Frontiers*, 185–186.

39. J. Resende Silva, *A Fronteira do sul*, 510, 670–671; Cesar, *O Contrabando*, 97; report "Transito de xarque nacional pelo porto de Montevideo," by Mario de Azevedo, Brazilian Consul General in Montevideo to the Brazilian Minister of Foreign Relations, Montevideo, 1 October 1928, Documentos avulsos lata 1928-C, APMT; report "Xarque brasileiro: Especulação em Montevideo/reacção em Cuba," by Ildefonso d'Abreu Albano, Brazilian Interim Commercial Attaché in Cuba and Central America to the Brazilian Minister of Foreign Relations, Havana, 29 September 1929 [should read 1928], Documentos avulsos lata 1929-C, APMT.

40. Report on the jerky market in Montevideo and Cuba [by the Brazilian Commercial Attaché in Montevideo] to the Brazilian Minister of Foreign Relations, n.p., December [1928], Documentos avulsos lata 1929-C, APMT; report "Transito de xarque nacional pelo porto de Montevideo," by Mario de Azevedo, Brazilian Consul General in Montevideo, to the Brazilian Minister of Foreign Affairs, Montevideo, 1 October 1928, Documentos avulsos lata 1928-C, APMT.

41. Report "O xarque uruguayo em Cuba," by the Brazilian Commercial Attaché in Uruguay, unidentified, to the Brazilian Minister of Foreign Relations, Montevideo, 12 December 1928, Documentos avulsos lata 1928-C, APMT; report "Xarque brasileiro em transito pelo Uruguay," by the Brazilian Commercial Attaché in Uruguay, unidentified, to the Brazilian Minister of Foreign Relations, Montevideo, 12 December 1928, Documentos avulsos lata 1928-C, APMT; Pesavento, *República velha gaúcha*, 269–271; Cesar, *O Contrabando*, 99; Nascimento, "As Charqueadas em Mato Grosso," 68–70.

42. Letter from Francisco Freire de Borja Sabeuma Garção, superintendent of the Cia. Nacional de Navegação a Vapor to its president, João Antonio Mendes Motta, Montevideo, 15 October 1888, Documentos avulsos lata 1888-F, APMT; Mato Grosso, *Mensagem do Presidente do Estado de Matto Grosso, Dr. Manoel José Murtinho à Assembléa Legislativa em sua 2ª legislatura aberta, 13 de maio de 1895* (Cuiabá: Typ. do Estado, 1895), 17.

43. Mato Grosso, *Relatório com que o exm. snr. Dr. João José Pedrosa, Presidente da Província de Matto Grosso, 22° legislatura da respectiva assembléa, no dia 1 de novembro de 1878* (Cuiabá: n.p., 1878), 38–39; *Relatório do Exm. Snr. Coronel Barão de Maracajú, Presidente da Província de Matto Grosso, 5 de Dezembro, 1879* (Cuiabá: n.p., 1879), 61; *Relatório do vice-presidente Dr. José Joaquim Ramos Ferreira devia apresentar à Assembléia Legislativa Províncial de Matto Grosso, 2ª legislatura de Setembro de 1887* (Cuiabá: n.p., 1887); Armen Mamigonian, "Inserção de Mato Grosso ao mercado nacional e a gênese de Corumbá" *Géosul* (Florianópolis) 1 (May 1986): 54; report from Antonio Canale, mayor of Miranda, to the President of Mato Grosso, Miranda, 31 December 1895, Documentos avulsos lata 1895-D, APMT.

44. Cardoso Ayala and Simon, *Album Graphico*, 67–68; report "Historical Sketch of the Lloyd Brasileiro Steamship Line," [by A. M. Gothchalk], US Consul General in Rio de Janeiro, Rio de Janeiro, 11 May 1917, Records of the US Shipping Board, Sub-classified General Files, ea. 1916–1936, file 153, Record Group 32, USNA; Brazil, Ministério da Tesouraria, letters from Lloyd Brasileiro to the Minister of the Treasury, Rio de Janeiro, 28 January 1893, 13 February 1894, 9 August 1895, 30 April 1896, "Companhia Lloyd Brasileiro, diversos, 1893–1897," IF¹, no. 159, Arquivo Nacional de Rio de Janeiro (hereafter ANRJ).

45. Report by José Alvarez Sánchez Surga, 1° Vice-Intendente of Nioác, to the Mato Grosso Secretario de Estado dos Negocios do Interior, Nioác, 19 January 1912, Documentos avulsos lata 1912-A, APMT; Virgílio Alves Corrêa, "Aos Fazendeiros," *Revista da Sociedade Matto-Grossense de Agricultura* (Cuiabá) 1 (May 1907): 28; Lisboa, *Oeste de S. Paulo*, 157–158; report "A Navegação Brasileira no Paraguay," from the Brazilian Minister in Paraguay, Dr. J. T. Nabuco de Gouvêa, to the Brazilian Minister of Foreign Relations, Asunción, October 1928, Documentos avulsos lata 1928-C, APMT. The Argentine gold peso was worth US$0.965 in 1928. "Exportação de Matto-Grosso, Decennio de 1926–1935," *Relatório . . . Manoel Ary Pires, 13 de junho de 1937*, annexo.

46. Cardoso Ayala and Simon, *Album Graphico*, 124–125; "Meza de Rendas de Corumbá, Registros de exportação, 1915, 1920, 1926," "Collectoria da Villa de Porto Murtinho, Impostos de exportação, 1919," and "Collectoria Estadoal do Porto Murtinho, Registro de exportações, 1925," Coletorias Avulsas, APMT.

47. "Exposição de motivos da Associação comercial aos representantes de Mato Grosso no Congresso Nacional, Corumbá, 10 de maio de 1923," lata 1923, APMT, cited in Lúcia Salsa Corrêa, "Corumbá: Um núcleo comercial na fronteira de Mato Grosso, 1870–1920" (MA thesis, Universidade de São Paulo, 1980), 135–140, annexo n. 1, 3; Antonio Carlos Simoens da Silva, *Cartas Mattogrossenses* (Rio de Janeiro: n.p., 1927), 28–29; "Notas," *A Cidade* (Corumbá) 6 (27 March 1923), 1. Part of the reason for such long journeys between Corumbá and Cuiabá was the need to tie up at night. Unlike service from Corumbá downstream, ships were not outfitted for night navigation.

48. "Noticias de Porto Murtinho, O Lloyd e o Commercio," *A Noticia* (Três Lagoas) 1 (17 April 1924): 2; report "A Navegação Brasileira no Paraguay," by Dr. J. T. Nabuco de Gouvêa, Brazilian Minister in Paraguay, to the Brazilian Minister of Foreign Relations, Asunción, October 1928, Documentos avulsos lata 1928-C, APMT; Mato Grosso, *Informações gerais do municipio de Corumbá* (Corumbá: n.p., 1932), 17–19; Virgílio Alves Corrêa Filho, *Pedro Celestino* (Rio de Janeiro: n.p., 1945), 242.

49. The motivators were particularly the conflict over Acre, where Brazilian settlers in the rich rubber-producing Bolivian territory sought support from Rio for political autonomy in the face of alleged arbitrary practices by Bolivian police and customs, and a heated war of words between Brazil and Argentina between 1906 and 1908 over the upgrading of naval vessels and regional power. The

conflict with Bolivia led to the annexation of Acre through the Treaty of Petrópolis in 1903.

50. Mato Grosso, *Relatório . . . João José Pedrosa . . .* 1878, 38–39; Demosthenes Martins, *História de Mato Grosso: Os fatos, os governos, a economia* (São Paulo: n.p., [1977]), 167–170; Roberto Etchepareborda, *Historia de las relaciones internacionales argentinas* (Buenos Aires: Editorial Pleamar, 1978), 81–85, 135–138, 142; Fernando de Azevedo, *Um Trem Corre para o Oeste: Estudo sobre a Noroeste e seu papel no sistema de viação nacional* (São Paulo: Livraria Martins Editora, 1950), 108–111, 144. See also Paulo Roberto Cimó Queiroz, *As curvas do trem e os meandros do poder: O Nascimento da Estrada de Ferro Noroeste do Brasil (1904–1908)* (Campo Grande: Ed. da UFMS, 1997). It could be argued that the Madeira-Mamoré Railroad in Acre, first attempted in the 1870s, was Brazil's first real "political highway." For a detailed study of the promise of a railway to Cuiabá that never materialized and the broader implications for the region, see Fernando Tadeu de Miranda Borges, *Esperando o trem: Sonhos e esperanças de Cuiabá* (Rio de Janeiro: Scortecci, 2005).

51. Brazil, Ministério dos Transportes, "Decreta aprova estudos para EF Noroeste do Brazil, e exposição de motivos, 23 de Outubro de 1913," IT¹, maço 202, 10.523, ANRJ; Azevedo, *Um Trem Corre para o Oeste,* 108–111, 249–250; Paulo Roberto Cimó Queiroz, *Uma ferrovia entre dois mundos: A EF Noroeste do Brasil na primeira metade do século 20* (Bauru, SP: EDUSC, 2004), chapter 2. Cimó Queiroz examines the development of the EFNOB in exceptional detail, amplifying and revising the conclusions of earlier authors.

52. Azevedo, *Um Trem Corre para o Oeste,* 108–109, 249–250.

53. C. R. Cameron, "Through Matto Grosso," *Bulletin of the Pan American Union* 66 (March 1932): 158–160; Azevedo, *Um Trem Corre para o Oeste,* 110–111.

54. Cimó Queiroz, *Uma ferrovia,* 321–328.

55. Brazil, Directoria Geral de Estatística, *Anuário estatística do Brasil, 1908–1912,* vol. 1, "População" (Rio de Janeiro: Typ. da Estatística, 1913), 327–328; Estado de Mato Grosso, Prefeitura Municipal de Campo Grande, *Relatório 1943* (Rio de Janeiro: Imprensa Nacional, 1944), 38; Brazil, Ministério da Agricultura, Commercio e Obras Públicas, Directoria Geral de Estatística, *Recenseamento realizado em 1 de Setembro de 1920, vol. 4, parte 1 "População,"* 408; Brazil, IBGE, *Recenseamento Geral do Brasil (1º de Setembro de 1940) . . . Censo Demográfico,* 51; Cardoso Ayala and Simon, *Album Graphico,* 409; Brazil, *Estudo dos factores . . . Campo Grande,* 39.

56. Report "Foreign Holdings in Mato Grosso," by Roger L. Heacock, US vice-consul in São Paulo, January 23, 1941, no. 478, Brazil, entry 5, box 20, Record Group 166, USNA; Antonio Carlos Simoens da Silva, *Cartas Matogrossenses,* 18; Corrêa Filho, *A propósito do Boi pantaneiro,* 52–54; Cardoso Ayala and Simon, *Album Graphico,* 292–294.

57. Mato Grosso, *Mensagem . . . Caetano Manoel de Faria e Albuquerque . . . 1916,* 35, 41. For a detailed discussion of these issues, see Cimó Queiroz, *Uma ferrovia,* chapter 4.

58. Arlindo de Andrade, *Erros da federação* (São Paulo: n.p., 1934), 89; Corrêa Filho, *A propósito do Boi pantaneiro*, 52–54; Mato Grosso, *Mensagem . . . 1924 . . . Pedro Celestino Corrêa da Costa, Presidente de Matto Grosso*, 75–76.

59. *Relatório do Município de Aquidauana, 1928* (São Paulo: n.p., 1929), 55–58; report by Jorge Bodstein Filho, president of the Aquidauana Municipal Council, to the President of Mato Grosso, Aquidauana, 14 February 1929, Documentos avulsos lata 1929-F, APMT; Nascimento, "As Charqueadas em Mato Grosso," 70; Cimó Queiroz, *Uma ferrovia*, 411–418.

60. Salsa Corrêa, "Corumbá: Um núcleo comercial," 123–124.

61. Arlindo de Andrade, *Erros da federação*, 72–73; Cimó Queiroz, *Uma ferrovia*, 399–411; Brazil, Ministério da Agricultura, Indústria e Commercio, Serviço de Inspecção e Fomento Agrícola, *Estudo dos Factores da Producção no Municípios Brasileiros e Condições economicos de cada um, Estado de Matto Grosso, Município de Corumbá* (Rio de Janeiro: Ministério da Agricultura, 1924), 17; US Department of State, Report "Rumored Plan for the Encouragement of Immigration of Cattle Raisers into the State of Matto Grosso," by John Hubner II, US Vice-Consul in São Paulo, São Paulo, 9 August 1938, Reports of the US Consuls in Brazil, 1910–1929, microfilm M-519, roll 27, no. 832.52 AM3/22, Record Group 59, USNA.

62. Azevedo, *Um Trem Corre para o Oeste*, 182–185, 191–196; Cimó Queiroz, *Uma ferrovia*, 347–358, 395–411; "O que nos contam as estatísticas da Estrada de Ferro Noroeste do Brasil," *Jornal do Commercio* (Campo Grande) 14 (30 May 1934): 1; Carlos Araujo, "A exportação dos bovinos," *Jornal do Commercio* 14 (24 May 1934): 1, 4; "A regularisação do trafego da EF Noroeste," *Jornal do Commercio* 14 (25 December 1934): 1, 4.

63. Azevedo, *Um Trem Corre para o Oeste*, 215–216; interview with Paulo C. Machado, Campo Grande, May 29, 1990; José de Lima Figueiredo, *A Noroeste do Brasil e a Brasil-Bolívia* (São Paulo: n.p., 1950), 104.

64. Salsa Corrêa, "Corumbá: Um núcleo comercial," 118–121; Virgílio Alves Corrêa Filho, *Pantanais Matogrossenses (Devassamento e Ocupação)* (Rio de Janeiro: IBGE, 1946), 139–144; Lúcia Salsa Corrêa, "Corumbá: O Comércio e o Casario do Porto (1870–1920)," in *Casario do Porto de Corumbá* (Campo Grande: Fundação de Cultura de Mato Grosso do Sul, 1985), 53–54; Gilberto Luiz Alves, "A Trajetória Histórica do Grande Comerciante dos Portos em Corumbá, 1857–1929 (A propósito das determinações econômico-sociais do Casario do Porto)," in *Casario do Porto*, 79–81; Cimó Queiroz, *Uma ferrovia*, 358–366.

65. Virgílio Corrêa Filho, *Fazendas de gado no pantanal Matogrossense*, Documentário da vida rural, no. 10 (Rio de Janeiro: Ministério da Agricultura, 1955), 42–43; Azevedo, *Um Trem Corre para o Oeste*, 118–119; José de Melo e Silva, *Fronteiras Guaranís* (São Paulo: Imprensa Metodista, 1939), 113–114; João Batista de Souza, *Evolução Histórica Sul Mato Grosso* (São Paulo: Organização Simões, [1961]), 133. The railroad continued its modest operations into the 1990s, though mostly for local transport and as an outlet for eastern Bolivia. In 2009 it was reorganized into Pantanal tourism and nicknamed Trem do Pantanal.

CHAPTER 4. LAND ACCESS

1. Brazil, Ministério da Agricultura, Commercio e Obras Públicas, Terras e Colonisação, Província de Mato Grosso, "Repartição Especial das Terras Públicas em Cuiabá, 28 de Agosto de 1861," caixa 1174 (2543), Terras e Colonização, Arquivo Nacional de Rio de Janeiro (hereafter ANRJ); Virgílio Alves Corrêa Filho, "Terras devolutas: Evolução do processo de adquiril as em Matto-Grosso," *Revista do Instituto Histórico de Matto-Grosso* 3 (1921): 65–69; Demosthenes Martins, *História de Mato Grosso: Os fatos, os governos, a economia* (São Paulo, n.p., [1977]), 166; Mato Grosso, *Relatório do Presidente da Província, Barão de Maracajú, 1° sessão, 23° legislatura, no dia 1° de outubro de 1880* (Cuiabá: Typ. Joaquim J. R. Calhão, 1880), 39–40.

2. Letter from District Judge in Santana do Paranaíba to president of Mato Grosso, Santana do Paranaíba, February 15, 1879, Documentos avulsos lata 1879-B, Arquivo Público do Estado de Mato Grosso, Cuiabá (hereafter APMT); Mato Grosso, Livros de correspondencia províncial, *Registro dos officios expedidos pela Presidencia da província às diversas autoridades do interior, annos 1879–1885*, book 492, various letters, 1879–1884, APMT.

3. Mato Grosso, *Registros dos officios . . . 1879–1885*, book 492, various letters, 1879–1884, APMT.

4. Corrêa Filho, "Terras devolutas," 65–67; Mato Grosso, *Registro de Correspondencia Geral com os Ministérios dos Negocios da Agricultura, 1885 à 1889*, book 179, various letters, APMT.

5. Rodolpho Endlich, "A criação do gado vaccum nas partes interiores da America do Sul," *Boletim da Agricultura* 4, no. 3 (1903): 128; Mato Grosso, *Registro de Correspondencia*, book 179, letters 12 (March 28, 1888), 30 (July 5, 1887), various other letters, APMT; Mato Grosso, *Correspondencia Official do Presidente da Província ao Ministério e Secretário de Estado dos Negocios da Agricultura, Commercio e Obras Públicas, 1881–1885*, book 167, letter of October 22, 1881, APMT; Mato Grosso, Livros de correspondencia províncial, *Registros dos officios expedidos pela Presidencia da província às Camaras Municipais, annos 1883–1889*, book 404, letter 48 (July 4, 1884), letters 74–84 (October 28, 1884), APMT; Mato Grosso, *Registro dos oficios expedidos pela Presidencia da Província aos Comandos das Fronteiras e Colonias Militares, 1881–1889*, book 166, December 6, 1888, APMT; Brazil, Ministério da Agricultura, Commercio e Obras Públicas, "Directoria da Agricultura, 1890–1892," letters of November 3, December 6, 11, 12, 28, 1888, GIFI, 5F, caixa 602, ANRJ. Although foreigners were denied requests for public lands by the imperial government, no mechanisms were put in place to limit the purchase of private land in border regions.

6. João Leite de Barros, "A Pecuária em Corumbá," in *Anuário de Corumbá, 1939* (Corumbá: n.p., 1939), 8–11; José de Barros Maciel, "A pecuária nos pantanaes de Matto Grosso," these apresentada ao 3° Congresso de Agricultura e Pecuária, 1922 (São Paulo: Imp. Metodista, 1922), 14–18; Brazil, Ministério da Agricultura,

Industria e Commercio, Directoria Geral de Estatística, *Synopse do recenseamento realizado em 1 de setembro de 1920; população pecuária* (Rio de Janeiro: Typographia da Estatística, 1922), 26; *Terra e Gente* (Rio de Janeiro) 1 (January 1956): 92–93; Gervásio Leite, *O gado na economia matogrossense* (Cuiabá: Escolas Profissionais Salesianos, 1942), 13; Bento de Souza Porto, "Programa de desenvolvimento da pecuária no Pantanal," in Estado de Mato Grosso, *Pantanal, Nova Fronteira Econômica, I Encontro do Prodepan (Programa de Desenvolvimento do Pantanal), Conferências, Proposições e Subsídios* (Corumbá: Secretária da Agricultura, 1974), 102–103. Nhecolândia as a separate unit was not conceived until the 1930s. Populations for the region before this date are extrapolated from those of the municipality of Corumbá, which covered more territory than the Nhecolândia of today. The bulk of cattle were found in Nhecolândia, however.

7. Corrêa Filho, *Fazendas de gado no pantanal Matogrossense*, Documentário da vida rural, no. 10 (Rio de Janeiro: Ministério da Agricultura, 1955), 20, 22; Octavio Domingues and Jorge de Abreu, *Viagem de estudos à Nhecolândia* (Rio de Janeiro: Instituto de Zootecnica, Ministério da Agricultura, 1949), 21; Brazil, Ministério da Agricultura, Indústria e Commercio, Directoria Geral de Estatística, *Valor das terras no Brazil, segundo o censo agrícola realizado em 1 de setembro de 1920* (Rio de Janeiro: Typ. da Estatística, 1924), 24–25; Brazil, IBGE, Recenseamento Geral do Brasil (1º de Setembro de 1940), Série Regional, Parte XXII, *Mato Grosso: Censo Demográfico, Censos Econômicos* (Rio de Janeiro: Serviço Gráfico do IBGE, 1952), 193; *Terra e Gente* 1 (January 1956), 92–93; Bento de Souza Porto, "Programa de desenvolvimento da pecuária no Pantanal," 103.

8. Brazil, *Mato Grosso: Censo Demografico, Censos Econômicos (1940)*, 193, 196–197; Brazil, *Valor das Terras no Brasil*, 24–25. The vast extent of the Mate Larangeira concession, over 1.4 million hectares, accounts for the difference, 47 percent of total private property.

9. Brazil, *Mato Grosso: Censo Demografico, Censos Econômicos (1940)*, 193, 196–197; Brazil, *Valor das Terras no Brasil*, 24–25; Afonso Simões Corrêa, "Pecuária de corte em Mato Grosso do Sul," report delivered at Encontros Regionais de Pecuária de Corte, Brasília, 27 November 1984, 1.

10. Corrêa Filho, "Terras devolutas," 70–72.

11. Corrêa Filho, "Terras devolutas," 72–74, 80–81. Some large ranches escaping penalties were Taboco (344,000 hectares), Firme (176,000 hectares), and Carandá (404,000 hectares) in the Pantanal, Passa-Tempo in Miranda (149,000 hectares), and Morro do Azeite in Paranaíba (171,000 hectares). Others forced to pay included Santa Virginia belonging to Mate Larangeira (68,000 hectares), Rio Branco in the Pantanal (384,000 hectares, reduced to 191,000 hectares), and Rio Negro in Aquidauana (95,000 hectares). It appears foreign ownership of Mate Larangeira and membership in opposition political parties by the others explains the variations.

12. Corrêa Filho, "Terras devolutas," 72, 80; Mato Grosso, *Mensagem apresen-*

tado à Assembléa Legislativa em 1º de Fevereiro de 1896 pelo Dr. Antonio Corrêa da Costa, Presidente do Estado (Cuiabá: Typ. do Estado, 1896), 22–23; report by João Pedro Gardés, director of the Mato Grosso land office, to the president of Mato Grosso, Cuiabá, December 31, 1897, Documentos avulsos lata 1898-A, APMT; S. Cardoso Ayala and Feliciano Simon, *Album Graphico do Estado de Matto-Grosso* (Corumbá/Hamburgo: n.p., 1914), 168–171.

13. Mato Grosso, *Mensagem apresentado à Assembléa Legislativa em 1º de Fevereiro de 1897 pelo Dr. Antonio Corrêa da Costa, Presidente do Estado* (Cuiabá: Typ. do Estado, 1897), 19–22.

14. Cardoso Ayala and Simon, *Album Graphico*, 100; Endlich, "A criação," 4 (1903): 128–129; A. Marques, *Matto Grosso: Seus recursos naturaes, seu futuro economico* (Rio de Janeiro: Papelaria Americana, 1923), 117–118; Miguel Arrojado Ribeiro Lisboa, *Oeste de S. Paulo, Sul de Mato-Grosso: Geologia, Indústria, Mineral, Clima, Vegetação, Solo Agrícola, Indústria Pastoril* (Rio de Janeiro: Typografia do Jornal do Commercio, 1909), 142–145; Arlindo de Andrade, *Erros da federação* (São Paulo: n.p., 1934), 87. Many of the prices per hectare given here have been calculated from information on the costs of land per league. Although the normal league in Brazil equaled 4,356 hectares, between the 1890s and 1914 in Mato Grosso a league was considered to be 3,600 hectares. This changed with standardization after World War I.

15. Letter from T. G. Chittenden, general manager of Brazil Land, Cattle and Packing Company, through the British Embassy in Rio de Janeiro, to the president of Brazil, São Paulo, August 9, 1920, Documentos avulsos lata 1920-C, APMT; Virgílio Corrêa Filho, *A propósito do Boi pantaneiro*, Monografias Cuiabanas, vol. 6 (Rio de Janeiro: Pongetti e Cia., 1926), 47 n. 59; Mato Grosso, *Mensagem pelo Cel. Pedro Celestino Corrêa da Costa, 1º Vice-Presidente do Estado, à Assembléa Legislativa, 13 de maio de 1911* (Cuiabá: Typ. Official, 1911), 18; letter from José Santiago, mayor of Campo Grande, to the president of Mato Grosso, Campo Grande, April 13, 1913, Documentos avulsos lata 1913-A, APMT.

16. Various messages (*mensagens*) from state presidents between 1920 and 1931. All presidential reports during this period included statistics on the legalization of properties, though records were incomplete and often uncertain, as many writers admitted.

17. Mato Grosso, *Mensagem à Assembléa Legislativa, 7 de Setembro de 1920 por D. Francisco de Aquino Corrêa, Bispo de Prusiade, Presidente do Estado* (Cuiabá: Typ. Official, 1920), 88; Mato Grosso, *Mensagem dirigido à Assembléa Legislativa em 13 de maio de 1924 pelo Cel. Pedro Celestino Corrêa da Costa, Presidente de Matto Grosso* (Cuiabá: Typ. Official, 1924), 51; Neill Macaulay, *The Prestes Column: Revolution in Brazil* (New York: New Viewpoints, 1974), 92–128, 230–234; Mato Grosso, *Mensagem apresentado pelo Presidente de Matto Grosso, Dr. Mario Corrêa, à Assembléa Legislativa, 13 de maio de 1929* (Cuiabá: Typ. Official, 1929), 126; Mato Grosso, *Mensagem apresentado à Assembléa Legislativa pelo Presidente de Mato Grosso, Dr. Annibal*

Toledo, 13 de maio de 1930 (Cuiabá: Imprensa Oficial, 1930), 80. In 1919, the year before the recession took effect, 977,000 hectares were granted provisional and definitive titles.

18. Brazil, Fundação Instituto Brasileiro de Geografia e Estatística, Séries Estatísticas Retrospectivas, vol. 1, *Repertório estatística do Brasil: Quadros retrospectivos*, Separata do Anuário Estatístico do Brasil, Ano V, 1939–1940 (Rio de Janeiro: IBGE, 1986), 57–58; Mato Grosso, Instituto Nacional de Estatística, Secretaria da Agricultura, Indústria, Comercio, Viação e Obras Públicas, Junta Executiva Regional de Estatística, *Sinopse estatística do Estado, no. 1*, Separata, com acréscimos, do Anuário Estatística do Brasil, Ano II, 1936 (Cuiabá: Tipografia A. Calhão, 1937), 98.

19. Otto Willi Ulrich, *Nos sertões do Rio Paraguay* (São Paulo: n.p., 1936), 87; Arlindo Sampaio Jorge, "A pecuária em Mato Grosso," *Revista dos Criadores* 19 (Janeiro 1948): 39.

20. Corrêa Filho, "Terras devolutas," 75; Mato Grosso, *O Município de Campo Grande, Estado de Matto Grosso* (n.p.: Publicação oficial, 1919), 77; Cartório do 2º Oficio, "Livros de Escripturas de compra e venta, 1914 a 1930" (Ponta Porã, Mato Grosso do Sul: n.p.).

21. Brazil, *Valor das Terras no Brasil*, 24; US Department of State, Report "Land Prices in Brazil," by A. Gaulin, US Consul in Rio de Janeiro, to Department of Agriculture, April 23, 1924, Brazil, entry 5, box 63, Record Group 166, United States National Archives (hereafter USNA); São Paulo, Secretaria da Agricultura, Commercio e Obras Públicas, *Os Municípios do Estado de São Paulo* (São Paulo: Serviço de Publicações, 1924), 199–200.

22. "Terras, compra e venda," *Jornal do Commercio* (Campo Grande) 14 (December 28, 1934): 4; "Propaganda de Matto Grosso e a zona Noroeste," *Almanaque Illustrado* (Três Lagoas) 2 (1929): 385–387; US Department of State, Division of the American Republics, "Rumored Plan for the Encouragement of Immigration of Cattle Raisers into the State of Matto Grosso," by John Hubner II, US Vice-Consul, São Paulo, August 9, 1938, no. 832.52 AM3/22, *Reports of the US Consuls in Brazil, 1910–1929*, microfilm M-519, roll 27, USNA; Antonio Carlos de Oliveira, *Economia pecuária do Brasil Central: Bovinos* (São Paulo: Departamento Estadual de Estatística de São Paulo, 1941), 9 n. 1.

23. Federico Rondon, "Mato Grosso Econômico," *Revista Brasileira dos Municípios* (Rio de Janeiro) 21 (1952): 130; Joe Foweraker, *The Struggle for Land: A Political Economy of the Pioneer Frontier in Brazil from 1930 to the Present Day* (New York: Cambridge University Press, 1981), 41, 46.

24. Antonio Carlos Simoens da Silva, *Cartas Mattogrossenses* (Rio de Janeiro: n.p., 1927), 48; tables of provisional and definitive land titles issued by the state, not signed, n.d., Documentos avulsos lata 1921-A, APMT; Antonio de Pádua Bertelli, *O Paraíso das espécias vivas: Pantanal de Mato Grosso* (São Paulo: Cerifa, 1984), 85. This work includes a translation of an unpublished report written in the early 1920s by the ranch administrator of Fazendas Francesas, Count Robert La Batut.

Most information on the ranch comes from his report. US Department of State, Report "Foreign Holdings in Mato Grosso," by Roger L. Heacock, US vice-consul, São Paulo, 23 January 1941, no. 479, Brazil, entry 5, box 20, Record Group 166, USNA; Cezar Benevides and Nanci Leonzo, *Miranda Estância: Ingleses, peões e caçadores no Pantanal mato-grossense* (Rio de Janeiro: Editora Fundação Getúlio Vargas, 1999), 28–30; J. Barbosa Rodrigues, *História de Mato Grosso do Sul* (São Paulo: n.p., 1985), 130–131. By comparison, in Texas the famous Matador and King ranches measured over 330,000 hectares (800,000 acres) and 520,000 hectares (1.25 million acres), respectively. Lewis Nordyke, *Great Roundup: The Story of Texas and Southwestern Cowmen* (New York: William Morrow & Company, 1955), 261; P. Rossi, "King Ranch, a maior fazenda de criar do mundo e a raça bovina Santa Gertrudis," *Revista de Indústria Animal* (São Paulo) 8 (January 1939): 163.

25. Mateo Martinic Beros, "La participación de capitales Británicos en el desarrollo económico del territory de Magallanes, 1880–1920," *Historia* (Santiago, Chile) 35 (2002): 299–321.

26. Charles A. Gauld, *The Last Titan: Percival Farquhar, American Entrepreneur in Latin America* (Stanford: Institute of Hispanic American and Luso-Brazilian Studies, 1964), 218, 226 n. 19; C. E. Perkins, "Report on Brazil Land, Cattle and Packing Company," exhibit I in *Brazil Railway Company Records*, 45–50, NB-15, Baker Library Historical Collections, Harvard Business School; Brazil Land, Cattle and Packing Company, São Paulo Office, "Accounts for the Receivers: Summarized Accounts Showing the Position of the Company as at 1st January 1915, and Subsequent Operations to 31st December 1918, as Appearing on the São Paulo Books," in *Brazil Railway Company Records*, appendix 2, Box 21, Baker Library Historical Collections, Harvard Business School; "O Caso da venda da Fazenda Paracatu," *O Jornal* (April 23, 1950): 2, in No. 205, Box 15, Percival Farquhar Papers, Manuscripts and Archives, Yale University Library; "Descalvados: Continúa sua luta de tradição e dinamismo," *Terra e Gente* 1 (January 1956): 102–103.

27. "Propaganda de Matto Grosso e zona Noroeste," *Almanaque Illustrado* (Três Lagoas) 2 (1929): 141–163, 266–325; Alexandre A. de Castro, ed., *Anuario de Matto Grosso 1930 1º Anno* (Corumbá: n.p., 1930), 113–115; Mato Grosso, *O Município de Campo Grande*, 63–64. By the 1950s, the Alves Ribeiro ranch, Taboco, had shrunk to 100,000 hectares. See "Fazenda Taboco," *Terra e Gente* 2 (January 1957): 47.

28. Brazil, Ministério da Agricultura, Commercio e Obras Públicas, Directoria Geral de Estatística, *Recenseamento realizado em 1 de Setembro de 1920, vol. 3, parte 1 "Agricultura"* (Rio de Janeiro: Typ. da Estatística, 1923), 150–153; Brazil, *Mato Grosso: Censo Demográfico, Censos Econômicos*, 196. I should reiterate that these numbers are not absolutes, given that the censuses of the day were famously flawed; not all properties were included, and frequently areas were little more than estimates, but the trends are clear.

29. Brazil, Ministério da Agricultura, *Recenseamento realizado em 1 de Setembro de 1920, vol. 3, parte 1, "Agricultura"* (Rio de Janeiro: Typ. da Estatística, 1926), 150–153; Brazil, *Mato Grosso: Censo Demográfico, Censos Econômicos*, 196–197; Afonso

Simões Corrêa, "Pecuária de corte em Mato Grosso do Sul," report delivered at Encontros Regionais de Pecuária de Corte, Brasília, 27 November 1984, 1. More study is needed on the role of smallholders in the history of Mato Grosso agriculture. On another note, the 1940 data deliberately omitted numbers for Cáceres in order to avoid "individualizing" the information. Since the major landowner in the municipality in 1920 was Brazil Land, with more than one million hectares at Descalvados, it seems this may have been an attempt to shield the company from criticism of foreign holdings.

30. Letter from José Pinheiro Machado, mayor of Aquidauana, to the president of Mato Grosso, Aquidauana, November 25, 1909, Documentos avulsos lata 1910-C, APMT; Miguel A. Palermo, *Subrogação dos direitos de propriedade da antiga e historica Fazenda do Apa perante os Poderes publicos e a opinião sensata do Estado de Matto-Grosso* (Cuiabá: n.p., 1914).

31. Palermo, *Subrogação*, 28–135; Matto Grosso, Comarca de Bela Vista, *Fazenda S. Raphael do Estrella: De Uma simples questão de facto—importantes questões de direito: Perambulos da invasão, Lesões ao direito de propriedade, por Ludgero Feital* (Rio de Janeiro: Typ. Revista do Tribunaes, 1921).

32. Various petitions concerning land to the president of Mato Grosso, Campo Grande and Nioác, 1920, Documentos avulsos lata 1920-B, APMT; Palermo, *Subrogação*; Matto Grosso, Comarca de Bela Vista, *Fazenda S. Raphael do Estrella*.

33. Letter from Elviro Mario Mancini for Cel. Alfredo Justino de Souza and others in Santana do Paranaíba, to the president of Mato Grosso, Santana do Paranaíba, January 3, 1913, Documentos avulsos lata 1913-A, APMT; letter from T. G. Chittenden, general manager of Brazil Land, Cattle and Packing Company, through the British Embassy in Rio de Janeiro, to the president of Mato Grosso, São Paulo, August 9, 1920, Documentos avulsos lata 1920-C, APMT; letters from Barnabé Antonio Gondim, Cuiabá police department, to the federal judge, Cuiabá, August 18, 21, 1921, Documentos avulsos lata 1921-B, APMT; Corrêa Filho, *A propósito do Boi pantaneiro*, 47–48.

34. Corrêa Filho, *Pedro Celestino* (Rio de Janeiro: n.p., 1945), 94–97; letter from Homero Baptista, Ministério dos Negocios da Fazenda, to the president of Mato Grosso, Rio de Janeiro, 5 April 1920, Documentos avulsos lata 1920-C, APMT; letter from Aristides Figueiredo, clerk, Secretário da Agricultura, to the Director Secretário do Governo de Mato Grosso, Cuiabá, 12 March 1921, Documentos avulsos lata 1921-C, APMT; "Destino Para O Nabileque," *Terra e Gente* 2 (January 1957): 78–82. The timing of the claim of such vast territory, only a few short years after a major diplomatic dispute between Brazil and Argentina, probably didn't help the Argentine company's position in Rio.

35. Letter from Oswaldo Aranha, Ministro da Justiça e Negocios Interiores, to the Federal Interventor in Mato Grosso, Rio de Janeiro, 31 August 1931, Documentos avulsos lata 1931-L4, APMT; "Destino Para O Nabileque," 78–82.

36. Corrêa Filho, *A propósito do Boi pantaneiro*, 46–48, 47 n. 58; Foweraker, *The Struggle for Land*, chapter 5.

37. Earl Richard Downes, "The Seeds of Influence: Brazil's 'Essentially Agricultural' Old Republic and the United States, 1910–1930" (Ph.D. diss. in History, University of Texas, 1986), 42–58. The SNA did not appear to have members from Mato Grosso until into the twentieth century. One of the results of the Sociedade's lobbying was the re-creation of the Ministry of Agriculture in 1909. Downes, *The Seeds of Influence*, 54–55; Paulo Roberto Cimó Queiroz, "Joaquim Murtinho, banqueiro: Notas sobre a experiência do Banco Rio e Mato Grosso (1891–1902)," *Estudos Históricos* (Rio de Janeiro) 23, no. 45 (janeiro-junho de 2010): 126.

38. Cimó Queiroz, "Joaquim Murtinho, banqueiro."

39. Gilberto Luiz Alves, "A Trajetória Histórica do Grande Comerciante dos Portos em Corumbá, 1857–1929 (A propósito das determinações econômico-sociais do Casario do Porto)," in *Casario do Porto de Corumbá*, ed. Valmir Batista Corrêa, Lúcia Salsa Corrêa, and Gilberto Luiz Alves (Campo Grande: Fundação de Cultura de Mato Grosso do Sul, 1985), 67–69; Gauld, *The Last Titan*, 220.

40. Paulo Coelho Machado, "Historia das ruas de Campo Grande, A rua 26 de Agosto: Artigos publicados no Jornal da Cidade, 1981/82" (Campo Grande: unpublished), 149–150.

41. Mato Grosso, *Mensagem dirigido pelo Dr. Caetano Manoel de Faria e Albuquerque, Presidente de Matto Grosso, à Assembléa Legislativa, 15 de maio de 1916* (Cuiabá: Typ. Official, 1916), 96–97; Alves, "A Trajetória," 73–74; Lúcia Salsa Corrêa, "Corumbá: O Comércio e o Casario do Porto (1870–1920)," in *Casario do Porto de Corumbá*, 53–54.

42. S. A. Maciel, "Em torno da Pecuária Matto-Grossense. Especialmente a do Pantanal," *Revista da Sociedade Rural Brasileira* 27 (setembro 1922): 498; A Feira de Gado de Três Lagoas, *Creação e installação* (São Paulo: n.p., 1922), 70; Marques, *Matto Grosso*, 172; Brazil, Ministério da Agricultura, Indústria e Commercio, Serviço de Inspecção e Fomento Agrícola, *Estudo dos Factores da Producção no Municípios Brasileiros e Condições economicos de cada um: Estado de Matto Grosso, Município de Corumbá* (Rio de Janeiro: Ministério da Agricultura, 1924), 21–22; Barros Maciel, "A pecuária nos pantanaes de Matto Grosso," 29–30.

43. Matto Grosso, *Mensagem dirigido à Assembléa Legislativa em 13 de maio de 1926, pelo Dr. Mario Corrêa da Costa, Presidente do Estado de Mato Grosso* (Cuiabá: Typ. Official, 1926), 17, 82–83; Brazil, Ministério da Agricultura, Indústria e Commercio, Serviço de inspecção e fomento agrícola, *Estudo dos factores da Producção nos Municípios Brasileiros e condições economicas de cada um: Estado de Matto Grosso, Município de Campo Grande* (Rio de Janeiro: Imprensa Nacional, 1929), 45–46; Sampaio Jorge, "A pecuária em Mato Grosso," 38–39.

44. For a detailed discussion of this issue for all of Brazil between 1910 and 1930, see Downes, *The Seeds of Influence*, chapter 5.

45. Lúcia Helena Gaeta Aleixo, "Mato Grosso: Trabalho escravo e trabalho livre, 1850–1888" (M.A. thesis, Pontífica Universidade Católica de São Paulo, 1980), 64 n. 20, 72–73; Joaquim Ferreira Moutinho, *Notícia sobre a província de Matto Grosso seguida d'um roteiro da viagem da sua capital á S. Paulo* (São Paulo: Typographia

de Henrique Schroeder, 1869), 135–138; Mato Grosso, Director Geral dos Indios, "Relatório de Henrique José Vieira, director geral dos Indios de Matto Grosso, datado de Cuyabá a 16 de Dezembro de 1853, dando conta do estado dos aldeiamentos dos Indios da mesma Província," Colleção Carvalho, CEHB no. 14.880, I-32, 14, 18, Seção de Manuscritos, Biblioteca Nacional, Rio de Janeiro (hereafter BNRJ).

46. Luiz d'Alincourt, "Resultado dos trabalhos e indagações statisticas da Província de Matto-Grosso," *Anais da Biblioteca Nacional* 8 (1880–1881): 46, 103–106; Joaquim Francisco Lopes, "Itinerario de Joaquim Francisco Lopes: Encarregado de explorar a melhor via de communicação entre a província de S. Paulo e a de Matto-Grosso pelo Baixo Paraguay," *Revista do Instituto Histórico e Geographico Brasileiro* 13 (1850): 329–334; João Augusto Caldas, *Memoria histórica sobre os indígenas da província de Matto-Grosso* (Rio de Janeiro: n.p., 1887), 19–20, 47 n. 39–42; Afonso d'Escragnolle Taunay, *Entre os nossos indios: Chanés, terenas, kinikinaus, guanás, laianas, guatós, guaycurús, caingangs* (São Paulo: n.p., 1931), 16, 19–21.

47. Roberto Cardoso de Oliveira, *Do Índio ao Bugre, O processo de assimilação dos Terêna* (Rio de Janeiro: Francisco Alves Editora, 1976), 59; Maria de Fátima Costa, "Indigenous Peoples of Brazil and the War of the Triple Alliance, 1864–1870," in *Military Struggle and Identity Formation in Latin America: Race, Nation, and Community During the Liberal Period*, ed. Nicola Foote and René D. Harder Horst (Gainesville: University Press of Florida, 2010), 159–174.

48. Mato Grosso, Report by Antonio Luiz Brandão, Cuiabá, March 13, 1872, "Relatório acerca dos aldeiamentos de índios da província de Matto Grosso pelo respectivo director geral," Colleção Carvalho, CEHB no. 14.882, I-32, 14, 20, Seção de Manuscritos, BNRJ; João Severiano da Fonseca, *Viagem ao Redor do Brasil, 1875–1878* (1880; Rio de Janeiro: Biblioteca do Exército Editora, 1986), 9–12. These population estimates likely were undercounts.

49. Mato Grosso, Report by Antonio Luiz Brandão, Cuiabá, March 13, 1872, "Relatório acerca dos aldeiamentos," BNRJ. For an overview of the history of native peoples in Brazil, see Manuela Carneiro da Cunha, "Introdução a uma história indígena," *História dos índios no Brasil*, 2nd ed., org. Manuela Carneiro da Cunha (São Paulo: Companhia das Letras, FAPESP, 1998): 9–24. This superb volume includes an essay on native peoples in southern Mato Grosso by Silvia M. Schmuziger Carvalho, "Chaco: Encruzilhada de povos e 'melting pot' cultural, suas relações com a bacia do Paraná e o Sul mato-grossense," 457–474.

50. Darcy Ribeiro, *Os Índios e a Civilização: A Integração das Populações Indígenas no Brasil Moderno* (Petrópolis: Vozes, 1982), 81–82; Roberto Cardoso de Oliveira, *Do Índio ao Bugre*, 59; Kalervo Oberg, *The Terena and the Caduveo of Southern Mato Grosso, Brazil*, publ. 9 (Washington, D.C.: Smithsonian Institution, Institute of Social Anthropology, 1949), 57–60; Antonio de Pádua Bertelli, *Os Fatos e os acontecidos com a poderosa e soberana Nação dos Índios Cavaleiros Guaycurús no Pantanal de Mato Grosso, entre os anos de 1526 até o ano de 1986* (São Paulo: Uyara, 1987), 165. "Caduveo" is the older spelling of "Kadiwéu." The name "Guaikuru" was most common before the war, after which "Kadiwéu" became the norm. The question of the re-

serve is vague, since there are no documents that prove this award, complicating Kadiwéu-settler relations over subsequent years.

51. Bertelli, *Os Fatos*, 154–155; "Colonisação," *A Opinião* (Corumbá) 1 (November 17, 1878): 1–2; letter from Thomaz Antonio de Miranda Loizaga, Director Geral dos Indios, to the President of Mato Grosso, Cuiabá, July 14, 1885, Documentos avulsos lata 1885-E, APMT.

52. Letter from Thomaz Antonio de Miranda Loizaga, July 14, 1885, Documentos avulsos lata 1885-E, APMT; letter from Teodoro Paes da Silva Rondon, Director dos Indios do Municipio de Miranda, to the General Director of Indians, Aquidauana, August 7, 1894, Documentos avulsos lata 1894-A, APMT; letter from Antonio J. Malheiros, Director dos Indios Cadiguios to the President of Mato Grosso, left bank of the Paraguay River, June 2, 1896, Documentos avulsos lata 1896-C, APMT; Emilio Rivasseau, *A Vida dos indios Guaycurús, quinze dias nas suas aldeias* (São Paulo: Companhia Editora Nacional, 1936), 64–68; letter from Antonio J. de Faria Albernaz, Director Geral dos Indios, to the President of Mato Grosso, Cuiabá, November 16, 1899, Documentos avulsos lata 1899-C, APMT; letter from Marianno Rostey, Director dos Indios Cadiuéos, to the President of Mato Grosso, Corumbá, November 21, 1899, Documentos avulsos lata 1899-C, APMT.

53. Letter from Marianno Rostey, Director dos Índios Cadiuéos, to the President of Mato Grosso, Corumbá, February 3, 1900, Documentos avulsos lata 1900-C, APMT.

54. Letter from Antonio Joaquim Malheiros to the President of Mato Grosso, Forte Olimpo, Paraguay, September 7, 1902, Documentos avulsos lata 1902-B, APMT; letters from Marianno Rostey, Director dos Índios Cadiuéos, to the President of Mato Grosso, Corumbá, September 5, 21, 1902, Documentos avulsos lata 1902-D, APMT; letter from Marianno Rostey, Director dos Índios Cadiuéos, to the President of Mato Grosso, Corumbá, July 27, 1903, Documentos avulsos lata 1903-A, APMT.

55. Contract between Marianno Rostey and the Mato Grosso government, Cuiabá, October 14, 1902, Documentos avulsos lata 1902-B, APMT; letter from the Director da Repartição de Terras, Minas e Colonisação, Evaristo Josetti, to the Director dos Índios Cadiuéos, Marianno Rostey, Cuiabá, March 24, 1906, Documentos avulsos lata 1906-A, APMT.

56. Fonseca, *Viagem*, 9–12; Cardoso Ayala and Simon, *Album Graphico do Estado de Matto Grosso*, EEUU do Brasil (Corumbá/Hamburg: n.p., 1914), 89; Oberg, *The Terena*, 3–4; Bertelli, *Os Fatos*, 156, 165.

57. Report from Nicolau Bueno Horta Barbosa, Auxiliar technico da Directoria dos Indios, to Coronel Antonino Menna Gonçalves, Interventor Federal em Mato Grosso, São Paulo, March 6, 1931, Documentos avulsos lata 1931–03, APMT; Bertelli, *Os Fatos*, 166–167. For more on the SPI in Mato Grosso, see chapter 5. For Rondon's contributions, see Todd A. Diacon, *Stringing Together a Nation: Cândido Mariano da Silva Rondon and the Construction of Modern Brazil, 1906–1930* (Durham NC: Duke University Press, 2004).

58. Oberg, *The Terena*, 4, 57–60; Centro Ecumênico de Documentação e Informação, Museu Nacional, Universidade Federal do Rio de Janeiro, *Terras Indígenas no Brasil* (Rio de Janeiro: CEDI, 1987), 63.

59. Fonseca, *Viagem*, 9–12; Mato Grosso, "Report by Antonio Luiz Brandão, Director Geral dos Indios Grosso, Cuiabá, March 13, 1872," BNRJ; Ribeiro, *Os Índios e a Civilização*, 83; Alfred Métraux, "Indians of the Gran Chaco, Ethnography of the Chaco," in *Handbook of South American Indians, vol. 1: The Marginal Tribes, Part 2*, ed. Julian H. Steward (New York: Cooper Square Publishers, 1963), 239, 306–307; Egon Schaden, *Aculturação Indígena: Ensaio sôbre fatôres e tendências da mudança cultural de tribas índias em contacto com o mundo dos brancos* (São Paulo: Pioneira Editôra, Universidade de São Paulo, 1969), 52; Frei Alfredo Sganzerla, *A História do Frei Mariano de Bagnaia* (Campo Grande: FUCMT, 1992), 173–214, 263–303.

60. Though it has circulated widely in the region, no evidence has been found for this latter conclusion.

61. Oberg, *The Terena*, 14; Schaden, *Aculturação Indígena*, 290 n. 6.

62. Cardoso de Oliveira, *Do Índio ao Bugre*, 64 n. 9; Oberg, *The Terena*, 4; Missão Rondon, *Apontamentos sobre os trabalhos realizados pelo Commissão de Linhas Telegraphicas Estrategicas de Matto-Grosso ao Amazonas sob a direcção de Coronel de Engenharia Cândido Mariano da Silva Rondon de 1907 a 1915* (Rio de Janeiro: Typ. do Jornal do Commercio, 1916), 54–57.

63. Letter from Antonio J. de Faria Albernaz, Director Geral dos Indios, to the President of Mato Grosso, Cuiabá, October 25, 1897, Documentos avulsos lata 1897-C, APMT; letter from Manoel Antonio de Barros, Director dos Indios em Miranda, to the President of Mato Grosso, Aquidauana, February 3, 1902, Documentos avulsos lata 1902-D, APMT; Cardoso de Oliveira, *Do Índio ao Bugre*, 75–77.

64. Cardoso de Oliveira, *Do Índio ao Bugre*, 71–88; Métraux, "Indians of the Gran Chaco," in *Handbook of South American Indians*, 240; letter from Antonio Martins Vianna Estigarribia, Interim Inspector for the Serviço de Protecção aos Indios, to the President of Mato Grosso, Cuiabá, November 30, 1922, Documentos avulsos lata 1922-C, APMT; report by Antonio Martins Vianna Estigarribia, Interim Inspector for the Serviço de Protecção aos Indios, to the President of Mato Grosso, Cuiabá, May 31, 1926, Documentos avulsos lata 1926-B, APMT; letter from Affonso Castro, Director General of the Ministério de Trabalho, Indústria e Commercio, to the Interventor Federal of Mato Grosso, Rio de Janeiro, March 12, 1931, Documentos avulsos lata 1931–03, APMT.

65. Brazil, Ministério da Agricultura, Conselho Nacional de Proteção aos Índios, Missão Rondon. *Relatório dos trabalhos realizados de 1900–1906 pela Comissão de Linhas Telegráficas do Estado de Mato Grosso, apresentado as autoridades do Ministério da Guerra pelo Major de Engenharia Cândido Mariano da Silva Rondon, como chefe da Comissão*, Publicação 69–70 (Rio de Janeiro: Imprensa Nacional, 1949), 81–82; Adriano Metello, *O Sul de Mato Grosso* (Rio de Janeiro: n.p., 1937), 8; Cardoso de Oliveira, *Do Índio ao Bugre*, 60, 71–72; Esther de Viveiros, *Rondon conta sua vida* (Rio de Janeiro: n.p., 1958), 202.

66. Missão Rondon, *Apontamentos sobre os trabalhos*, 57–59; Renato Alves Ribeiro, *Taboco, 150 anos: Balaio de Recordações* (Campo Grande: n.p., 1984), 73–84; Ribeiro, *Os Índios e a Civilização*, 84–88.

CHAPTER 5. COWBOYS, HANDS, AND NATIVE PEOPLES

1. For a good comparison of cowboys and social relations in the Americas, though with only scattered reference to Brazil, see Richard W. Slatta, *Cowboys of the Americas* (New Haven: Yale University Press, 1990), especially chapter 4, and Edward Larocque Tinker, *Horsemen of the Americas and the Literature They Inspired*, 2nd ed. (Austin: University of Texas Press, 1967).

2. Virgílio Alves Corrêa Filho, *Pantanais Matogrossenses (Devassamento e Ocupação)* (Rio de Janeiro: IBGE, 1946), 123–124; Antonio de Pádua Bertelli, *O Paraíso das espécias vivas: Pantanal de Mato Grosso* (São Paulo: Cerifa, 1984), 199–200; Adriano Metello, *O Sul de Mato Grosso* (Rio de Janeiro: n.p., 1937), 8; Francisco Antônio Pimenta Bueno, *Memória justificativa dos trabalhos de que foi encarregado à Provincia de Matto Grosso, segundo as instrucções do Ministério da Agricultura de 27 de maio de 1879* (Rio de Janeiro: Typ. Nacional, 1880), 85.

3. Franz Van Dionant, *Le Rio Paraguay et l'État Brésilien de Matto-Grosso* (Brussels: L'Imprimerie Nouvelle, 1907), 161–162; Bertelli, *O Paraíso*, 202; W. Jaime Molins, *Paraguay: Crónicas americanas*, 2nd ed. (Buenos Aires: n.p., 1916), 186; Virgílio Corrêa Filho, *Á sombra dos hervaes matogrossenses*, Monografias Cuiabanas, vol. 4 (São Paulo: São Paulo Editora, 1925), 15; Virgílio Alves Corrêa Filho, *Ervais do Brasil e ervateiros*, Documentário da vida rural, no. 12 (Rio de Janeiro: Ministério da Agricultura, Serviço de Informação Agrícola, 1957), 30–35; Irene Arad, "La ganadería en el Paraguay: Período 1870–1900," *Revista Paraguaya de Sociología* 10 (September–December 1973): 192–195; Paulo Coelho Machado, *Pelas Ruas de Campo Grande, Vol. 1: A Rua Velha* (Campo Grande: Tribunal de Justiça de Mato Grosso do Sul, 1990), 62–65, 80. For a more detailed discussion of the Paraguayan presence in Mato Grosso, see Robert Wilcox, "Paraguayans and the Making of the Brazilian Far West, 1870–1935," *The Americas* 49, no. 4 (April 1993): 479–512.

4. J. Barbosa Rodrigues, *História de Mato Grosso do Sul* (São Paulo: n.p., 1985), 135–146; Corrêa Filho, *Ervais do Brasil*, 50–53; report by Hugo Heyn, Mate Larangeira superintendent, to the president of Mato Grosso, Concepción, Paraguay, November 11, 1897, Documentos avulsos lata 1897-C, Arquivo Público do Estado de Mato Grosso, Cuiabá (hereafter APMT); Miguel Arrojado Ribeiro Lisboa, *Oeste de S. Paulo, Sul de Mato-Grosso: Geologia, Indústria, Mineral, Clima, Vegetação, Solo Agrícola, Indústria Pastoril* (Rio de Janeiro: Typografia do Jornal do Commercio, 1909), 141–145; Metello, *O Sul de Mato Grosso*, 8.

5. Lúcia Helena Gaeta Aleixo, "Mato Grosso: Trabalho escravo e trabalho livre, 1850–1888" (M.A. thesis, Pontífica Universidade Católica de São Paulo, 1980), 24–26, 67, 108; Joaquim Ferreira Moutinho, *Notícia sobre a província de Matto Grosso*

seguida d'um roteiro da viagem da sua capital á S. Paulo (São Paulo: Typographia de Henrique Schroeder, 1869), 34–35; Corrêa Filho, *Pantanais Matogrossenses*, 123–124.

6. Rodolpho Endlich, "A criação do gado vaccum nas partes interiores da America do Sul," *Boletim da Agricultura* (São Paulo) 4, no. 4 (1903): 183–184.

7. Endlich, "A criação," 4 (1903): 183–184; Virgílio Alves Corrêa Filho, *Fazendas de gado no pantanal Matogrossense*, Documentário da vida rural, no. 10 (Rio de Janeiro: Ministério da Agricultura, 1955), 25. Farofa is a coarse manioc flour, commonly toasted with fat and bacon, and customary in Brazilian cuisine.

8. Endlich, "A criação," 4 (1903): 183–186; José de Barros Maciel, "A pecuária nos pantanaes de Matto Grosso," these apresentada ao 3° Congresso de Agricultura e Pecuária, 1922 (São Paulo: Imprenta Metodista, 1922), 18–19; J. Lucídio N. Rondon, *Tipos e aspectos do Pantanal* (São Paulo: n.p., 1972), 77–80. Today, castration normally does not require a knife. The testes cords are crushed with a special type of pliers, a safer and more efficient method that allegedly causes less discomfort for the animal.

9. Machado, *A Rua Velha*, 85–86; Roy Nash, *The Conquest of Brazil* (New York: AMS Press, 1926), 260; M. Cavalcanti Proença, *No têrmo de Cuiabá* (Rio de Janeiro: Biblioteca de Divulgação Cultural, Série A-XVI, 1958), 72–73; Theodore Roosevelt, *Nas selvas do Brasil* (Rio de Janeiro: n.p., 1948), 117. An influential rancher from the south of the state, Coelho Machado was Secretary of Agriculture for Mato Grosso in the early 1970s, a time when the nation and state promoted development at all costs. His opinions reflected the elite perspective on the region and its populations.

10. Proença, *No têrmo de Cuiabá*, 72, 83; Rondon, *Tipos e aspectos do Pantanal*, 50–51, 65; Bertelli, *O Paraíso*, 196–197.

11. Van Dionant, *Le Rio Paraguay*, 161–162; Bertelli, *O Paraíso*, 200–202; Rondon, *Tipos e aspectos do Pantanal*, 151, 154; Metello, *O Sul de Mato Grosso*, 8; Corrêa Filho, *Pantanais Matogrossenses*, 116.

12. Bertelli, *O Paraíso*, 199–200; Corrêa Filho, *Pantanais Matogrossenses*, 114; Brazil, Ministério da Agricultura, Indústria e Commercio, Serviço de Inspecção e Fomento Agrícola, *Aspectos da economia rural brasileira* (Rio de Janeiro: Ministério da Agricultura, 1922), 975; Maurício Coelho Vieira, "A Pecuária," in *Grande Região Centro-oeste*, Vol. 2 of *Geografia do Brasil*, org. Marília Velloso Galvão, 196–199, Publ. 16, Conselho Nacional de Geografia (Rio de Janeiro: IBGE, 1960), 209–211.

13. Aleixo, "Mato Grosso: Trabalho escravo," 72–73, 75.

14. Endlich, "A criação," 4 (1903): 183; Mato Grosso, *Mensagem apresentado à Assembléa Legislativa em 1° de Fevereiro de 1896 pelo Dr. Antonio Corrêa da Costa, Presidente do Estado* (Cuiabá: Typ. do Estado, 1896), 19–20; Lisboa, *Oeste de S. Paulo*, 145. A devaluation of the milreis on the international market was responsible for the low dollar value of wages in 1898. Bertelli, *O Paraíso*, 201–202; Brazil, Ministério da Agricultura, Indústria e Commercio, Serviço de inspecção e fomento agrícola, *Estudo dos factores da Producção nos Municípios Brasileiros e condições economicas*

de cada um: Estado de Matto Grosso, Município de Campo Grande (Rio de Janeiro: Imprensa Nacional, 1929), 41.

15. Brazil, *Aspectos da economia rural brasileira*, 973–975; US Department of Agriculture, Report "The Cattle Industry of Paraguay," September 15, 1922, Records of the Foreign Agricultural Service, Narrative Agricultural Reports, 1904–1954, Paraguay, entry 5, box 429, Record Group 166, United States National Archives (hereafter USNA); Campo Grande, Matto Grosso, *O Município de Campo Grande em 1922*, Relatório do anno de 1922, Apresentado á Camara Municipal pelo Intendente Geral Dr. Arlindo de Andrade Gomes (São Paulo: Cia. Melhoramentos, 1923), 93; Barros Maciel, "A pecuária nos pantanaes de Matto Grosso," 18–19.

16. Campo Grande, *O Município de Campo Grande em 1922*, 92–93.

17. *A Situação* (Corumbá) 6 (January 16, 1873): 2; report by the interim Police Chief of Mato Grosso to the President of Mato Grosso, Corumbá, February 16, 1878, and passed on to the Ministry of Justice in Rio de Janeiro, "Oficios dos Presidentes de Matto Grosso, 1877–1878," IJ¹, no. 689, Arquivo Nacional de Rio de Janeiro (hereafter ANRJ). In this context, illegal slavery refers to the practice of enslaving individuals who had not been born into slavery, such as native peoples and in this case Paraguayan prisoners of war.

18. Genaro Romero, *Repatriación* (Asunción: n.p., 1913), 25; Endlich, "A criação," 4 (1903): 183–184; F. C. Hoehne, "O Grande pantanal de Matto Grosso," *Boletim de Agricultura* (São Paulo) 37 (1936): 467; Corrêa Filho, *Pantanais Matogrossenses*, 124; Fazenda Taboco, various Livros Razão, Livros Diários, 1927–1940, Private archive of Renato Ribeiro, Campo Grande, Mato Grosso do Sul.

19. Bertelli, *O Paraíso*, 87, 200–202; Brazil, *Estudo dos factores . . . Campo Grande*, 42.

20. Eduardo Alfonso Cadavid Garcia, *Análise técnico-econômico da pecuária bovina do Pantanal: Sub-região da Nhecolândia e dos Paiaguás*, circular técnica 15 (Corumbá: EMBRAPA-CPAP, March 1985), 48–49; interview with Dr. Cassio Leite de Barros, Corumbá, September 1990; Abílio Leite de Barros, "O Homen pantaneiro," paper given at conference Pantanal Alerta Brasil, São Paulo, December 17, 1985, 6–10; Abílio Leite de Barros, *Gente pantaneira: Crônicas de sua história* (Rio de Janeiro: Lacerda Editores, 1998), 174–179; interview with Dr. Arnildo Pott, EMBRAPA-CPAP, Corumbá, June 7, 1990.

21. Garcia, *Análise técnico-econômica*, 48–49; Leite de Barros, "O Homen pantaneiro," 8; Paulo Coelho Machado, "Historia das ruas de Campo Grande, A rua 26 de Agosto: Artigos publicados no Jornal da Cidade, 1981/82" (Campo Grande: unpublished), 62; Bertelli, *O Paraíso*, 99.

22. Valmir Batista Corrêa, *Coronéis e bandidos em Mato Grosso (1889–1943)*, 2nd ed. (Campo Grande: Editora UFMS, 2006), 77–179. Another contribution is Lúcia Salsa Corrêa, *História e fronteira: O Sul de Mato Grosso, 1870–1920* (Campo Grande: Editora UCDB, 1999).

23. Batista Corrêa, *Coronéis*; Brazil, *Estudo dos factores . . . Campo Grande*, 41.

24. Batista Corrêa, *Coronéis*, 37. The verse is from Matogrossense poet Hé-

lio Serejo: "Quatrero veio de longe, dos confins do território paraguaio. Surgiu por força de uma sangrenta revolução guarani." Quoted in Batista Corrêa, *Coronéis*, 202.

25. Brazil, *Estudo dos factores* . . . *Campo Grande*, 45; Bertelli, *O Paraíso*, 197; Carlos Vandoni de Barros, *Nhecolândia* (n.p., 1934), 30–31; Leite de Barros, "O Homen pantaneiro," 6.

26. Rondon, *Tipos e aspectos do Pantanal*, 47–48.

27. Endlich, "A criação," 4 (1903): 183; Vandoni de Barros, *Nhecolândia*, 23; "O Gado da Nhecolândia," *O Observador económico e financeiro* (São Paulo) 10 (October 1945): 125–126; Bertelli, *O Paraíso*, 197–199.

28. Renato Alves Ribeiro, *Taboco, 150 anos: Balaio de Recordações* (Campo Grande: n.p., 1984), 33–38, 121, 160; Bertelli, *O Paraíso*, 197; Rondon, *Tipos e aspectos do Pantanal*, 87–90; Eudes Fernando Leite, *Marchas na história: Comitivas e peões-boiadeiros no Pantanal* (Campo Grande: Editora Universidade Federal de Mato Grosso do Sul, 2003), 112–114.

29. Brazil, *Estudo dos factores* . . . *Campo Grande*, 40–41.

30. Paulo Coelho Machado, *Parceria pecuária: Seus Contratos e Modalidades, inclusive o FICA de Mato Grosso* (São Paulo: Editora Saraiva, 1972), 47–59. For a more recent example of parceria that involves the Kadiwéu, see Alain Moreau, "Parceria Pecuária em Terra Indígena: a Novidade Kadiwéu," in *Povos Indígenas no Brasil 500, 1996–2000* (São Paulo: Instituto Socioambiental, 2000), 749–753.

31. Machado, *Parceria pecuária*, 47–49; Cartório do 2º Oficio, "Livros de Escripturas de compra e venta, 1914 a 1930," (Ponta Porã, Mato Grosso do Sul: n.p.), book 2: 35b-36, book 4: 56–58, book 5: 160–161, book 13: 54b–55b.

32. Afonso d'Escragnolle Taunay, ed. *Cartas da campanha de Matto Grosso (1865 a 1866) por o Visconde de Taunay* (São Paulo: n.p., [1942]), 142–143; Corrêa Filho, *Pantanais Matogrossenses*, 115; Proença, *No têrmo de Cuiabá*, 125.

33. Dióres Santos Abreu, "Communicações entre o sul de Mato Grosso de o sudoeste de São Paulo. O comércio de gado," *Revista de História* (São Paulo) 53 (1976): 197.

34. Affonso Costa, *Questões economicas, Fatores da nossa riqueza: Entraves á producção* (Rio de Janeiro: n.p., 1918), 108 n. 5; Matto Grosso, *O Estado de Matto-Grosso (Brazil): Notas e apontamentos uteis aos immigrantes e industrias europeus* (Cuiabá: Typ. Official, 1898), 21; Endlich, "A criação," 4, no. 6 (1903): 271; Machado, *A Rua Velha*, 78; US Department of State, Division of Latin American Affairs, report "Matto Grosso and Its Finances," by US Consul C. R. Cameron, São Paulo, 24 December 1927, No. 832.51 M43/2, p. 27, Reports of US Consuls in Brazil, 1910–1929, microfilm M-519, roll 27, Record Group 59, USNA; US Department of Agriculture, "Markets for Pure Bred Live Stock in South America," preliminary report by David Harrel and H. P. Morgan, Records of the Bureau of Agricultural Economics, General Correspondence of the Bureau of Markets and Bureau of Agricultural Economics, 1912–1952, Brazil, No. 39 [1918], 6, 10–11, Record Group 83, USNA; Nash, *The Conquest of Brazil*, 262; Campo Grande, *O Município de Campo Grande em 1922*, 43. With time, cattle trails improved and the drives became more efficient. In the 1950s, it

took forty days to drive animals from the Pantanal or the Vacaria to São Paulo, a significant reduction from previous decades, though still a considerable length of time. See Maurício Coelho Vieira, "A Pecuária," in *Grande Região Centro-oeste*, 216; and Econ J. Monserrat and Carlos A. Gonçalves, *Observações sobre a pecuária no Brasil Central: Relatório de viagem apresentado ao Ilmo. Sr. Dr. Manoel Corrêa Soares, D.D., Presidente do Instituto Sul Rio Grandense de Carnes, em 8 de agosto de 1953* ([Porto Alegre]: n.p., 1954), 54.

35. Matto Grosso, *O Estado de Matto-Grosso*, 21; Endlich, "A criação," 4, no. 6 (1903): 271; Machado, *A Rua Velha*, 78; "Memoria de la Sociedad Ganadera del Paraguay, Ejercicio 1927–1928," presentado por la Comisión Directiva a la Asamblea General Ordinaria convocada para el 15 de Noviembre de 1928 (Asunción: La Colmena, 1928), 5–6.

36. José Jorge da Silva, "Communicado: O commercio de gado," *Correio da Tarde* (Rio de Janeiro) 4 (December 15, 1858): 2, Doc. Arm. 5, Gav. 3, No. 42, Instituto Histórico e Geográfico Brasileiro (hereafter IHGB); Harrel and Morgan, "Markets for Pure Bred Live Stock," Brazil, No. 39 [1918], 6, 10–11, Record Group 83, USNA; Cameron, "Mato Grosso and Its Finances," Reports of US Consuls, 27, USNA; "A creação em Matto Grosso," *Brasil Agrícola* (Rio de Janeiro) 1 (December 1916): 363; Costa, *Questões economicas*, 108 n. 5; Campo Grande, *O Município de Campo Grande em 1922*, 43; Dolor F. Andrade, *Mato Grosso e a sua pecuária* (São Paulo: Universidade de São Paulo, 1936), 9.

37. Andrade, *Mato Grosso*, 6; Durval Garcia de Menezes, *O Indubrasil: Conferencia pronunciada em 2 de Maio de 1937 no recinto da 3ª Exposição Agro-pecuária de Uberaba, em homenagem á Sociedade Rural do Triangulo Mineiro* (Rio de Janeiro: Sociedade Rural do Triangulo Mineiro, 1937), 24; Antonio Carlos de Oliveira, "Economia pecuária do Brasil Central, Bovinos: III Parte," *Boletim do Departamento Estadual de Estatística* 2 (February 1942): 47, 61.

38. Machado, *A Rua Velha*, 78–81; Ribeiro, *Taboco, 150 anos*, 32–33.

39. Rondon, *Tipos e aspectos do Pantanal*, 95–100; Brazil, *Estudo dos factores . . . Campo Grande*, 34; José Alípio Goulart, *Brasil do boi e do couro*, vol. 1 (Rio de Janeiro: Edições GRD, 1965), 43–44; Corrêa Filho, *Pantanais Matogrossenses*, 117.

40. Machado, *A Rua Velha*, 79–80; Charles A. Gauld, *The Last Titan: Percival Farquhar, American Entrepreneur in Latin America* (Stanford: Institute of Hispanic American and Luso-Brazilian Studies, 1964), 220.

41. Monserrat and Gonçalves, *Observações*, 29; EMBRAPA, CPAP, *Programa nacional de pesquisa do pantanal* (Corumbá: EMBRAPA, 1990), 39; Machado, *A Rua Velha*, 79; Rondon, *Tipos e aspectos do Pantanal*, 95–100; Leite, *Marchas na história*, 106–107, 158.

42. Leite, *Marchas na história*, 153–154, 159, 176; Monserrat and Gonçalves, *Observações*, 29; Proença, *No têrmo de Cuiabá*, 122–125. Curiously, I have found virtually no mention of dogs in accounts of ranching in Mato Grosso. Dogs certainly lived on ranches, and no doubt they worked with cattle. Passing references are made to their presence on drives, but descriptions and pictures seldom include

them. It is unclear whether this was because dogs were not significant partici-pants or because they were taken for granted and no one felt compelled to men-tion them.

43. Rondon, *Tipos e aspectos do Pantanal*, 95–100; Menezes, *O Indubrasil*, 24; Coelho Vieira, "A Pecuária," in *Grande Região Centro-oeste*, 216–217; author's per-sonal observation.

44. Report from José Victorino de Sousa, Capitão Director da Colonia Mili-tar de Miranda, to the President of Mato Grosso, Miranda, August 28, 1880, Docu-mentos avulsos lata 1880-C, APMT; report from João Caetano Teixeira Muzzi, Ca-pitão Director da Colonia Militar do Brilhante, to the President of Mato Grosso, Colonia Brilhante, April 27, 1879, Documentos avulsos lata 1879-C, APMT; Lis-boa, *Oeste de S. Paulo*, 163–164; Darcy Ribeiro, *Os Índios e a Civilização: A Integração das Populações Indígenas no Brasil Moderno* (Petrópolis, RJ: Vozes, 1982), 89–90; Ma-ria Inés Ladeira, "As Demarcações Guarani, a Caminho da Terra Sem Mal," in *Po-vos Indígenas no Brasil 500*, 782–785.

45. Letter from Antonio Martins Vianna Estigarribia, interim inspector for the Serviço de Protecção aos Indios in Mato Grosso, to the President of Mato Grosso, Documentos avulsos lata 1924-A, APMT; report from Antonio Martins Vianna Estigarribia, interim inspector for the Serviço de Protecção aos Indios in Mato Grosso, to the President of Mato Grosso, Documentos avulsos lata 1926-B, APMT; "Protecção aos selvicolas," *O Progresso* (Ponta Porã) 4 (August 5, 1923): 2; "Edital. Delegacia de Policia," *O Progresso* 8 (May 1, 1927): 4; Egon Schaden, *Acul-turação Indígena: Ensaio sôbre fatôres e tendências da mudança cultural de tribas índias em contacto com o mundo dos brancos* (São Paulo: Pioneira Editôra, Universidade de São Paulo, 1969), 36–40; Centro Ecumênico de Documentação e Informação, Museu Nacional, Universidade Federal do Rio de Janeiro, *Terras Indígenas no Brasil* (Rio de Janeiro: CEDI, 1987), 43–113. Smaller bands of Guarani also have survived in-side Paraguay, along the Brazilian border, and in São Paulo, Rio Grande do Sul, and Paraná. Statistics from Instituto Socioambiental's *Povos Indígenas no Brasil* indi-cated that in 1998 there were more than 25,000 Guarani in Mato Grosso do Sul, in several locations. *Povos Indígenas no Brasil 500*, 11, 741–743.

46. Antonio de Pádua Bertelli, *Os Fatos e os acontecidos com a poderosa e soberana Nação dos Índios Cavaleiros Guaycurús no Pantanal de Mato Grosso, entre os anos de 1526 até o ano de 1986* (São Paulo: Uyara, 1987), 157, 165; Claude Lévi-Strauss, *Tristes Tro-piques*, translated by John Russell (New York: Atheneum, 1969), 149–150, 153, 162, 177–178; Ribeiro, *Os Índios*, 82–83.

47. Centro Ecumênico, *Terras Indígenas no Brasil*, 63; Moreau, "Parceria Pecu-ária," 749–753, and 11.

48. Roberto Cardoso de Oliveira, *Do Índio ao Bugre, O processo de assimila-ção dos Terêna* (Rio de Janeiro: Francisco Alves Editora, 1976), 61, 67 n. 34; Lisboa, *Oeste de S. Paulo*, 163–164; Aleixo, "Mato Grosso: Trabalho escravo," 72–73; Bra-zil, Ministério da Agricultura, Conselho Nacional de Proteção aos Índios, Missão Rondon, *Relatório dos trabalhos realizados de 1900–1906 pela Comissão de Linhas Tele-*

gráficas do Estado de Mato Grosso, apresentado as autoridades do Ministério da Guerra pelo Major de Engenharia Cândido Mariano da Silva Rondon, como chefe da Comissão, publicação 1, 69–70 (Rio de Janeiro: Imprensa Nacional, 1949), 83–84.

49. Kalervo Oberg, *The Terena and the Caduveo of Southern Mato Grosso, Brazil,* publ. 9 (Washington, D.C.: Smithsonian Institution, Institute of Social Anthropology, 1949), 4. It is highly ironic that by the 1960s, corruption, exploitation, and patronage within the SPI had created conditions that brought accusations that the SPI was perpetrating genocide on native peoples, especially in the Amazon, leading to its dissolution in 1967 and the creation of today's Fundação Nacional do Índio (FUNAI). The SPI's creation, then, could be considered yet another parallel to later conditions in the country.

50. Chapter 1, Article 6, of the Brazilian Civil Code of 1916 determined that "savages" (*silvícolas*) were considered "relatively unfit" (*incapazes, relativamente a certos actos*), along with citizens between 16 and 21 and those who were considered financially and otherwise irresponsible (*pródigos*). See Brazil, Senado Federal, *Código Civil: Quadro Comparativo 1916/2002* (Brasília: Secretaria Especial de Editoração e Publicações, 2003), 7–8.

51. Letter from Estevão Alves Corrêa, Intendente Geral do Município de Aquidauana, to the President of Mato Grosso, Aquidauana, February 2, 1917, Documentos avulsos lata 1917-B, APMT; letter from Antonio Martins Vianna Estigarribia, Interim Inspector for the SPI, to the President of Mato Grosso, Cuiabá, November 30, 1922, Documentos avulsos lata 1922-C, APMT.

52. Oberg, *The Terena,* 14–15, 36–37; letter from Affonso Castro, General Director of the Ministério de Trabalho, Indústria e Commercio, to the President of Mato Grosso, Rio de Janeiro, March 12, 1931, Documentos avulsos lata 1931–03, APMT.

53. Schaden, *Aculturação Indígena,* 291; Oberg, *The Terena,* 14–15, 34–38, 46–47; Cardoso de Oliveira, *Do Índio ao Bugre,* 77–80.

54. Oberg, *The Terena,* 38; Centro Ecumênico, *Terras Indígenas no Brasil,* 43–91; *Povos Indígenas no Brasil 500,* 14.

CHAPTER 6. THE DYNAMICS OF THE MUNDANE

1. Mato Grosso, *Relatório apresentado à Assembléa Legislativa da Provincia de Matto-Grosso, 4 de outubro de 1872, pelo Presidente, Tenente Coronel Dr. Francisco José Cardozo Junior* (Rio de Janeiro: Typ. Official, 1873), 86–87, 91. All presidential reports are found on microfilm in the Núcleo de Documentação e Informação Histórica Regional, Universidade Federal de Mato Grosso, Cuiabá (NDIHR), and reports to 1930 are online at the Center for Research Libraries Global Resources Network, http://www-apps.crl.edu/brazil/provincial/mato_grosso. Joaquim Ferreira Moutinho, *Notícia sobre a província de Matto Grosso seguida d'um roteiro da viagem da sua capital à S. Paulo* (São Paulo: Typographia de Henrique Schroeder, 1869),

31–35; Virgílio Corrêa Filho, *A propósito do Boi pantaneiro*, Monografias Cuiabanas, vol. 6 (Rio de Janeiro: Pongetti e Cia., 1926), 39–40.

2. João Severiano da Fonseca, *Viagem ao Redor do Brasil, 1875–1878*, vol. 1 (1880; Rio de Janeiro: Biblioteca do Exército Editora, 1986), 165–167; Mato Grosso, "Industria pastoril," *Relatório do vice-presidente Dr. José Joaquim Ramos Ferreira devia apresentar à Assembléa Legislativa Provincial de Matto Grosso, 2ª sessão da 26ª legislatura de Setembro de 1887* (Cuiabá: n.p., 1887).

3. David McCreery, *Frontier Goiás, 1822–1889* (Stanford: Stanford University Press, 2006), chapter 5; Carlos Pereira de Sá Fortes, *Industria pastoril*, relatório apresentado à commissão fundamental do Congresso Agrícola, Commercial e Industrial de Minas (Belo Horizonte: n.p., 1903), 11–12, 72–81; Eduardo Cotrim, *A Fazenda Moderna: Guia do Criador de Gado Bovino no Brasil* (Brussels: Typ. V. Verteneuil et L. Desmet, 1913), 145.

4. Virgílio Alves Corrêa, "Aos Fazendeiros," *Revista da Sociedade Matto-Grossense de Agricultura* (Cuiabá) 1 (May 1907): 13–20, 28–31.

5. Barão de Melgaço, "Apontamentos para o diccionario chorographico da província de Matto-grosso," *Revista do Instituto Histórico e Geographico Brasileiro* 47 (1884): 348; Luiz d'Alincourt, "Resultado dos trabalhos e indagações statisticas da Província de Matto-Grosso," *Anais da Biblioteca Nacional* 3 (1877–1878): 256.

6. Ana Primavesi, *Manejo Ecológico de Pastagens Em Regiões Tropicais e Subtropicais* (São Paulo: Nobel, 1986), 45–46, 113; John C. Tothill, "Uso de Fogo," in *Ecologia e manejo de pastagens nativas na área de sistemas de produção de carne*, A. L. Gardner et al., mimeo Fol. 002 (Campo Grande: EMBRAPA-CNPGC, January 23–26, 1978), 11–15; Jean Koechlin, "Végétation et mise en valeur dans le sud du mato Grosso," in *Le Bassin Moyen du Paraná Brésilien: L'Homme et son Milieu*, ed. Raymond Pebayle et al. (Bordeaux: Centre d'Études de Géographie Tropicale, CNRS, 1978), 126–132; William Sanford and Elizabeth Wangari, "Tropical grasslands: Dynamics and utilization," *Nature and Resources* 21 (July–September 1985): 13; Dominique Gillon, "The Fire Problem in Tropical Savannas," in *Tropical Savannas: Ecosystems of the World, 13*, ed. François Bourlière (Amsterdam: Elsevier Scientific Publishing Co., 1983), 617–627; quote from Fausto Vieira de Campos, *Retrato de Mato Grosso* (São Paulo: n.p., 1960), 69.

7. Letter from Alfredo Taunay, Matto Grosso, 9 December 1865, in *Cartas da campanha de Matto Grosso (1865 a 1866) por o Visconde de Taunay*, ed. Afonso D'Escragnolle Taunay (São Paulo: n.p., [1942]), 135–136.

8. Joaquim Carlos Travassos, *Monographias agrícolas* (Rio de Janeiro: Typographia Altina, 1903), 68.

9. Franz Van Dionant, *Le Rio Paraguay et l'État Brésilien de Matto-Grosso* (Brussels: L'Imprimerie Nouvelle, 1907), 76.

10. J. Barbosa Rodrigues, *Palmae mattogrossenses* (Rio de Janeiro: n.p., 1898), X–XI, Colección Enrique Solano López, No. 637, Biblioteca Nacional de Asunción (hereafter BNA); Moises S. Bertoni, *Condiciones generales de la vida organica y division territorial* (Puerto Bertoni, Paraguay: n.p., 1918), 44–45; A. Marques, *Matto*

Grosso: Seus recursos naturaes, seu futuro economico (Rio de Janeiro: Papelaria Americana, 1923), 63–64; Antonio Carlos Simoens da Silva, *Cartas Mattogrossenses* (Rio de Janeiro: n.p., 1927), 17–18; Antonio C. Allem and José F. M. Valls, *Recursos forrageiros nativos do Pantanal mato-grossense* (Brasília: EMBRAPA, 1987), 191; various letters from Alfredo D'E. Taunay, in *Cartas da campanha de Matto Grosso (1865 a 1866)*, ed. Afonso D'E. Taunay, 63, 73, 108–110, 132.

11. *Sexto Congresso Agrícola do Estado de São Paulo, 15 a 18 de dezembro de 1912* (São Paulo: n.p., [1912]), 50; Tothill, "Uso de Fogo," 13; C. R. Cameron, "Through Matto Grosso," *Bulletin of the Pan American Union* 66 (March 1932): 166–167.

12. Rodolpho Endlich, "A criação do gado vaccum nas partes interiores da America do Sul," *Boletim da Agricultura* 4, no. 1 (1903): 27–28 (from a series of articles that appeared in 1902 and 1903, in seven continuous issues); Cotrim, *A Fazenda Moderna*, 90.

13. Primavesi, *Manejo*, 62–64, 75–77, 113; Allem and Valls, *Recursos forrageiros*, 223–226.

14. Luiz D'Alincourt, "Resultado," *Anais da Biblioteca Nacional* 3 (1877–1878): 256; Fonseca, *Viagem ao Redor do Brasil* 1, 169; Renato Alves Ribeiro, *Taboco, 150 anos: Balaio de Recordações* (Campo Grande: n.p., 1984), 27, 102; Allem and Valls, *Recursos forrageiros*, 189–190; José de Barros, *Lembranças para os meus filhos e descendentes* (São Paulo: n.p., 1987), 45.

15. Octavio Domingues and Jorge de Abreu, *Viagem de estudos à Nhecolândia*, Publ. 3 (Rio de Janeiro: Ministério da Agricultura, Instituto de Zootecnia, Dezembro 1949), 11; Arnildo Pott, *Pastagens no Pantanal* (Corumbá: EMBRAPA, CPAP, 1988), 25; Simoens da Silva, *Cartas Mattogrossenses*, 72; Carolina Joana da Silva, "Nota prévia sobre o significado biológico dos termos usados no Pantanal Matogrossense I, Batume e Diquada," *Universidade, Revista da Universidade Federal de Mato Grosso* 4 (May–August 1984): 30–35; "Ocorrência do Fenômeno Natural 'dequada' no Pantanal," Website *Ambiente Brasil* http://ambientes.ambientebrasil .com.br/agua/artigos_agua_doce/ocorrencia_do_fenomeno_natural_%E2%80%9C dequada%E2%80%9D_no_pantanal.html (accessed September 2015); F. Iglesias, "O Toque," *Brasil Agrícola* (Rio de Janeiro) 1 (November 1916): 346.

16. José de Melo e Silva, *Fronteiras Guaranís* (São Paulo: Imprensa Metodista, 1939), 146–147; Domingues and Abreu, *Viagem de estudos à Nhecolândia*, 21.

17. Cotrim, *A Fazenda Moderna*, 145.

18. Mato Grosso, *Mensagem dirigido pelo Dr. Caetano Manoel de Faria e Albuquerque, Presidente de Matto Grosso, à Assembléa Legislativa, 15 de maio de 1916* (Cuiabá: Typ. Official, 1916), 15–18, 89–97; A Feira de Gado de Três Lagoas, *Creação e installação* (São Paulo: n.p., 1922), 70.

19. Mato Grosso, *Mensagem pelo Dr. Joaquim A. da Costa Marques, Presidente do Estado à Assembléa Legislativa, 13 de maio de 1913* (Cuiabá: Typ. Official, 1913), 33–34.

20. Mato Grosso, *Mensagem pelo Dr. Joaquim A. da Costa Marques . . . 1913*, 74; São Paulo, Secretaria de Estado dos Negocios da Agricultura, Commercio e Obras Públicas, *Almanach para o anno de 1917* (São Paulo: Secretaria de Estado, 1917), 40,

56, 113, 122; Geraldo Leme da Rocha and Ademir Giacomo Pietrosanto, "O Instituto de Zootecnia e a agropecuária Brasileira," *Instituto de Zootecnia, Boletim Técnico* (Nova Odessa, SP) 20 (1986): 3–16; A Feira de Gado de Três Lagoas, *Creação e installação*, 7–10, 23; Pedro Celestino Corrêa da Costa, "Crise da pecuária," *Revista da Sociedade Rural do Brasil* (São Paulo) 27 (Setembro 1922): 523. Though not a major cattle producer at the time, São Paulo was in the forefront of most agricultural innovation in Brazil and had created not only schools and experimental farms but also zootechnical posts for the care and study of animals. For São Paulo's overall role, see Warren Dean, "The Green Wave of Coffee: Beginnings of Tropical Agricultural Research in Brazil (1885–1900)," *Hispanic American Historical Review* 69, no. 1 (February 1989): 91–115.

21. Mato Grosso, *Mensagem dirigido pelo Dr. Caetano Manoel de Faria e Albuquerque . . . 1916*, 23–25; *Mensagem à Assembléa Legislativa, 7 de setembro de 1920 por D. Francisco de Aquino Corrêa, Bispo de Prusiade, Presidente do Estado* (Cuiabá: Typ. Official, 1920), 85–87, 96–104; *Mensagem do Presidente de Mato Grosso, Dom Francisco de Aquino Corrêa, 7 de setembro de 1919* (Cuiabá: Typ. Official, 1919), 16–17; *Mensagem dirigido à Assembléa Legislativa em 13 de maio de 1926, pelo Dr. Mario Corrêa da Costa, Presidente do Estado de Mato Grosso* (Cuiabá: Typ. Official, 1926), 81–82; *Mensagem à Assembléa Legislativa, 13 de maio de 1927, por Mario Corrêa, Presidente do Estado de Mato Grosso* (Cuiabá: Typ. Official, 1927), 155; letter from the director of the Fazenda Modelo de Criação, A. Teixeira Vianna, to the president of Mato Grosso, Campo Grande, 2 August 1925, Documentos avulsos lata 1925-B, Arquivo Público do Estado de Mato Grosso (hereafter APMT); Mato Grosso, *Mensagem apresentado à Assembléa Legislativa pelo Presidente de Mato Grosso, Dr. Annibal Toledo, 13 de maio de 1930* (Cuiabá: Imprensa Oficial, 1930), 24.

22. One argument made elsewhere is that national governments concentrated on the inadequacy of regional ranchers to modernize in order to emphasize the importance of the growing national state and to limit the power of landed elites. If that was the case in Mato Grosso, it appears to have failed until the Vargas dictatorship, though one could argue it applied once the Estado Novo was established. See Shawn Van Ausdal, "Reimagining the Tropical Beef Frontier and the Nation in Early Twentieth-Century Colombia," in *Trading Environments: Frontiers, Commercial Knowledge and Environmental Transformation, 1820–1990*, ed. Gordon Winder and Andreas Dix, chapter 8 (London: Routledge, 2015).

23. Arlindo de Andrade, *Erros da federação* (São Paulo: n.p., 1934), 77; Dolor F. Andrade, *Mato Grosso e a sua pecuária* (São Paulo: Universidade de São Paulo, 1936), 8; Maurício Coelho Vieira, "A Pecuária," in *Grande Região Centro-oeste*, Vol. 2 of *Geografia do Brasil*, org. Marília Velloso Galvão, 206, Publ. 16, Conselho Nacional de Geografia (Rio de Janeiro: IBGE, 1960); "História," EMBRAPA Gado de Corte, https://www.embrapa.br/gado-de-corte/historia (accessed September 2015).

24. Bruno Garcia, "A Cia. Feira de Gado, tem nos beneficiado? Não," *A Noticia* (Três Lagoas) 5 (August 6, 1925): 1; A.G., "Com a Feira," *A Noticia* 5 (August 13, 1925): 6; Mato Grosso, *Mensagem dirigido . . . 1926 . . . Mario Corrêa da Costa*, 98–100.

25. Letter from the mayor of Três Lagoas, Bruno Garcia, to the Mato Grosso Secretary of Agriculture, Três Lagoas, 1 March 1930, Documentos avulsos lata 1930-C, APMT.

26. "Posto de Assistencia Veterinaria," *Gazeta de Commercio* (Três Lagoas) 7 (January 16, 1927): 1; Domingues and Abreu, *Viagem de estudos à Nhecolândia*, 32–33; report by Dr. Arthur Moses to the 2º Congresso Brasileiro de Expansão Economica, "Secção III, Indústria Animal, . . . Setembro-Outubro 1919," Documentos avulsos lata 1919-D, APMT; Estado de Mato Grosso, "Relatorio da Secretaria da Agricultura, Industria, Commercio, Viação e Obras Publicas, referente ao exercicio de 1921," by Carlos Gomes Borralha, Documentos avulsos lata 1921-C, APMT; Mato Grosso, *Mensagem apresentado . . . Annibal Toledo . . . 1930*, 18–24.

27. Simoens da Silva, *Cartas Mattogrossenses*, 51.

28. Endlich, "A criação," *Boletim da Agricultura* 4, no. 2 (1903): 82–84, and no. 3 (1903): 128–129; Miguel Arrojado Ribeiro Lisboa, *Oeste de S. Paulo, Sul de Mato-Grosso: Geologia, Indústria, Mineral, Clima, Vegetação, Solo Agrícola, Indústria Pastoril* (Rio de Janeiro: Typografia do Jornal do Commercio, 1909), 141–150.

29. Roy Nash, *The Conquest of Brazil* (New York: AMS Press, 1926), 255; "Primeiro Congresso de Pecuária," *Correio Paulistano* (São Paulo) September 19, 1916: 2; Andrade, *Mato Grosso e a sua pecuária*, 11; Econ J. Monserrat and Carlos A. Gonçalves, *Observações sobre a pecuária no Brasil Central: Relatório de viagem apresentado ao Ilmo. Sr. Dr. Manoel Corrêa Soares, D.D., Presidente do Instituto Sul Rio Grandense de Carnes, em 8 de agosto de 1953* ([Porto Alegre]: n.p., 1954), 53; Maurício Coelho Vieira, "A Pecuária," in *Grande Região Centro-oeste*, 196–199.

30. Letter from José Jorge da Silva, "Communicado: O commercio de gado," *Correio da Tarde* (Rio de Janeiro) 4 (December 15, 1858): 2, Doc. Arm. 5, Gav. 3, No. 42, Instituto Histórico e Geográfico Brasileiro (hereafter IHGB); F. M. Draenfert, "Valor forrageiro de plantas brasileiras," *Revista Agricola* (São Paulo) 1 (outubro 1895): 68–70; J. Carlos Travassos, *Industria pastoril*, Fasciculo 2 (Rio de Janeiro: Sociedade Nacional de Agricultura, 1898), 29–30; Sandro Dutra Silva, Rosemeire Aparecida Mateus, Vivian da Silva Braz, and Josana de Castro Peixoto, "A Fronteira do Gado e a *Melinis minutiflora* P. Beauv. (Poaceae): A História Ambiental e as Paisagens Campestres do Cerrado Goiano no Século XIX," *Sustentabilidade em Debate* (Brasília) 6, no. 2 (mai/ago 2015): 22.

31. Campo Grande, Matto Grosso, *O Município de Campo Grande em 1922*, Relatório do anno de 1922, Apresentado á Camara Municipal pelo Intendente Geral Dr. Arlindo de Andrade Gomes (São Paulo: Cia. Melhoramentos, 1923), 48–49, 57, 65–77; Brazil, Ministério da Agricultura, Indústria e Commercio, Serviço de inspecção e fomento agrícola, *Estudo dos factores da Producção nos Municípios Brasileiros e condições economicas de cada um: Estado de Matto Grosso, Município de Campo Grande* (Rio de Janeiro: Imprensa Nacional, 1929), 13–14, 32; Dutra Silva et al., "A Fronteira do Gado," 17–32; Cotrim, *A Fazenda Moderna*, 90–109; Monserrat and Gonçalves, *Observações*, 18–19.

32. Roberto Cochrane Simonsen, *The Meat and Cattle Industry of Brazil: Its Importance to Anglo-Brazilian Commerce* (London: Industrial Publicity Service, 1919), 12–13; Maurício Coelho Vieira, "A Pecuária," in *Grande Região Centro-oeste*, 191–193, 195–197; Raymond Pebayle and Jean Koechlin, "Les fronts pionniers du Mato Grosso méridional: Approche géographique et écologique," in *Le Bassin Moyen*, 149.

33. Alves Corrêa, "Aos Fazendeiros," 28–31; Endlich, "A criação," *Boletim da Agricultura* 4, no. 3 (1903): 131–132; Domingues and Abreu, *Viagem de estudos à Nhecolândia*, 13, 21–22; José de Barros Maciel, "A pecuária nos pantanaes de Matto Grosso," these apresentada ao 3° Congresso de Agricultura e Pecuária, 1922 (São Paulo: Imprenta Metodista, 1922), 27.

34. Corrêa Filho, *A propósito do Boi pantaneiro*, 46–52; Carlos Vandoni de Barros, *Nhecolândia* (n.p., 1934), 32–33; "O Gado de Nhecolândia," *O Observador econômico e financeiro* (São Paulo) 10 (October 1945): 124.

35. Francisco Fortes de Pinho, "As Cêrcas," *Cadernos AABB* (Rio de Janeiro) 23 (1957): 10–23.

36. Delegacia Fiscal do Tesouro Nacional em Mato Grosso, Alfândega de Corumbá, Capatazia, Guias de Importação, 1895–1917, microfilm rolls 41–43, NDIHR; Delegacia . . . , Mesa de Renda Alfandegada de Porto Murtinho, Guias de Importação, 1908–1944, microfilm rolls 44–45, NDIHR; Delegacia . . . , Mesa de Renda Alfandegada de Port Esperança, Guias de Importação, 1918–1931, microfilm roll 44, NDIHR.

37. Fortes de Pinho, "As Cêrcas," 65–66; Lisboa, *Oeste de S. Paulo*, 142–145; Sociedade Nacional de Agricultura, *O Rebanho bovino brasileiro e a exportação de carnes* (Rio de Janeiro: n.p., 1916), 5–12; Antonio de Pádua Bertelli, *O Paraíso das espécies vivas: Pantanal de Mato Grosso* (São Paulo: Cerifa, 1984), 240–241; Antonio Carlos de Oliveira, *Economia pecuária do Brasil Central: Bovinos* (São Paulo: Departamento Estadual de Estatística de São Paulo, 1941), 9 n. 1; Andrade, *Mato Grosso e a sua pecuária*, 10–11; Domingues and Abreu, *Viagem de estudos à Nhecolândia*, 21–22.

38. Fonseca, *Viagem ao Redor do Brasil*, 166–167; Virgílio Alves Corrêa Filho, *Fazendas de gado no pantanal Matogrossense*, Documentário da vida rural, no. 10 (Rio de Janeiro: Ministério da Agricultura, 1955), 25–27.

39. Endlich, "A criação," *Boletim da Agricultura* 4, no. 2 (1903): 83–84, and no. 6 (1903): 283–285; Lisboa, *Oeste de S. Paulo*, 143–144, 147.

40. Andrade, *Mato Grosso e a sua pecuária*, 10–11. By comparison, the average birthing rate in the United States from 1930 to the 1990s was 88 percent. David Aadland, "The Economics of Cattle Supply," working paper (December 1999), 15. http://citeseerx.ist.psu.edu/viewdoc/download?doi=10.1.1.196.5782&rep=rep1&type=pdf (accessed September 2015). The Argentine figures are lower and comparable to Mato Grosso, roughly 75 percent in the 1930s. Lovell S. Jarvis, "Supply Response in the Cattle Industry: The Argentine Case," special report (Giannini Foundation of Agricultural Economics, University of California, August 1986), Appendix V, 104.

41. Endlich, "A criação," *Boletim da Agricultura* 3, no. 12 (1902): [748], and 4, no. 6 (1903): 184; Lisboa, *Oeste de S. Paulo*, 147.

42. Lisboa, *Oeste de S. Paulo*, 146; Endlich, "A criação," *Boletim da Agricultura* 4, no. 6 (1903): 276; Jacomo Vicenzi, *Paraiso Verde: Impressões de uma viagem a Matto Grosso em 1918* (n.p.), 57; Andrade, *Mato Grosso e a sua pecuária*, 10–11; Monserrat and Gonçalves, *Observações*, 57; Octavio Domingues, *O Zebu, sua reprodução e multiplicação dirigida* (São Paulo: Nobel, 1971), 58. A study in the Pantanal in the 1980s revealed calf mortality of 21 percent, considerably higher than the rates cited here. The difference may have something to do with either better data collection or the location of the study, which was conducted largely in Nhecolândia, a region that had been degraded gradually over preceding years by overpasturing. See Eduardo Alfonso Cadavid Garcia, *Análise técnico-econômico da pecuária bovina do Pantanal: Sub-região da Nhecolândia e dos Paiaguás*, circular técnica 15 (Corumbá: EMBRAPA-CPAP, March 1985), 56.

43. Endlich, "A criação," *Boletim da Agricultura* 4, no. 4 (1903): 183–186; Fortes de Pinho, "As Cêrcas," 25.

44. Virgílio Alves Corrêa Filho, *Pantanais Matogrossenses (Devassamento e Ocupação)* (Rio de Janeiro: IBGE, 1946), 29; General Cândido Rondon, *Matto-Grosso, O que elle nos offerece e o que espera de nós. Conferenci realizada a 31 de Julho de 1920, pelo Exmo. Snr. General Cândido Mariano da Silva Rondon, perante a Sociedade Rural Brasileira no cidade de S. Paulo* (São Paulo: n.p., 1920), 9–10; F. Iglesias, "O Toque," *Brasil Agrícola* (Rio de Janeiro) 1 (November 1916): 345–346.

45. João Leite de Barros, "A Pecuária em Corumbá," *Anuário de Corumbá, 1939* (Corumbá: n.p., 1939), 8–11; Allem and Valls, *Recursos forrageiros*, 46–47; Nash, *The Conquest of Brazil*, 252.

46. Lisboa, *Oeste de S. Paulo*, 145; Corrêa Filho, *A propósito do Boi pantaneiro*, 38 n. 47; Mato Grosso, *Mensagem pelo Dr. Joaquim A. da Costa Marques . . . 1913*, 22–23.

47. Barros Maciel, "A pecuária nos pantanaes de Matto Grosso," 21; Alberto Maranhão, *These no. 30: Sal.* Conferencia nacional de pecuária (Rio de Janeiro: n.p., 1917), 2–3; Andrade, *Mato Grosso e a sua pecuária*, 11; Oliveira, *Economia pecuária do Brasil Central: Bovinos*, 9 n. 1.

48. Endlich, "A criação," *Boletim da Agricultura* 3 (1902): 818; *Commercial Almanach "Matto-Grossense"* (São Paulo: n.p., 1916), 180; Antonio da Silva Neves, *Primeira conferencia nacional de pecuária: Origem provavel das diversas raças que povoam o territorio patrio, alimentação racional, hygiene animal* (São Paulo: n.p., 1918), 90–91; "Propaganda de Matto Grosso e a zona Noroeste," *Almanaque Illustrado* (Três Lagoas) 2 (1929): 348; Primavesi, *Manejo*, 55–56.

49. Ribeiro, *Taboco, 150 anos*, 110–111; J. Lucídio N. Rondon, *Tipos e aspectos do Pantanal* (São Paulo: n.p., 1972), 79; "O Gado de Nhecolândia," *O Observador económico e financeiro* (São Paulo) 10 (Outubro 1945): 125: "Sal é o melhor vaqueiro."

50. Rondon, *Tipos e aspectos do Pantanal*, 139–140; Ribeiro, *Taboco, 150 anos*, 110; Cadavid Garcia, *Análise técnico-econômica*, 53; Afonso Simões Corrêa, "Pecuária de

corte em Mato Grosso do Sul," report delivered at Encontros Regionais de Pecuária de Corte, Brasília, 27 November 1984, 11.

51. Moutinho, *Noticia*, 31–35; Barros, *Lembranças*, 30; Endlich, "A criação,"
Boletim da Agricultura 4, no. 6 (1903): 279–280, and no. 5 (1903): 221; Paulo Coelho
Machado, "Historia das ruas de Campo Grande, A rua 26 de Agosto: Artigos publicados no Jornal da Cidade, 1981/82" (Campo Grande: unpublished), 67; Van Dionant, *Le Rio Paraguay*, 127–128.

52. Geoffrey P. West, ed., *Black's Veterinary Dictionary* (London: Adam &
Charles Black, 1979), 87–788; Marc Desquesnes, Philippe Holzmuller, De-Hua
Lai, Alan Dargantes, Zhao-Rong Lun, and Sathaporn Jittaplapong, "*Trypanosoma evansi* and Surra: A Review and Perspectives on Origin, History, Distribution, Taxonomy, Morphology, Hosts, and Pathogenic Effects," *BioMed Research
International* 2013, Hindawi Publishing Corp., Open Access; personal interview
with Dr. Arnildo Pott, EMBRAPA-CPAP, Corumbá, June 7, 1990; Ribeiro, *Taboco, 150 anos*, 29–30; Bertelli, *O Paraíso*, 87; Corrêa Filho, *Fazendas de gado*, 44–
45. Surra is well-known among horses, cattle, and camels in Africa and Asia. The
connection between outbreaks in both continents was not made until recently,
however.

53. Rondon, *Matto-Grosso*, 14; C. Pereira and W. F. de Almeida, "Observações
sobre parasitologia humana e veterinária em Mato Grosso," *Memorial do Instituto
Oswaldo Cruz* 36 (1941): 302–309.

54. Mato Grosso, *Relatório do Exm. Snr. Coronel Barão de Maracajú, Presidente
da Província de Matto Grosso, 5 de Dezembro, 1879* (Cuiabá: n.p., 1879), 43–45; *Relatorio do Presidente Gustavo Galvão, 3 de maio de 1881* (Cuiabá:, n.p., 1881), 29; Mato
Grosso, *O Estado de Matto-Grosso (Brazil): Notas e apontamentos uteis aos immigrantes
e industrias europeus* (Cuiabá: Typ. Official, 1898), 23; Corrêa Filho, *Fazendas de gado*,
45; Bertelli, *O Paraíso*, 225. Endlich for one believed that surra was transmitted by
intestinal worms; others thought leeches were the culprits. See Endlich, "A criação," *Boletim da Agricultura* 4 (1903): 283 n. 1, and Barros, *Lembranças*, 73. The anon-
ymous Paraguayan was actually the team of Drs. Miguel Elmassian and Luis Enrique Mignone, who made their original discovery of insect vectors in 1901 and
later published a widely distributed paper in France on the subject: M. Elmassian
and E. Mignone, "Sur le Mal de caderas ou flagellose parésiante des équidés sud-
américaines," *Annales de l'Institut Pasteur* (Paris) 17, no. 4 (Avril 1903): 241–267. At
roughly the same time, the highly respected Brazilian researcher Adolfo Lutz confirmed the same in Marajó. See Cicero Neiva, "Adolfo Lutz e a Medicina Veterinária," *Revista do Instituto Adolfo Lutz* 15, no. 1–2 (1955): 1–7.

55. Mato Grosso, *Mensagem dirigido pelo Dr. Joaquim A. da Costa Marques, Presidente do Estado de Mato Grosso, à Assembléa Legislativa, 13 de maio de 1915* (Cuiabá:
Typ. Official, 1915), 47–52; *Mensagem dirigido pelo Dr. Caetano Manoel de Faria e Albuquerque . . . 1916*, 28–31.

56. Andrade, *Mato Grosso e a sua pecuária*, 11; Ribeiro, *Taboco, 150 anos*, 29–30;

Terra e Gente (Rio de Janeiro) 1 (January 1956): 93. Though there have been few recorded outbreaks of the disease since the development of effective prophylactic drugs, Dr. Paulo C. Machado, the late rancher and former Secretary of Agriculture for Mato Grosso, emphasized that without continued prophylactic action the disease could still be devastating. Interview with Paulo C. Machado, Campo Grande, 22 May 1990; also Paulo Coelho Machado, *Pelas Ruas de Campo Grande, Vol. 1: A Rua Velha* (Campo Grande: Tribunal de Justiça de Mato Grosso do Sul, 1990), 87.

57. Tesouraria da Fazenda Nacional em Mato Grosso and Delegacia Fiscal do Tesouro Nacional em Mato Grosso, Alfândega de Corumbá, Capatazia, Guias de Importação, 1884–1895 and 1903–1904, microfilm rolls 8–41, NDIHR; Bertelli, *O Paraíso*, 223–225; Ribeiro, *Taboco, 150 anos*, 30; José de Barros Netto, *A vontade natural e o Pantanal da Nhecolândia* (São Paulo: Editora Alfa-Omega, 2001), 70; Lisboa, *Oeste de S. Paulo*, 144, 149.

58. Machado, *A Rua Velha*, 86–87; Brazil, Ministério da Agricultura, Indústria e Commercio, Directoria Geral de Estatística, *Estimativa do Gado existente no Brazil em 1916* (Rio de Janeiro: Typ. da Estatística, 1917), 66; Ministério da Agricultura, Indústria e Commercio, Directoria Geral da Estatística, *Synopse do recenseamento realizado em 1 de setembro de 1920: População pecuária* (Rio de Janeiro: Typographia da Estatística, 1922), 26; Brazil, Ministério da Agricultura, Indústria e Commercio, Serviço de inspecção e fomento agrícola, *Estudo dos factores . . . Campo Grande*, 36; Brazil, Mato Grosso, Instituto Nacional de Estatística, Diretoria de Estatística e Publicidade, *Sinopse estatística do Estado, no. 2*, Ano III, 1937, Separata, com acréscimos (Cuiabá: Imprensa official, 1938), 42; Ribeiro, *Taboco, 150 anos*, 29–30.

59. Bertelli, *O Paraíso*, 223; Estevão Alves Corrêa Filho, "O Cavalo pantaneiro," *Revista Médica Veterinária* (São Paulo) 8 (March 1973): 296–305; Octavio Domingues, *Contribuição ao estudo do cavalo pantaneiro* (Rio de Janeiro: Instituto de Zootecnia, publicação 17, abril 1957), 5–17; Domingues and Abreu, *Viagem de estudos à Nhecolândia*, 18–19.

60. Endlich, "A criação," *Boletim da Agricultura* 4 (1903): 281; West, *Black's Veterinary Dictionary*, 35–36, 84–85; Bertelli, *O Paraíso*, 225–226; Nelson Catunda, "Pecuária, sua organização e desenvolvimento," *Livro do Centenario da Camara dos Deputados, 1826–1926* (Rio de Janeiro: Camara dos Deputados, 1926), 444; Travassos, *Monographias agrícolas*, 335–337; US Department of Agriculture, "Markets for Pure Bred Live Stock in South America," by David Harrel and H. P. Morgan, Records of the Bureau of Agricultural Economics, General Correspondence of the Bureau of Markets and Bureau of Agricultural Economics, 1912–1952, Brazil, No. 39 [1918], 5–6, 10–11, Record Group 83, USNA; Brazil, Ministério da Agricultura, *Manual Técnico Para Criação de Gado de Corte em Mato Grosso* (Campo Grande: Embratur, 1978), 21–33; EMBRAPA-CPAP, *Programa nacional de pesquisa do pantanal* (Corumbá: EMBRAPA, 1990), 33; personal observation, May 1990.

61. Fred Brown, "The history of research in foot-and-mouth disease," in

Foot-and-Mouth Disease, ed. David J. Rowlands, 3–7 (Amsterdam: Elsevier, 2003); Paul Sutmoller, Simon S. Barteling, Raul Casas Olascoaga, and Keith J. Sumption, "Control and eradication of foot-and-mouth disease," in *Foot-and-Mouth Disease*, ed. David J. Rowlands, 104; Lourenço Granato, *A febre aphtosa* (São Paulo: n.p., 1913), 6–7; Eduardo Cotrim, *Indústria Pecuária: Problemas da indústria pecuária no Repúbica Argentina e estudo comparativo com o Brasil* (Rio de Janeiro: Typ. Serviço Estatístico, 1912), 163–164; US Department of State, Division of Latin American Affairs, "Mato Grosso and Its Finances," by C. R. Cameron, US Consul, São Paulo, 24 December 1927, *Reports of US Consuls in Brazil, 1910–1929*, microfilm M-519, roll 27, letter no. 832.51M43/2, USNA; Cadavid Garcia, *Análise técnico-econômica*, 57–58; Alberto Alves Santiago, *O Zebu na Índia, no Brasil e no mundo* (Campinas, SP: Instituto Campineiro de Ensino Agrícola, 1985), 717.

62. Barros, *Lembranças*, 63; Simões Corrêa, "Pecuária de corte em Mato Grosso do Sul," 11; EMBRAPA, *Programa nacional*, 33. Foot-and-mouth disease comprises seven serotypes worldwide, of which three are found in the Americas.

63. Paul Sutmoller and Raul Casas Olascoaga, "The successful control and eradication of foot-and-mouth disease epidemics in South America in 2001," presented to a meeting of the Temporary Committee on Foot-and-Mouth Disease of the European Parliament, 2 September 2002, Strasbourg, 1–3. By the 2010s, regular vaccination was providing sufficient protection to permit some regions of South America, particularly Uruguay, to market their beef to North America. In August 2015, a Brazilian beef import rule was made final, in which the United States allowed entry of chilled or frozen beef from vaccinated animals from a number of states in Brazil, including Mato Grosso and Mato Grosso do Sul. For details, see: United States, Federal Register, "Importation of Beef from a Region in Brazil," https://www.federalregister.gov/articles/2015/07/02/2015-16337/importation-of-beef-from-a-region-in-brazil#table_of_contents.

64. Santiago, *O Zebu na Índia*, 659–661, 717; "O Fim do zebú no México," *Revista dos Criadores* (São Paulo) 17 (July 1946): 22; "Liberado o Zebú no México," *Revista dos Criadores* 17 (November 1946): 22; "Zebú no México," *Revista dos Criadores* 18 (January 1947): 18, 20; Manuel A. Machado, *Aftosa: A Historical Survey of Foot-and-Mouth Disease and Inter-American Relations* (Albany: State University of New York, 1969), 37–42, 45. For their positions on the rancorous debate in Mexico, see Guillermo Quesada Bravo, *La Verdad sobre el Ganado Cebú Brasileño, la fiebre Aftosa y la cuarentena en la Isla de Sacrificios, Veracruz* (Mexico, D.F.: n.p., 1946), and Ing. Marte R. Gómez, *La Verdad sobre los Cebus: Conjeturas sobre la Aftosa* (Mexico, D.F.: n.p., 1948). P. Rossi, "King Ranch, a maior fazenda de criar do mundo e a raça bovina Santa Gertrudis," *Revista de Indústria Animal* (São Paulo) 2 (January 1939): 163–167. The United States also imported more than 200 zebu bulls and cows from Brazil through Mexico between 1923 and 1925. Santiago argues that these animals helped establish the American Brahman, though to be fair most of the breed was developed in the United States itself. Santiago, *O Zebu na Índia*, 106–107, 611–647.

65. Robert Southey, *History of Brazil, Part the Third* (1870; New York: Franklin, 1970), 376; "Cattle Ranching in Paraguay," *Paraguay* (Asunción) 1 (February 28, 1913): 14; letter from José Jorge da Silva, "Communicado": 2, IHGB; Endlich, "A criação," *Boletim da Agricultura* 4 (1903): 282.

66. West, *Black's Veterinary Dictionary*, 589–590. Lopes Oliveira Filho, *Combate ao berne* (São Paulo: n.p., 1922), 4–7.

67. Nash, *The Conquest of Brazil*, 253.

68. Luiz D'Alincourt, *Anais da Biblioteca Nacional* 3 (1877–1878): 256–257; West, *Black's Veterinary Dictionary*, 588–591; Brazil, Ministério da Agricultura, *Manual Técnico*, 33; Lopes Oliveira Filho, *Combate ao berne*, 9–10, 23; Travassos, *Monographias agrícolas*, 360; Domingues, *O Zebu*, 32–34; Cameron, "Through Matto Grosso," 166–167; Bertelli, *O Paraíso*, 255; Monserrat and Gonçalves, *Observações*, 52; Allem and Valls, *Recursos forrageiros*, 12.

69. Mato Grosso, *Mensagem apresentada pelo Governador do Estado, Fernando Corrêa da Costa, por ocasião da abertura da sessão legislativa de 1955* (Cuiabá: Imprensa Oficial, 1955), 88–90.

70. West, *Black's Veterinary Dictionary*, 565, 596. The tick was previously named *Boophilus microplus* or *B. argentinus*.

71. Domingues, *O Zebu*, 24–32; *Sexto Congresso Agricola do Estado de São Paulo, 15 a 18 de dezembro de 1912* (São Paulo: n.p., [1912]), 41–42; Monserrat and Gonçalves, *Observações*, 52. Most observers today conclude that the tick and zebu likely evolved together on the Indian subcontinent, thus permitting the zebu to develop resistance.

72. US Department of State, "Mato Grosso and Its Finances," Reports of US Consuls in Brazil, microfilm M-519, roll 27, 28, Record Group 59, USNA; Bertelli, *O Paraíso*, 225–226; Luis Freire Esteves and Juan C. González Peña, *El Paraguay Constitucional, 1870–1920* (Buenos Aires: Empresa Gráfica del Paraguay G. Peña y Cía., 1921), 222; Brazil, Ministério da Agricultura, *Manual Técnico*, 32–33.

73. Corrêa Filho, *Fazendas de gado*, 12–13; Corrêa Filho, *Pantanais Mato-grossenses*, 70. On some ranches, professional jaguar hunters were employed year-round.

74. Ribeiro, *Taboco, 150 anos*, 199–200; "O Gado da Nhecolândia," 126; Bertelli, *O Paraíso*, 253–254.

75. Bertelli, *O Paraíso*, 245–254; Pereira da Cunha, *Descrições de viagens* (n.p.), 90; Otto Willi Ulrich, *Nos sertões do Rio Paraguay* (São Paulo: n.p., 1936), 99–102; interview with Dr. Arnildo Pott, EMBRAPA-CPAP, Corumbá, 7 June 1990. Bertelli also reported that young heron chicks were slaughtered for their feathers and meat as early as the 1860s. Bertelli, *O Paraíso*, 281.

76. Annibal Amorim, *Viagens pelo Brasil: do Rio ao Acre, aspectos da Amazônia, do Rio a Matto Grosso* (Rio de Janeiro: Garnier, 1917), 474.

77. Cleber José Rodrigues Alho, "Manejo da fauna silvestre," in *Anais do 1° Simpósio sobre Recursos Naturais e Sócio-Econômicos do Pantanal*, Corumbá, Mato

Grosso do Sul, 28 de novembro a 4 de dezembro de 1984 (Brasília: EMBRAPA-CPAP, 1986), 190–192.

78. Ribeiro, *Taboco, 150 anos*, 200–205; *Programa nacional de pesquisa do pantanal* (Corumbá: EMBRAPA-CPAP, 1990), 33; Alho, "Manejo da fauna silvestre," 194.

79. Ribeiro, *Taboco, 150 anos*, 200–205; interview with Dr. Cassio Leite de Barros, Corumbá, September 1990. This lacuna has been partially relieved by a chapter on the commodification of wildlife in the dissertation of Jason B. Kauffman, "The Unknown Lands: Nature, Knowledge, and Society in the Pantanal of Brazil and Bolivia," University of North Carolina, Chapel Hill, 2015, chapter 5.

80. Antonio Correa do Couto, "Dissertação sobre o actual governo da república do Paraguay" (Rio de Janeiro: Typ. do Imperial Instituto Artístico, 1865), 98–99.

81. Van Dionant, *Le Rio Paraguay*, 77; Gillon, "The Fire Problem," in *Tropical Savannas*, 628–636; Juhani Ojasti, "Ungulates and Large Rodents of South America," in *Tropical Savannas*, 431, 436–437.

82. Travassos, *Monographias agrícolas*, 54–56; Durval Garcia de Menezes, *O Indubrasil: Conferencia pronunciada em 2 de Maio de 1937 no recinto da 3ª Exposição Agro-pecuária de Uberaba, em homenagem á Sociedade Rural do Triangulo Mineiro* (Rio de Janeiro: Sociedade Rural do Triangulo Mineiro, 1937), 24; Primavesi, *Manejo*, 140; Pebayle and Koechlin, "Les fronts pionniers," in *Le Bassin Moyen du Paraná Brésilien*, 162.

CHAPTER 7. NATIONAL BREEDS AND HINDU IDOLS

1. For recent studies on the origins and genetics of cattle, especially tropical cattle, see David Caramelli, "The Origins of Domesticated Cattle," *Human Evolution* 21, no. 2 (2006): 107–122; Daniel G. Bradley, Ronan T. Loftus, Patrick Cunningham, and David E. MacHugh, "Genetics and Domestic Cattle Origins," *Evolutionary Anthropology* 6, no. 3 (1998): 79–86; W. J. A. Payne and J. Hodges, *Tropical cattle: Origins, breeds and breeding policies* (London: Wiley-Blackwell, 1998). For a provocative and entertaining study of animals, breeding, and history, see Harriet Ritvo, *Noble Cows and Hybrid Zebras: Essays on Animals and History* (Charlottesville: University of Virginia Press, 2010).

2. Alberto Alves Santiago, *O Zebu na Índia, no Brasil e no mundo* (Campinas, SP: Instituto Campineiro de Ensino Agrícola, 1985), 8, 12–15; William Norris, ed., *The Heritage Illustrated Dictionary of the English Language* (New York: Heritage Publishing, 1975), 1488; Oklahoma State University, Department of Animal Science, *Breeds of Livestock*, online: http://www.ansi.okstate.edu/breeds/cattle (accessed September 2015). Apparently the term "zebu" was first used at the Paris Agricultural Fair of 1752 to refer to *Bos indicus*. Tibetan "Zen" or "Zeba" some-

times has been translated as "hump of the camel." This was first put forward in the late nineteenth century by H. A. Jäschke. See H. A. Jäschke, *A Tibetan-English Dictionary* (London: Secretary of State for India, 1881), 489. The Hobson-Jobson Anglo-Indian glossary argues that if there is a subcontinent origin, it may be the Ladak word "zobo," referring to a hybrid: http://dsalsrvo2.uchicago.edu/cgi-bin /philologic/contextualize.pl?p.2.hobson.1219108 (accessed September 2015).

3. S. Cardoso Ayala and Feliciano Simon, *Album Graphico do Estado de Matto-Grosso*, EEUU do Brazil (Corumbá/Hamburg: n.p., 1914), 288–289; Eduardo Cotrim, *A Fazenda Moderna: Guia do Criador de Gado Bovino no Brasil* (Brussels: Typ. V. Verteneuil et L. Desmet, 1913), 135; Manuel Paulino Cavalcanti, *Raças de carne* (Rio de Janeiro: Ministério da Agricultura, Indústria e Commercio, 1928), 33–34.

4. Miguel Arrojado Ribeiro Lisboa, *Oeste de S. Paulo, Sul de Mato-Grosso: Geologia, Indústria Mineral, Clima, Vegetação, Solo Agrícola, Indústria Pastoril* (Rio de Janeiro: Typografia do Jornal do Commercio, 1909), 136–137; J. Lucídio N. Rondon, *Tipos e aspectos do Pantanal* (São Paulo, n.p., 1972), 58–59; Cotrim, *A Fazenda Moderna*, 136–145; Dolor F. Andrade, *Mato Grosso e a sua pecuária* (São Paulo: Universidade de São Paulo, 1936), 7; Octavio Domingues and Jorge de Abreu, *A pecuária nos pantanaes de Matto Grosso*, these apresentada ao Terceiro Congresso de Agricultura e Pecuária, 1922 (São Paulo: n.p., 1922), 17; Maria Cristina Medeiros Mazza, Carlos Alberto da Silva Mazza, José Robson Bezerra Sereno, Sandra Aparecida Santos, and Aiesca Oliveira Pellegrin, *Etnobiologia e conservação do bovino pantaneiro* (Corumbá: EMBRAPA-CPAP, 1994), chapter 3.

5. Antonio Carlos de Oliveira, "Economia pecuária do Brasil Central, Bovinos: I Parte," *Boletim do Departamento Estadual de Estatística* (São Paulo) 10 (Outubro 1941): 65; José de Barros Maciel, *A pecuária nos pantanaes de Matto Grosso*, these apresentado ao 3º Congresso de Agricultura e Pecuária, 1922 (São Paulo: Imprenta Metodista, 1922), 29; Ricardo Ernesto Ferreira de Carvalho, *Indústria pastoril: Promptuário de noções geraes e especias de zootecnia* (São Paulo: n.p., 1906), 162; interview with Dr. Cassio Leite de Barros, Corumbá, Mato Grosso do Sul, September 1990; Medeiros Mazza et al., *Etnobiologia e conservação do bovino pantaneiro*, chapter 4.

6. Rodolpho Endlich, "A criação do gado vaccum nas partes interiores da America do Sul," *Boletim da Agricultura* 3, no. 12 (1902): 742–744; Lisboa, *Oeste de S. Paulo*, 137–139; Cotrim, *A Fazenda Moderna*, 136–145.

7. Endlich, "A criação," *Boletim da Agricultura* 3, no. 12 (1902): 745–746; Lisboa, *Oeste de S. Paulo*, 140; Cotrim, *A Fazenda Moderna*, 136–145; Carvalho, *Indústria pastoril*, 159–160.

8. Mato Grosso, *Relatório do vice-presidente Dr. José Joaquim Ramos Ferreira devia apresentar à Assembléa Legislativa Provincial de Matto Grosso, 2ª sessão da 26ª legislatura de Setembro de 1887* (Cuiabá: n.p., 1887); Lisboa, *Oeste de S. Paulo*, 140; Joaquim Carlos Travassos, *Indústria pastoril*, Fasciculo 2 (Rio de Janeiro: Sociedade Nacional de Agricultura, 1898), 35–36; Fernando Ruffier, *Dos Meios de melhorar as raças nacionaes*, Primeira Conferencia Nacional de Pecuária, these 12 (Rio de Janeiro: n.p.,

1917), 58–59, 65–66. For a revealing discussion of the experience in Rio Grande do Sul just a few short years previously, see Stephen Bell, *Campanha Gaúcha: A Brazilian Ranching System, 1850–1920* (Stanford: Stanford University Press, 1998), 99–117.

9. Santiago, *O Zebu na Índia*, 176.

10. Santiago, *O Zebu na Índia*, 114–115, 168–171; José Jorge da Silva, "Communicado: O commercio de gado," *Correio da Tarde* (Rio de Janeiro) 4 (December 15, 1858): 2, Doc. Arm. 5, Gav. 3, No. 42, Instituto Histórico e Geográfico Brasileiro (hereafter IHGB).

11. Santiago, *O Zebu na Índia*, 169–171; Maria Antonia Borges Lopes and Eliane M. Marquez Rezende, *ABCZ, 50 Anos de História e Estorias* (Uberaba, Minas Gerais: Associação Brasileira de Criadores de Zebú, 1984), 17–19, 22–23. Prices charged Triângulo ranchers by the importers were high, between 100 and 400 milreis (US$55 to $220) in 1875, depending on the age, quality, and sex of the animal.

12. Lopes and Rezende, *ABCZ*, 31; Santiago, *O Zebu na Índia*, 119–134, 143, 168–171.

13. Virgílio Corrêa Filho, *A propósito do Boi pantaneiro*, Monografias Cuiabanas, vol. 6 (Rio de Janeiro: Pongetti e Cia., 1926), 48–50; Fernando Ruffier, *Guerra ao Zebú, um pouco de agua fria . . .* (Castro, Paraná: n.p., 1919), 7–10; Paulo de Moraes Barros, *O Sul de Matto Grosso e a pecuária* (n.p., 1922), 12, 17–21; Matto Grosso, Prefeitura Municipal de Campo Grande, *Relatório 1943* (Rio de Janeiro: Imprensa Nacional, 1944), 36; Mato Grosso, Município de Campo Grande, *Relatório do anno de 1922, Apresentado á Camara Municipal pelo Intendente Geral Dr. Arlindo de Andrade Gomes* (São Paulo: Companhia Melhoramentos, [1922]), 15–16; John Mackenzie, "A story of the Brazil Land Co.," n.d., Matador Land and Cattle Company Records, 1874–1960, Headquarters Collection Records, 1881–1960, box 13, Brazil Land, Cattle and Packing Company, Southwest Collection/Special Collections Library, Texas Tech University, Lubbock (hereafter SWC/SCL-TTU).

14. Andrade, *Mato Grosso e a sua pecuária*, 6; Endlich, "A criação," *Boletim da Agricultura* 3 (1902): 744–745; Paulo Coelho Machado, *Pelas Ruas de Campo Grande, Vol. 1: A Rua Velha* (Campo Grande: Tribunal de Justiça de Mato Grosso do Sul, 1990), 93–95; Arsenio López Decoud, *Album Gráfico de la República del Paraguay, 1811–1911* (Buenos Aires: n.p., 1911), CIII–CIX.

15. Lisboa, *Oeste de S. Paulo*, 152–153; Endlich, "A criação," *Boletim da Agricultura* 3 (1902): 745; Machado, *A Rua Velha*, 93; Antonio da Silva Neves, *Primeira conferencia nacional de pecuária: Origem provavel das diversas raças que povoam o territorio patrio, alimentação racional, hygiene animal* (São Paulo: n.p., 1918), 50–52; Campo Grande, Matto Grosso, *O município de Campo Grande em 1922*, Relatório do anno de 1922, Apresentado á Camara Municipal pelo Intendente Geral Dr. Arlindo de Andrade Gomes (São Paulo: Cia. Melhoramentos, 1923), 43.

16. Lisboa, *Oeste de S. Paulo*, 139.

17. Ruffier, *Dos Meios*, 3, 53–57; Travassos, *Indústria pastoril*, 34; Eduardo A. Torres Cotrim, "Contribuição para o estudo das vantagens ou desvantagens da introducção do sangue do gado Zebú nas nossas manadas," in *Inquerito sobre o gado*

zebú (Rio de Janeiro: Sociedade Nacional de Agricultura, 1907), 87–89; Arary Prudente Corrêa, "A raça caracú e a pecuária do Brasil central," *Gado Caracú, Orgão da Associação Herd Book Caracú* (São Paulo) 2 (janeiro 1937): 9–10. Today, the caracu is raised in limited numbers to keep its gene pool available, although it has no significant role in commerce.

18. Lopes and Rezende, *ABCZ*, 34. Pereira Barreto was also a planter and public health expert in São Paulo. He was a major figure in the movement for agricultural improvement in the state and was at one time Secretary of Agriculture. His contribution to São Paulo's economic development was recognized by naming a town after him.

19. Lisboa, *Oeste de S. Paulo*, 154; Cotrim, *A Fazenda Moderna*, 135; Cotrim, "Contribuição," 71–92. Similar ideas were expressed across tropical America. See Shawn Van Ausdal, "Reimagining the Tropical Beef Frontier and the Nation in Early Twentieth-Century Colombia," in *Trading Environments: Frontiers, Commercial Knowledge and Environmental Transformation, 1820–1990*, ed. Gordon Winder and Andreas Dix (London: Routledge, 2015), chapter 8. For a more extensive discussion of the environmental impact of zebu in Central Brazil, see Robert W. Wilcox, "Zebu's Elbows: Cattle Breeding and the Environment in Central Brazil, 1890–1960," in *Territories, Commodities and Knowledges: Latin American Environmental Histories in the Nineteenth and Twentieth Centuries*, ed. Christian Brannstrom, 218–246 (London: Institute for the Study of the Americas, 2004), 218–246.

20. Carlos Pereira de Sá Fortes, *Industria pastoril*, relatório apresentado à commissão fundamental do Congresso Agrícola, Commercial e Industrial de Minas (Belo Horizonte: n.p., 1903), 4–7, 11–12, 18–24; Lopes and Rezende, *ABCZ*, 35–36.

21. Joaquim Carlos Travassos, *Monographias agrícolas* (Rio de Janeiro: Typographia Altina, 1903), 257–296, 321–323, 330–332; Robert Wallace, *India in 1887* (Edinburgh: Oliver and Boyd, 1888), 52–54, 99–100, 312–314.

22. Santiago, *O Zebu na Índia*, 18–25; Octavio Domingues, *O Zebu, sua reprodução e multiplicação dirigida* (São Paulo: Nobel, 1971), 14–18.

23. Santiago, *O Zebu na Índia*, 168–171; Lopes and Rezende, *ABCZ*, 27–28; Alexandre Barbosa da Silva, *O Zebú na India e no Brasil* (Rio de Janeiro: n.p., 1947), 71–72.

24. Corrêa Filho, *A propósito do Boi pantaneiro*, 44–46; "A creação em Matto Grosso," *Brasil Agrícola* (Rio de Janeiro) 1 (December 1916): 362–363; General Cândido Rondon, *Matto-Grosso, O que elle nos offerece e o que espera do nós. Conferencia realizada a 31 de Julho de 1920, pelo Exmo. Snr. General Cândido Mariano da Silva Rondon, perante a Sociedade Rural Brasileira no cidade de S. Paulo* (São Paulo: n.p., 1920), 19; José de Barros, *Lembranças para os meus filhos e descendentes* (São Paulo: n.p., 1987), 63; "Congresso de Pecuária," *Correio Paulistano* (São Paulo) September 23, 1916: 3–4.

25. The concept of heterosis (hybrid vigor), still debated among geneticists, surfaced in the nineteenth century and entered the scientific lexicon in 1914. See James F. Crow, "90 Years Ago: The Beginning of Hybrid Maize," *Genetics* 148 (March 1998): 923–928. Ruffier's observation paralleled the ideas in genetics then

beginning to emerge, though it is unclear whether he was aware of the most recent research. The concept was not referred to by Brazilian ranchers until many decades later, and even the scientific literature of the time is silent.

26. Ruffier, *Dos Meios*, 39–42, 58–59, 65–66, 72–78; Neves, *Primeira conferencia nacional de pecuária*, 58–59, 63–68. For a brief recent discussion of the controversy, as well as references to imported grasses and fencing wire, see Ricardo Ferreira Ribeiro, "The Ox from the Four Corners of the World: The Historic Origins of the Brazilian Beef Industry," *Agrarian South: Journal of Political Economy* 1, no. 3 (December 2012): 315–340.

27. Ruffier, *Guerra ao Zebú*, 7–10; Domingues, *O Zebu*, 40, 43. The Brazilian preference for lean meat is best illustrated in accounts of consumers turning up their noses to imported Argentine beef, complaining that "you buy meat and receive fat and tallow." See Domingues, *O Zebu*. Van Ausdal argues that such controversy in Colombia was more about providing beef to the masses, as a way to improve their "race," than about boosting exports. Perhaps that idea was implicit, but it was not expressed by Brazilian observers, who were obsessed with economic expansion. The ready availability of dried beef may have superseded such an approach. Van Ausdal, "Reimagining the Tropical Beef Frontier."

28. Ruffier, *Guerra ao Zebú*, 18–28. Two clarifications should be made concerning Ruffier's arguments. It appears the British importation of Australian beef probably included few if any zebu. According to Frank O'Loghlen, the first organized import of zebu into Australia occurred in 1933, fifteen years after the 1918 ban. This is not to say that there were no zebu in Australia in 1918, but their numbers were certainly insufficient to play a notable part in meat exports, particularly to finicky Britain. Frank O'Loghlen, ed., *Beef Cattle in Australia, 1956* (Sydney: F. H. Johnston Publishing, 1956), 79–80. See also Santiago, *O Zebu na Índia*, 675–676. In addition, at the time there were few areas in the world raising zebu for beef export besides Brazil. It is possible Egypt exported some beef to Britain at the time, but the amount would have been minuscule in comparison with imports from other, more traditional producers. The ban on zebu was directed exclusively at Brazil. Ruffier's argument about poor preparation was relevant, though, since it went to the heart of the matter.

29. Ruffier, *Guerra ao Zebú*, 18–28. There was a more serious challenge to Argentina from Britain in 1900, when Argentine meat was again temporarily banned, this time because of a British outbreak of foot-and-mouth traced to imported Argentine beef. The ban provoked Argentina to enact its first animal sanitary law in 1903, subsequently followed by other Latin American nations. See David Sheinin, "Defying infection: Argentine foot-and-mouth disease policy, 1900–1930," *Canadian Journal of History* 29, no. 3 (December 1994): 503. Ironically, one group that quietly supported a halt in imports from India was Mineiro zebu breeders. The reason was obvious.

30. Santiago, *O Zebu na Índia*, 143–145, 169–170.

31. West, *Black's Veterinary Dictionary*, 166–168. Southern Africa suffered a

devastating epidemic among native cattle in the late nineteenth century, which facilitated European occupation of the lands of native Africans. See Pule Phoofolo, "Epidemics and Revolutions: The Rinderpest Epidemic in Late Nineteenth-Century Southern Africa," *Past and Present* 138 (February 1993): 112–143.

32. US Department of State, "Imports of Zebu Cattle Suspended," Trade report 292, January 7, 1922, Records of the Foreign Agricultural Service, Narrative Agricultural Reports, 1904–1954, Brazil, entry 5, box 64, Record Group 166, United States National Archives (hereafter USNA); Mato Grosso, *Mensagem dirigido à Assembléa Legislativa, 7 de setembro de 1921, por D. Francisco de Aquino Corrêa, Bispo de Prusiade, Presidente do Estado* (Cuiabá: Typ. Official, 1921), 71, 97; Santiago, *O Zebu na Índia*, 143–145; Brazil, "Relatorio da Secretaria da Agricultura, Industria, Commercio, Viação e Obras Publicas, referente ao exercicio de 1921, Estado de Matto Grosso," 10–20, Documentos avulsos lata 1921-C, Arquivo Público do Estado de Mato Grosso (hereafter APMT).

33. Oswaldo Affonso Borges, *O Zebú do Brasil* (Uberaba, Minas Gerais: Sociedade Rural do Triângulo Mineiro, 1947), 199; Alberto Alves Santiago, *O Nelore: Origem, Formação e Evolução do Rebanho* (São Paulo: n.p., 1958), 125–130.

34. Santiago, *O Zebu na Índia*, 169–170; "Propaganda de Matto Grosso e a zona Noroeste," *Almanaque Illustrado* (Três Lagoas) 2 (1929): 243–246.

35. "Uma opinião sobre o gado zebú," *Gazeta de Commercio* (Três Lagoas) 7 (July 27, 1927): 4. Vestey had three frigoríficos in Brazil, at Barretos and Santos in São Paulo and at Mendes near Rio de Janeiro.

36. "O inquerito sobre o gado zebú: 'Precisamos julgar definitivamente o gado zebú,'" *Rural: Orgam da Sociedade Rural de Cuiabá* 1 (August 1930): 1–5.

37. Andrade, *Mato Grosso e a sua pecuária*, 7–8; Gervásio Leite, *O gado na economia matogrossense* (Cuiabá: Escolas Profissionais Salesianos, 1942), 9–11; Antonio Carlos de Oliveira, "Economia pecuária do Brasil Central, Bovinos," 184–185. There was some debate over whether Mato Grosso had a higher percentage of zebu blood in its herds than other regions. Durval Menezes argued in 1940 that Goiás produced more beef per animal because there was a higher concentration of zebu in that state than in Mato Grosso. Durval Garcia de Menezes, "O Zebú: Riqueza Paulista," *O Zebú* (Uberaba) 1 (August 1940): 21.

38. Gastão de Oliveira, "Intensifica: Se cada vez mais a criação de Gado Caracú nos pantanais de Mato Grosso," *Gado Caracú* (São Paulo) 10 (1945): 31–35; interview with Renato Alves Ribeiro, Campo Grande, May 29, 1990; "O Gado da Nhecolândia," *O Observador económico e financeiro* (São Paulo) 10 (October 1945): 124; Octavio Domingues and Jorge de Abreu, *Viagem de estudos à Nhecolândia* (Rio de Janeiro: Instituto de Zootecnica, Ministério da Agricultura, 1949), 15–18.

39. Santiago, *O Zebu na Índia*, 72–73, 81, 86. Two recent excellent Brazilian dissertations have addressed the issues of zebu breeding and genetics from distinct perspectives. See Joana Medrado, "Do Pastoreio a Pecuária: A invenção da modernização rural nos sertões do Brasil Central," Tese de doutorado, Programa de Pós-Graduação em História da Universidade Federal Fluminense, Niterói, 2013;

and Natacha Simei Leal, "Nome aos Bois. Zebus e zebueiros em uma pecuária de elite," Tese de doutorado, Programa de Pós-Graduação em Antropologia Social da Universidade de São Paulo, São Paulo, 2014.

40. Durval Garcia de Menezes, *O Indubrasil: Conferencia pronunciada em 2 de Maio de 1937 no recinto da 3ª Exposição Agro-pecuária de Uberaba, em homenagem á Sociedade Rural do Triangulo Mineiro* (Rio de Janeiro: Sociedade Rural do Triangulo Mineiro, 1937), 8; Santiago, *O Zebu na Índia*, 468.

41. Menezes, *O Indubrasil*, 16–19; Santiago, *O Nelore*, 125–131.

42. Menezes, "O Zebú: Riqueza Paulista," 15.

43. Menezes, "O Zebú: Riqueza Paulista," 18; Lopes and Rezende, *ABCZ*, 63–65; Santiago, *O Zebu na Índia*, 467, 470, 475.

44. Brazil, Ministério da Agricultura, *Manual Técnico Para Criação de Gado de Corte em Mato Grosso* (Campo Grande: Embratur, 1978), 2–3; Santiago, *O Zebu na Índia*, 170–171, 273, 354, 440.

45. Santiago, *O Zebu na Índia*, 106–107. There have been a few scattered exports, but Santiago's prediction has not gained legs in India. Brazilian genetic material, including embryos, has found a modest market in China. As noted in the Introduction, in recent years Indian beef exports have increased dramatically, referred to as the Pink Revolution. Most writers argue that beef slaughtered in India is primarily buffalo (carabeef, not considered by many to be beef), but some observe that a significant amount of the meat is sourced from beef cows. See Sena Desai Gopal, "Selling the Sacred Cow: India's Contentious Beef Industry," *Atlantic*, February 12, 2015, http://www.theatlantic.com/business/archive/2015/02/selling-the-sacred-cow-indias-contentious-beef-industry/385359/ (accessed June 2015).

46. Lopes and Rezende, *ABCZ*, 35–36; Oscar Lisboa, "Congresso de Pecuária," *Correio Paulistano* (São Paulo) September 26, 1916: 4. The joke is that the jogo de bicho is a semilegal lottery popular throughout Brazil that uses animal (*bicho*, in Portuguese) representations.

CONCLUSION. TRANSFORMATION AND CONTINUITY

1. The works cited in the Introduction reveal the extent of this attention. The most recent and welcome addition is Jeffrey Hoelle's *Rainforest Cowboys: The Rise of Ranching and Cattle Culture in Western Amazonia* (Austin: University of Texas Press, 2015), an anthropological approach that introduces the nexus of economy and environment through the cultural experiences of small ranchers and rubber tappers in Acre.

2. Brazil, IBGE, SIDRA, Banco de Dados Agregados, *Pecuária*, 2015, http://www.sidra.ibge.gov.br/bda/pecua/default.asp?t=2&z=t&o=24&u1=1&u3=1&u4=1&u5=1&u6=1&u7=1&u2=35 (accessed June 2015); Brazil, IBGE, "Efetivos da Pecuária em 1920, 1940 e 1950, Segundo as Regiões Fisiográficas e as Unidades da Federação,"

Série Nacional, *Vol. 2: Censo Agrícola do Brasil, 1950* (Rio de Janeiro: IBGE, 1956), 128; Associação Brasileira das Indústrias Exportadoras de Carne, online report, "Exportações Brasileiras de Carne Bovina, jan.-out. 2012," 2–5, http://www.abiec .com.br/download/Relatorioexportacao2012_jan_out.pdf (accessed June 2015); Brazil, IBGE, "Efetivo dos rebanhos de grande porte em 31.12, segundo as Grandes Regiões e as Unidades da Federação—2011," table 3 in *Produção da Pecuária Municipal*, 2011, vol. 39 (Rio de Janeiro: IBGE, 2011): 1–63, ftp://ftp.ibge.gov.br /Producao_Pecuaria/Producao_da_Pecuaria_Municipal/2011/tabelas_pdf/tab03 .pdf (accessed January 2013).

3. Brazil, IBGE, Estatísticas do Século XX, *Econômicas*, 2015, http:// seculoxx.ibge.gov.br/images/seculoxx/economia/atividade_economica/setoriais /agropecuaria/6_04ab_agro1920_96.xls (accessed June 2015).

4. Brazil, IBGE, SIDRA, Banco de Dados Agregados, *Censo Demográfico e Contagem da População*, 2015, http://www.sidra.ibge.gov.br/bda/tabela/listabl.asp ?z=cd&o=2&i=P&c=200 (accessed June 2015); Brazil, Banco de Dados Agregados, *Pecuária*, 2015. These statistics also indicate that in Amazonia, cattle populations doubled between 1995 and 2006, from 17 million head to 32.5 million. For the Mato Grossos, cattle numbers skyrocketed, from 3.5 million head in 1950 to 33 million head in 1995, 20 million in Mato Grosso do Sul alone. By 2011, that population had surged to 51 million, almost 30 million in Mato Grosso alone. ftp://ftp.ibge.gov.br /Producao_Pecuaria/Producao_da_Pecuaria_Municipal/2011/tabelas_pdf/tab03 .pdf (accessed January 2013).

5. Silvio R. Duncan Baretta and John Markoff, "Civilization and Barbarism: Cattle Frontiers in Latin America," *Comparative Studies in Society and History* 20 (October 1978): 587–620; Alistair Hennessy, *The Frontier in Latin American History* (London: Edward Arnold, 1978), 82–89.

6. Gomercindo Rodrigues and Linda Rabben, *Walking the Forest with Chico Mendes: Struggle for Justice in the Amazon* (Austin: University of Texas Press, 2007). This is the most complete full-length biography of Chico Mendes and the rubber tapper struggle yet published.

7. There have been many news stories of murders of activists like Chico Mendes in recent years, including of an American nun, Dorothy Stang, in 2005. Of course, it is often ordinary Brazilians who have suffered the most through such violence, many of them local environmental activists. For example, see Jon Lee Anderson, "Murder in the Amazon," *The New Yorker*, June 15, 2011.

8. Valmir Batista Corrêa, *Coronéis e bandidos em Mato Grosso, 1889–1943*, 2nd ed. (Campo Grande: Editora UFMS, 2006); Joe Foweraker, *The Struggle for Land: A Political Economy of the Pioneer Frontier in Brazil from 1930 to the Present Day* (New York: Cambridge University Press, 1981), chapters 4 through 6.

9. At the same time, in recent years these communities have suffered from several incidents of teenage suicide due to the lack of opportunities and future for the younger generation, as well as the conflicts over land. They are a symbol of the ongoing predicament of native peoples living in proximity to nonaboriginal

communities throughout Brazil. This distressing plight of the Guarani in Dourados has been reported worldwide, including in English. Most recently, see Amnesty International, "Pushed into Poverty," *Wire* 40, no. 4 (August–September 2010): 2–5, and Wyre Davis, "Brazil's Guarani-Kaiowa tribe allege genocide over land disputes," *BBC News* (September 8, 2015). http://www.bbc.com/news/world -latin-america-34183280 (accessed September 9, 2015). Also, Survival International, http://www.survivalinternational.org/tribes/guarani (accessed June 23, 2014); the Amnesty International action file, http://www.amnesty.org/en/library/info /AMR19/016/2009/en (accessed June 23, 2014), and Rubem Thomaz de Almeida, "A 'Entrada' no Tekoha," in *Povos Indígenas no Brasil 500, 1996–2000* (São Paulo: Instituto Socioambiental, 2000), 745–748.

10. Recent figures indicate that the GINI coefficient (which measures income gap) for Brazil has hovered around 53 for the past ten years, despite an ambitious governmental program to alleviate extremes. Few countries have a higher coefficient than Brazil. The higher the number, the greater the gap (Denmark, by contrast, is 27). World Bank Data, http://data.worldbank.org/indicator/SI.POV.GINI (accessed June 2015).

11. For more on EMBRAPA, see the website https://www.embrapa.br/en /home.

12. Afonso Simões Corrêa, "Pecuária de corte em Mato Grosso do Sul," paper presented to Encontros Regionais de Pecuária de Corte, Brasília, 27 November 1987, 3; Raymond Pebayle and Jean Koechlin, "Les fronts pionniers du Mato Grosso méridional: Approche géographique et écologique," in *Le Bassin Moyen du Paraná Brésilien: L'Homme et son Milieu*, ed. Raymond Pebayle et al., 156–166 (Bordeaux: Centre d'Études de Géographie Tropicale, CNRS, 1978).

13. Ana Primavesi, *Manejo Ecológico de Pastagens Em Regiões Tropicais e Subtropicais* (São Paulo: Nobel, 1986), 22–29, 41–42, 59, 83–85, 101. Primavesi's statistic for planted pasture in the United States has not changed since the 1980s, confirmed by USDA information. See USDA Natural Resources Conservation Service, "Pasture Resources": http://www.nrcs.usda.gov/wps/portal/nrcs/main/national /landuse/rangepasture/pasture (accessed September 2015); Simões Corrêa, "Pecuária de corte em Mato Grosso do Sul," 1–8.

14. R. M. Boddey, R. Macedo, R. M. Tarré, E. Ferreira, O. C. de Oliveira, C. de P. Rezende, R. B. Cantarutti, J. M. Pereira, B. J. R. Alves, S. Urquiaga, "Nitrogen cycling in *Brachiaria* pastures: The key to understanding the process of pasture decline," *Agriculture, Ecosystems and Environment* 103, no. 2 (July 2004): 389–403; Sergio Margulis, *Causes of Deforestation of the Brazilian Amazon*, World Bank Working Paper 22 (Washington, D.C.: World Bank, 2004), 6, 10; Êrika B. Fernandes Cruvinel, Mercedes M. da C. Bustamante, Alessandra R. Kozovits, Richard G. Zepp, "Soil emissions of NO, N_2O and CO_2 from croplands in the savanna region of central Brazil," *Agriculture, Ecosystems and Environment* 144, no. 1 (November 2011): 29–40; O. C. de Oliveira, I. P. de Oliveira, B. J. R. Alves, S. Urquiaga, and R. M. Boddey, "Chemical and biological indicators of decline/degradation of

Brachiaria pastures in the Brazilian Cerrado," *Agriculture, Ecosystems and Environment* 103 (2004): 289–300; interviews with Dr. Alfonso Simões Corrêa, EMBRAPA-CNPGC, Campo Grande, 24 May 1990, and Dr. Paulo Coelho Machado, Campo Grande, 22 May 1990; A. Pott, A. K. M. Oliveira, G. A. Damasceno-Junior, J. S. V. Silva, "Plant diversity of the Pantanal wetland," *Brazilian Journal of Biology* 71, no. 1 suppl. (April 2011), 268 (table 1).

15. Ilsyane do Rocio Kmitta, "Descortinando os pantanais: A construção de um paraíso às avessas entre o limite das águas e dos homens," Ph.D. thesis, Universidade Federal da Grande Dourados, Dourados, Mato Grosso do Sul, 2016, 113. Many thanks to Ilsyane for generously allowing me to cite her work before its finalization. Antonio C. Allem and José F. M. Valls, *Recursos forrageiros nativos do Pantanal mato-grossense* (Brasília: EMBRAPA, 1987), 189–190, 232; Pebayle and Koechlin, "Les fronts pionniers," 148–149; Vitor Del'Alamo Guarda and Renato Del'Alamo Guarda, "Brazilian Tropical Grassland Ecosystems: Distribution and Research Advances," *American Journal of Plant Sciences* 5 (2014): 924–932, http://dx.doi.org/10.4236/ajps.2014.57105 (accessed August 2015). Pensacolo and Townsville lucerne were imported from Australia.

16. A. da S. Mariante, M. do S. M. Albuquerque, A. A. do Egito, and C. McManus, "Advances in the Brazilian animal genetic resources conservation programme," *Animal Genetic Resources Information* 25 (1999): 107–121.

17. Carlos A. Klink and Ricardo B. Machado, "Conservation of the Brazilian Cerrado," *Conservation Biology* 19, no. 3 (June 2005): 700–707.

18. David Kaimowitz and Joyotee Smith, "Soybean Technology and the Loss of Natural Vegetation in Brazil and Bolivia," in *Agricultural Technologies and Tropical Deforestation*, ed. A. Angelsen and D. Kaimowitz, 195–211 (Wallingford, Oxon, UK: CABI Publishing, 2001).

19. William H. Fisher, "Surrogate Money, Technology, and the Expansion of Savanna Soybeans in Brazil," in *Rethinking Environmental History: World-System History and Global Environmental Change*, ed. Alf Hornborg, J. R. McNeill, and Joan Martinez-Alier, 345–360 (Lanham, MD: Altamira Press, 2007); Mariana Soares Domingues and Célio Bermann, "O arco de desflorestamento na Amazônia: Da pecuária à soja," *Ambiente e Sociedade* (São Paulo) 15, no. 2 (maio-agosto 2012): 1–22.

20. Christian Brannstrom, "South America's Neoliberal Agricultural Frontiers: Places of Environmental Sacrifice or Conservation Opportunity," *AMBIO: A Journal of the Human Environment* 38, no. 3 (2009): 141–149.

21. Brazil, IBGE, Diretoria de Pesquisas, Coordinação de Agropecuária, *Resultados da Produção Agrícola Municipal 2011* (Brasil: IBGE, 26 de outubro 2012), table 6, Principais Municípios produtores de cana-de-açucar em 2011.

Glossary

agregado. Tenant or semipermanent ranch hand; often in charge of a *retiro*, or line camp

alqueire. Unit of area equal to 2.42 hectares, or measure of capacity equal to 36.27 liters (roughly one US bushel)

angico. *Piptadenia* sp.; tree with high concentration of tannin used in artisanal hide tanning

arroba. 15 kilograms (33 pounds); common weight used for bagged dried goods in Brazil into the twentieth century

baía. Sweet-water lagoon found throughout the Pantanal

bandeiras. Expeditions in search of gold and slaves during the Colonial period, originating in São Paulo into the interior of Brazil

barba de bode. *Aristida pallens*; coarse invader grass species of poor nutritional value for cattle

barreiro. Natural salt lick found primarily in the Pantanal

berne. Infestation of bot fly (*Gasterophilus* spp.) or warble fly (*Hypoderma bovis* or *H. lineatum*); common in the cerrado, causing lesions and irritation to cattle

boiadeiro. Cattle drover; the drive is called a boiada

bombachas. Baggy trousers typical of the gauchos of Rio Grande do Sul and the Río de la Plata

braça. Linear measure of 2.2 meters

braquiária. *Brachiaria brizantha*, *B. decumbens*, and *B. humidicola*; most common grasses planted for ranching in Brazil today

camarada. Ranch hand; usually refers to ranch workers who do not handle cattle, as opposed to cowboys

campo. Field; pasture; open country

campo de demonstração. Experimental farm or ranch

campo limpo. Pasture land of native grasses found in southern Mato Grosso do Sul; also called the Vacaria

capataz. Ranch foreman

capybara. *Hydrochoerus hydrochaeris*; largest rodent on earth; originally found throughout lowland South America, today common in the Pantanal

caracu. Breed of cattle descended from original animals imported from Portugal; at one time promoted by animal scientists in São Paulo for its rusticity and strength; today bred to preserve gene pool and for scientific study

carandá. *Copernicia alba*; common palm tree in Pantanal often used for fence posts

carona. *Elionurus muticus*; ubiquitous coarse invader grass in much of Mato Grosso; poor nutritional value for cattle; appears native to region

carrapato. Cattle tick (*Rhipicephalus microplus*, formerly *Boophilus microplus* or *B. argentinus*); cause of Texas tick fever; the fever is not common in Mato Grosso

cavalo pantaneiro. Horse bred in the Poconé region of the northern Pantanal exclusively for use in the Pantanal

Cerrado. Semiarid savanna region of Central Brazil, extending over parts of the states of Mato Grosso, Mato Grosso do Sul, Goiás, and Minas Gerais; also known as Planalto

Chaco. Semiarid geographical region stretching west and northwest from the Paraguay River across Paraguay into Argentina and Bolivia

charque. Dried beef or jerky, sometimes spelled *xarque*

charqueada. Beef jerky factory, sometimes called *xarqueada* or *saladeiro*

chino. Term given one of several breeds of cattle imported into Brazil and adapted to local conditions; name may have come from the Spanish-American slang for mestizo

colonião. *Panicum maximum*; African-origin grass common in pasture planting of the past; so widespread that at one time it was considered native to Brazil; high nutritional value for cattle

conto. 1,000 milreis

cordilheiras. Hillocks

coronel. Literally colonel (*coronéis* [pl]); regional political boss or cattle baron; originally an honorary rank in the imperial national guard

correntino. A breed of cattle developed in Argentina and Paraguay, descended from original Spanish breeds

crioulo. Mixed-blood breed of cattle most common in Brazil until the twentieth century

EMBRAPA. Empresa Brasileira de Pesquisa Agropecuária; federal agricultural research corporation created in 1973

erva mate. *Ilex paraguariensis*; Paraguayan tea, *yerba mate* in Spanish; a regional tea most common in Argentina, Brazil, Paraguay, and Uruguay, made from the leaves of a native holly tree

estancia. Large ranch

farofa. Dried and toasted manioc flour, common as a side dish for most Brazilian dishes

fazenda. Large rural property, farm, or plantation; in this study the term refers principally to a ranch, large or small

fazendas reais. Royal ranches established in the late eighteenth century

febre aftosa. Foot-and-mouth disease; viral; common in Mato Grosso but of low virulence

feira de gado. Cattle fair or marketplace

Fluminense. A native of Rio de Janeiro state

franqueiro. Mestizo cattle breed developed in Brazil, possibly in the São Paulo city of Franca

friagem. Cold front from Antarctica that occasionally occurs in June to September

frigorífico. Packing plant

Gaúcho. A native of Rio Grande do Sul; also refers to the cowboy of that state

gordura. *Melinis multiflora*; African-origin grass common in pasture planting; so widespread that at one time it was considered native to Brazil; good nutritional value for cattle

grilagem. Land grabbing, claim-jumping; perpetrated by a *grileiro*

Guaraní. Native language of Paraguay; also spoken by Guarani natives

hectare. 2.47 acres

Indubrasil. Breed of zebu developed in Brazil by crossing Gir and Guzerat cattle; see also *zebu*

interventor. State governor appointed by decree by the federal government

jaraguá. *Hyparrhenia rufa*; grass of African origin; common in pasture planting throughout tropical Latin America until replaced by brachiaria; so widespread that at one time it was considered native to Brazil; high nutritional value and resistant to fire

juiz comissário. Municipal property judge

legua. League, unit of land measurement equal to 6,600 meters (linear) or 4,356 hectares (area); in Mato Grosso and much of Brazil in the nineteenth century and into the twentieth, 3,600 hectares

mal das cadeiras. *Trypanosoma evansi* or *T. equinum* (surra, equine anemia); also known as peste de cadeiras; anemic disease of horses and mules caused by a protozoan bacterium that causes loss of ability to support hindquarters; thought to be transmitted by biting flies and possibly vampire bats

manqueira. Blackleg or symptomatic carbuncle; cattle disease common in Mato Grosso causing weakening of motor abilities of young animals; sometimes called mancha negra, mancha branca, or manchilla

mantas. Flanks of a steer used for *charque*, especially in Argentina and Uruguay

Matogrossense. Native of Mato Grosso state; Sulmatogrossense is a native of Mato Grosso do Sul state

milreis. Brazil's currency during the period under study, written as 1$000; a conto was equivalent to 1,000 milreis (1:000$000)

mimosos. *Hemarthria altissima, Axonopus purpusii, Reimarochloa brasiliensis, Heteropogon villosus*, and other species; most common native grasses in Mato Grosso, particularly in the Pantanal and Vacaria; very high nutritional value and well-adapted to seasonal flooding

Mineiro. Native of Minas Gerais state

mutirão. Sometimes *muxirão*; cooperative labor among rural residents of Brazil, similar to North American house or barn raising

pangola. *Digitaria decumbens*; African-origin grass also used in pasture planting

pantanais. Wetlands; in this case, the various subdivisions of the Pantanal

Pantanal. Seasonal fluvial floodplain of Mato Grosso and Mato Grosso do Sul; one of the world's great natural bird sanctuaries

Pantaneiro. Native of the Pantanal; native cattle and horses of the Pantanal

paratudo. *Tabebuia caraiba*; tree common throughout the Pantanal

parceria. Partnership; share contract involving the care of cattle for a period of time in return for a percentage of production or calves; common in Mato Grosso from the 1920s

Paulista. A native of São Paulo state

peão. Landless rural worker, *camarada*; probably originally from the Spanish *peón*

Planalto. Geographical name for the Cerrado; Brazilian highlands

posse. Rural landholding, often squatted; though usually small, in Mato Grosso the term also refers to unregistered ranches

postas. Meat from legs and other extremities used for *charque*, especially in Mato Grosso and Rio Grande do Sul

Repartição de Terras. State Land Office

retiro. Line camp; outlying area of ranch with corrals and buildings where cattle are pastured, handled, and rounded up for sale; sometimes called a posto; usually run by an *agregado*

salina. Saltwater lagoon in the Pantanal; usually found in the higher elevations and in conjunction with *barreiros*

serra. Hill chain, highland

sertanejo. Resident of the sertão; descendant of original Portuguese cattle adapted to the interior of Brazil

sertão. Interior of Brazil; sometimes referring to the arid interior of the northeast; backlands, as used in the English translation of Euclides da Cunha's study *Os Sertões* (*Rebellion in the Backlands*) and João Guimaraes Rosa's novel *Grande Sertão: Veredas* (*The Devil to Pay in the Backlands*)

sesmaria. Colonial and imperial land grant of 13,068 square hectares

tereré. Cold mate tea commonly consumed in Paraguay and the Pantanal

toque. Debilitating condition suffered by cattle when they ingest too much ash after a pasture is burned or if they rely too much on barreiros or salinas for dietary salt

Triângulo Mineiro. Westernmost region of Minas Gerais bordering Mato Grosso; the "triangle" name refers to its geographical shape

tristeza. Texas tick fever; caused by the parasite *Babesia bigemina*

tucura. Local term given to original cattle breed common to the Pantanal, the pantaneiro

Vacaria. Region of southern Mato Grosso do Sul where there is an abundance of nutritious native grasses; also called campo limpo

vaqueiro. Cowboy

zebu. *Bos taurus indicus* or *Bos indicus*; species of cattle native to the Indian subcontinent and imported into Brazil in the late nineteenth century; most common variety in Brazil is Nelore; a related animal in North America is the Brahman

Index

Page numbers in italics refer to figures and tables.